Vault Reports Career Guide to High Tech

The Team

VAULT REPORTS, INC.

80 Fifth Avenue
11th Floor
New York, NY 10011
(212) 366-4212
Fax: (212) 366-6117
www.VaultReports.com

Vault Reports Career Guide to High Tech

MARCY LERNER AND NIKKI SCOTT

Library of Congress CIP Data is available.

ISBN 1-58131-043-9

Printed in the United States of America

ACKNOWLEDGEMENTS

Warm thanks to:

Hernie, Glenn Fischer, Carol and Bart Fischer, Lee Black, Jay Oyakawa, Ed Somekh, Todd Kelleher, Bruce Bland, Celeste and Noelle, Rob Copeland, Muriel and Stephanie, Michael Kalt, Ravi Mhatre, Tom Phillips, Bryan Finkel, Geoff Baum, Gary Mueller, Ted Liang, Brian Fischer, Glen and Dorothy Wilkins, Sarah Griffith, Russ Dubner, Kirsten Fragodt, Aldith and Robert Scott-Asselbergs, C., J. Nilsson, Geoff Vitale, Dana Evans, Olympia, Marky Mark, Katie T., Rachel, Jen, and The Garden Tavern (8th Avenue between 30th & 31st).

And to:

Amy Wegenaar, Angela Tong, Kofo Anifalaje, Alex Apelbaum, Al Gatling, Kevin Salgado, Candice Mortimer, Samir Shah, Thomas Lee, Joan Lucas, Kelly Guerrier, Sylvia Kovac, and Austin Shau

And Artists:

Robert Schipano and Jake Wallace

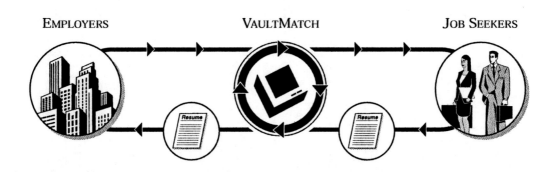

Contents

Contents

Profiles

Articles

A Guide to this Guide

If you're wondering what all those snazzy icons in our company entries are for, or how we developed our information, read on. Here's a guide to the information you'll find in each entry of our book.

The Meters

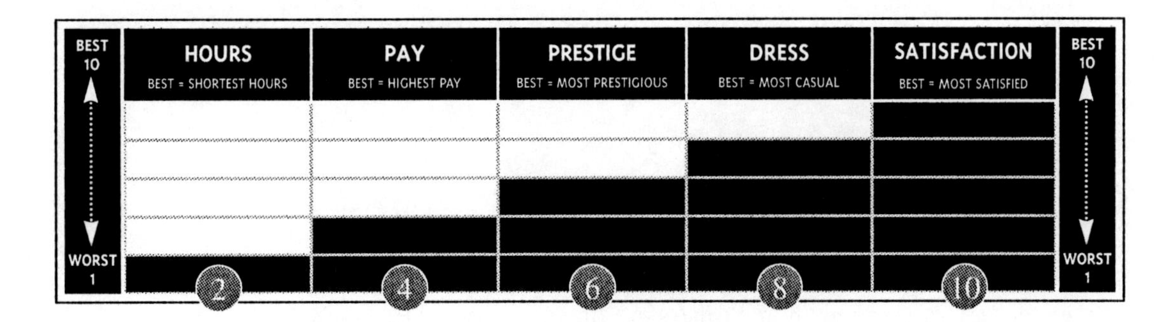

The Vault Reports Meters are scored on a 1-10 scale, with 10 being the highest. The scores are based upon how a particular company compares to the average employer (including industries other than high tech). A company with a score of "5" in pay signifies that the company has an average salary. In determining our ratings, we looked at both company data and our own surveys and benchmarking of companies (derived from years of research).

The meters attempt to give an at-glance look at a company. However, keep in mind that the issues they measure are complex, and a closer look at the entry is necessary for a fuller view.

Work Hours: The shorter the workweek, the higher the score. We did not adjust these scores based on the company location or to account for the long hours in the investment banking industry in general. Working 70 hours a week may be a vacation at a software company on deadline but it's still a mighty long workweek.

Pay: A company's pay score is based on home office pay, with bonuses (and likelihood of achieving them) factored in. We did not factor in cost of living or stock options.

Prestige: The higher the score, the more prestigious the company.

Dress: The higher the score, the more casual the dress code. A score of 10 means shorts and ripped T-shirts; a score of 1 means strict business formal.

Satisfaction: Employee satisfaction is based on our extensive interviews and surveys of employees at each company.

The Graphs

The information given in the graphs for our companies are drawn from a variety of sources. Much of the basic financial and employment data for public companies (net revenues, number of employees, etc.) comes from business information provider Hoover's Online and company annual reports. In some cases, the companies themselves supplied data.

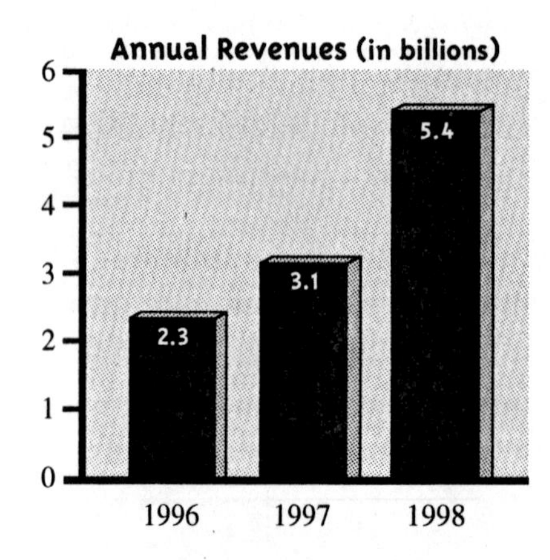

THE ENTRIES

Each of our entries is broken into three sections: The Scoop, Getting Hired, and Our Survey Says.

The Scoop: This section includes information about the company's history, strategy and organization, recent business performance, and other points of interest.

Getting Hired: This section includes information about the company's hiring process, concentrating on recruiting for junior-level employees and interviews.

Our Survey Says: This section is based on quotes and comments from employees at the company, and covers topics such as company culture, pay, perks, social life, as well as many others.

Introduction

A HIGH VELOCITY WORLD – HIGH TECH 1998

The high tech world is a dynamic one that moves at "Internet speed" – faster than a speeding e-mail. In 1998, the industry was marked by just as many ups and downs as Wall Street (where high tech has been causing much of the ruckus) or Hollywood. Computer makers, chip manufacturers, software makers and Internet companies have been acquiring one another, merging, and forging partnerships like crazy. After picking up software makers in 1997, IBM has moved into the chip-making business; AOL just purchased browser giant Netscape; and Microsoft is embroiled in an antitrust case that will go down in history.

THE CRAZE ON WALL STREET

Wall Street circa 1998 was marked by IPO mania. Internet and software companies made a huge splash on NASDAQ, as the market boomed and Wall Street traders lived in fear of missing out on the next Microsoft or Dell. As of December 1998, Yahoo! market cap was higher than that of well-known firms like JP Morgan, 3Com and General Mills (roughly $20 billion). New media companies Ticketmaster Online-CitySearch, EarthWeb and Theglobe.com racked up 187.5 percent, 240 percent, and 606 percent gains, respectively, when they went public. It's not hard to believe that the relatively unknown portal Xoom.com saw its stock price jump 146 percent on the first day. And it's a special kind of mania that allows a company to tout itself as the next Amazon and see a stock jump from 50 cents to $17 in a week – especially when that company, PinkMonkey.com, has annual revenues of about $100. (No, that's not a typo.)

EVERYBODY WANTS TO BE A PORTAL

A big trend in 1998 was the rush among Internet firms to become a portal, an entry to the Internet. Though portals used to be simple search engine sites, the frantic rush to attract "eyeballs" and advertising revenues has caused companies like Yahoo, Metacrawler and Excite to hawk everything from horoscopes and stock quotes to free e-mail and online address books. The goal – to ensnare surfers before they go somewhere else. In fact, the hottest portal offering of late is the web community – which encourages consumers to stick around the site instead of using it as a means to move off. Communities allow Net heads with similar interests (new moms, gourmet cooks, Metallica fans) to congregate online, post messages, even build their own sites. The web community could become an advertising mecca – the community attracts a recurring audience with a super-specific set of interests. Someone who habitually surfs through sites on Nepal and Brazil, for example, might be an attractive target for travel marketers.

THE GOVERNMENT STEPS IN

In a trial of crucial importance to the high-tech world, the Justice Department filed an antitrust suit against Microsoft in May 1998. The government alleges that, by bundling its Internet Explorer with its operating systems (essentially giving it away) near-monopolist Microsoft had unfairly leveraged its control of the operating systems market into a devastating advantage over its browser rival, Netscape, thus thwarting competition in the software market. (Netscape had worked with Sun on promoting Java, a computer language which can run on any operating system – a direct threat to Microsoft.) Before Microsoft's bundling action in 1996, Netscape had had nearly 90 percent of the browser market; a year later, the two companies were nearly even. But in the fast-paced high-tech industry, events threaten to overtake the antitrust trial.

THE AOL SURPRISE

In November 1998, America Online, perhaps the only successful pay-for-content provider on the Web, aside from some financial magazines (such as TheStreet.com and a wide variety of pornographic sites), bid to acquire Netscape Communications for $4.2 billion. The deal is widely seen as weakening the government's allegation that Microsoft has squashed competition in the software industry. (An alternative explanation is that Netscape had been so weakened by Microsoft that it was forced to merge.) But Microsoft still faces other threats – from the "open source" movement. Netscape, prior to its acquisition, had released its browser's source code, thus essentially making the browser a public resource. Sun has also loosened its control of Java software code in order to promote the language as an alternative to Windows. Sun and Oracle plan to work together on a new computer that does not need an operating system – an initiative that would directly compete with Microsoft's heavy-duty Windows NT system.

In the hard-wired world

Meanwhile, PC ownership has become increasingly attractive to consumers as prices drop. Packard Bell has introduced PCs that retail for under $600 dollars. These PCs use slower, older generations of microchips. (Intel's effort to introduce the Merced, its latest model of microchip co-produced with Hewlett-Packard, has run into trouble and will not be introduced until 2000.) Major hardware companies like Dell and Gateway have introduced a leasing program for home computer consumers, with payments as low as $40 a month, which should ensure that even more computers wend their way into American homes. New toys on the hardware market are increasing the appeal of PCs as well. The DVD-ROM, a disk which can store up to 15 gigabytes of info, is hitting the consumer market, as are cable modems – modems that use coaxial cable lines, which have wider bandwidth, and are up to 10 times faster than the 56K modems now used. Research company Dataquest estimates that cable modem shipments will exceed 2.4 million by 2002.

E-commerce arrives

Is content on the Web there to attract customers? Or is the content of the future Web to be e-commerce? Christmas 1998 was the first holiday for which shoppers hit the Web in force, and e-retailers rang up a very merry season indeed. Preliminary indications are that online holiday sales hit $5 billion, more than double the $2.5 billion analysts had anticipated.

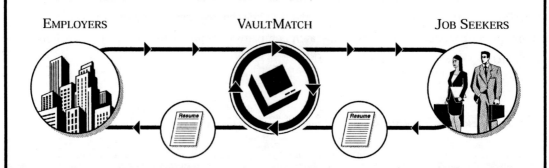

WHERE THE TECH IS

Up to a year ago, people still saw Silicon Valley as the epicenter of the high-tech boom. Home to more than 7000 high-tech firms including Netscape and Intel, the Valley is also the home of the venture capital firms that provide the funding to fuel high-tech ambitions. But there are other established "Silicon Cities," and new ones are cropping up faster than spam on an AOL account. One stalwart is Seattle – home to Amazon.com, RealNetworks and Microsoft. Then there's Austin, Texas, called "Silicon Gulch," home to hardware powerhouses Dell and Compaq, chipmaker Advanced Micro Devices, and the up-and-coming software maker Trilogy. New York City, often regarded as the content capital of the Internet, has been wittily dubbed "Silicon Alley." Once simply seen as the e-version of underemployment for New York's young designers and writers, the Alley is now host to a maturing set of new media extensions of traditional media, such as advertising, publishing, and marketing. In New York, content – and commerce – is king.

Look far east of New York, and you'll find one of the most vibrant high-tech industries in the world. Second only to the U.S. for the number of high-tech firms within its borders, Israel is home to top manufacturers of network security, data communications and Internet products. Young Israeli techies get their feet wet during mandatory military service – where they are exposed to some of the most sophisticated technologies in the world. Budding entrepreneurs benefit from government-sponsored technology incubators and a growing number of local and foreign venture capitalists.

Other growing tech havens include Boise, Idaho (home to promising hardware firm Micron Technology), North Carolina's Research Triangle, NC (actually three cities: Raleigh, Durham and Chapel Hill); and Atlanta, GA, which hosted the 1996 Summer Olympics. The city is now wired with more fiber-optic cable than any other city in the world.

Silicon Wadi

It's the birthplace of some of the world's hottest software, Internet products and electronic devices. Though no larger than New Jersey, this country is second only to the U.S. when it comes to the number of high-tech firms within its borders. It's Israel – a country people tend to identify with kibbutzim and political strife – and today it boasts one of the most vibrant high-tech industries on the world, with strengths in network security, data communications and Internet products. Many compare Israel to the burgeoning Silicon Valley of the early 1980s. The hub of the country's high-tech revolution is sunny Sharon Valley, which, with its red-roofed stucco houses and abundant palm trees, is strangely reminiscent of its predecessor in California. Dubbed Silicon Wadi – a reference to the dried up riverbeds that crisscross the area – the Middle Eastern version taps the talent of army-trained computer wunderkinds. The country's mandatory military service exposes young citizens to some of the most sophisticated technologies around, knowledge easily applicable to commercial products. Israel's strength in technology was sparked in part by France's 1967 embargo on weapons exports to Israel. Forced to focus energy on its defense industry, the country developed a strong tradition of high-tech innovation.

Noticing an increase in entrepreneurialism among its young citizens, the Israeli government set up a number of high-tech incubators in 1991. Since then it has funneled more than $125 million into about 300 projects, more than half of which have evolved into independent companies. Young companies have also benefited from venture capital firms – which have blossomed as much in the Holy Land as they have in Silicon Valley. In 1991, there was one VC firm in Israel, with $35 million to invest. As of 1998 there were 70 firms in Israel, with combined funds totaling about $700 million. There are the foreign investors as well: the U.S. is the largest VC investor in Israel, and acquisitive Japan (known for its weak software industry) comes in second.

Among the most successful Israeli start-up stories is that of Mirabilis, which recently sold its ICQ (I seek you) software to America Online for $287 million. Its four founders are the youngest and most celebrated entrepreneurs in the country. With ICQ, users can alert a list of friends when they log on to the Internet, then exchange messages in real time or participate in chat rooms. The cool part is that ICG became the Web's chat software of choice with nary an advertisement. The phenomenon has spread to more than 12 million users, strictly "by word of mouse." Other thriving software companies are GEO-Interactive, which makes software that compresses and sends video, animation and sound over the Internet; and CheckPoint, which sells software that protects corporate computer systems from hackers.

Industry Segments

Technology firms can be best placed into five major categories: hardware, semiconductors, software, Internet and networking. This doesn't mean that high-tech companies work in isolation; Intel, for example, frequently co-markets its microprocessors with computer hardware firms (which must feature the trademark "Intel Inside" tune clip), and Compaq pre-loads software onto its PCs. Dell may cobble together Microsoft software, a Netscape browser, 3Com modems and Intel microprocessors to create its computers.

Hardware

The computer hardware market is enormous – and getting bigger all the time. In 1997, it was estimated to be a $500 billion business, and growing 15 to 20 percent yearly. Sales of personal computers alone totaled $60 billion in 1996 and $75 billion in 1997. The drop in PC prices has made computer ownership more attractive to home consumers, many of whom are now buying, second, lower-powered computers for their children.

Computer hardware companies put together the terminals that house the software that run the computer. They assemble the terminals, microprocessors (also called "microchips," or just "chips"), hard drives, keyboards and CD-ROMs that compose the desktop and laptop computers we know and use. Other products: workstations (more powerful computers typically used for graphics, scientific computing, complex financial models, and other applications requiring high performance and memory) and servers (a powerful computer that is accessed by other computers, called clients, and which supplies data, software, or services to a network). "Fat servers" do most of the processing, leaving little or none that must be done by the client.

Once viewed as a major threat to the PC industry, network computers have failed to take off. Oracle CEO Larry Ellison first floated the concept of these cheap, stripped-down terminals that would access data and applications stored on central servers. The idea was that NCs would reduce the need for memory, disk drives, and other hardware currently needed to run computers. But the 'stripped down' idea didn't prove viable. A big strike against it was the significant decline in PC prices in 1998. Even the cheapest PC has more to offer than a machine with no memory and no disk drive. Hardware makers are now offering up variations on the NC theme – ranging from digital set-top boxes to Internet-based PCs for home and corporate users.

But there are growth possibilities for the hardware industry too. Chips and computers don't have to be confined to desktops. A major growth area targeted by both hardware and software firms is embedded computers – computing chips everywhere from "smart TV" boxes to "dashboard computers" in automobiles and planes to "smart cards" – cards that can carry data and serve as a sort of replacement for pocket change. Another new addition to the hardware scene is the DVD. The DVD is an optical storage system that can store up to 15 gigabytes of information. Currently used in computers and as a VCR substitute, the DVD can access CDs imprinted with movies, multimedia applications, music CDs – you name it – all with much crisper graphics and sound. Its major drawback, for the time being, is that the DVD can't be rewritten and reprogrammed (so you can't tape that irresistible Ally McBeal rerun) – but these problems are expected to be solved in the next year or two.

Major hardware companies

- Compaq
- Dell Computer
- Gateway
- Hewlett-Packard
- Sun Microsystems

Cheap PCs

Years ago people saved up and waited patiently for the latest, most powerful computer to come out each year. Then they dutifully emptied their bank accounts into the retailers' pockets and prayed for a little time before the next software development rendered their PCs obsolete. But that was then. Today, "affordable computer" doesn't have to mean "last year's version." Production prices for personal computers have been falling over the past two years, and PC manufacturers are taking advantage of the opportunity to expand their customer bases. Families who would otherwise do without can now afford home computers, and those who could afford the newest technology are more apt to buy a budget model for the kids to use. The new "Cheap PCs" are usually priced at $1000 or less. You may not be able to run a business on these babies, but you'll certainly be getting lots of power for your penny.

The biggest force driving down PC prices has been the financial crisis that began in late 1997 in Asia, where chip manufacturers are cutting prices for semiconductors and other key components in an effort to keep products from sitting in their warehouses. Production costs are also dropping as PC makers revamp their product design and the manufacturing process – many have managed to shorten production time and reduce the cost of materials. The decline in computer prices also reflects the relatively slow development in the software industry. In the past, software and hardware technology evolved at similar rates. However there's not much new software that requires the processing power available with the latest machines. Even Microsoft's Windows 98 runs fine on the less expensive microprocessors. For these reasons, the industry has seen a massive decline in memory prices over the last six months. Even Intel, the industry leader, was forced to create a line of lower-priced chips, called Celeron. Dell uses the new chip in its version of the low-cost PC – the Dell Dimension, which sells for about $1400.

The best part about this trend is that the benefits are so widespread. More consumers are able to afford computers, and the larger market base helps manufacturers cut costs – and everyone from modem makers to mousepad manufacturers enjoys increasing demand. In addition, Internet usage increases, a boon to the growing number of companies advertising on the Web.

Of course there is the fear in that sales of these so-called "sub-zeroes" will cannibalize business from higher-priced models, but few in the industry are truly worried. By the end of 1997, almost 40 percent of the PCs sold in the U.S. cost less than $1000, but about 80

Cheap PCs, cont'd

percent of those purchasing them were first-time buyers. And the fact is, there will always be customers who want 'serious machines' – for multimedia, graphics and other such programs. Besides, manufacturers are confident that evolution in the software market will force the budget-PC owners to upgrade later.

Compaq introduced the first of the brood in February 1997. Instead of Intel's famous (and pricey) Pentium processor, its $999 Presario is powered by a less expensive chip from Cyrix Corporation. The other catch: that $999 price did not include a monitor. By the end of that year, HP introduced its Pavilion for $1000, which boasted the same power and included a monitor. In 1998, the choices grew, and the quality of the machines vastly improved. At retailers like CompUSA and Staples, you can get a Packard Bell or Hewlett Packard desktop computer (300MHz) for around $900 (after rebates), and both include a monitor and a color printer. The most recent Compaq Presario (with a monitor) is still $999, as is the Everex AGI, and the Packard Bell S700, which comes with a Pentium CPU, a 5.1 Gigabyte hard drive.

What you really get for $999

So how much computer are you really getting for these low, low prices? A year ago, you could get a good machine, but really just the basics – a 233MHz processor (probably not a Pentium), 32MB of RAM, a graphics card and (hopefully) a modem. In many cases the monitor was not included. Today, a $1000 PC will include can a Pentium processor, 32 MB of memory, a 24-speed CD-ROM drive, 15-inch monitor, and a 56K modem.

These systems run only 20 to 25 percent slower than the fastest machines on the market, which is perfectly fine if you're just using your computer for word processing and surfing the net. You can use the money you save to get a good backup drive and some speakers. Industry insiders say the only things you'll have problems with are programs like Photoshop and Quark, animation, multimedia and 3D modeling software.

Most $999 specials are currently sold through retail outlets. Places like CompUSA and OfficeMax offer some pretty good deals on desktop machines, and many include both a monitor and printer. Direct-sales PC makers like Gateway and Dell maintain that their customers don't want bargain computers and intend to avoid the trend for as long as possible. Apple's newsy iMac currently retails for $1299, but early discounting has pushed its price down to the three figure range as well.

Vault Reports Profile

Michael Dell, CEO: Dell Computer

Though he's the youngest Fortune 500 CEO in history, the fact is, his face is not as easily recognizable as Bill's or Steve's. Nevertheless, Michael Dell's story, at least, is one oft-told around the entrepreneurial campfire. His anonymity is actually pretty understandable considering his business strategy – he doesn't sell his computers in stores, so the only people that know about him are the people who really know computers. This direct-sales approach may have seemed strange at first, but Dell's success is proof that sometimes the quiet ones are smarter than you think.

Dell began his company during his freshman year at the University of Texas. He invested $1000 in discounted PC parts he bought from a retailer and used them to upgrade some old PCs. During his first month of business, Dell made $180,000 in sales. Needless to say, Dell opted for money over completing his sophomore year, and Dell has been known to comment that it took his parents two years to accept his decision to leave school. Since day one, Dell has stuck to a unique set of business strategies. First, the company sells direct to the consumer, sidestepping the usual markups implemented by the retailer. Second, each PC is made-to-order, and the company only keeps eight to 12 days worth of inventory on hand at any time, often purchasing components when the prices are low. Corporate customers (which make up the majority of its business) save time and money since Dell configures the computers and loads all the software right in the factory. Among home users, Dell established itself as the sophisticated buyer's brand – offering high performance computers at reasonable (but realistic) prices. Dell sweetens the deal with top-ranked customer service and free, unlimited technical support.

Michael Dell, 33, is now worth $7 billion – much more than Bill Gates was at that age. Though he's quite often the youngest person in the room at most of his meetings, Dell is not the arrogant Gen-X computer geek one might expect. He's a brilliant businessman, seemingly unfazed by the pressures of his position at the helm of his multibillion-dollar business. Not only was he smart enough to identify the need for a company like Dell, he was humble enough to bring in help when it was needed. After rocky times in the early 1990s – a failed attempt to sell through retail stores; an messy first attempt at the laptop computer, and perennial customer support issues – Dell was wise enough to look for strong management to get things in order. He brought in Mort Topfer as COO and Tom Meridith as CFO, and they quickly got things in order. Today, Dell ranks third among the world's computer makers (behind IBM and Compaq), and controls 7 percent of the PC market. Dell and Gateway are the only two major PC makers that are profitable at the moment, and Dell is the darling of Wall Street. Industry observers say the company's stock is likely to be named the decade's best performing stock on the S&P 500.

Is Apple Back?

In the mid-1980s, 'Apple' was synonymous with home computers. But after popularizing the personal computer, the Cupertino, CA company fell to the wayside as consumers flocked to worship at the shrine of the Wintel. IBM, Microsoft, and seemingly just about everyone else in the high-tech industry effectively let the air out of Apple's tires. By 1997, the company was in dire straits. Its market share was less than 3 percent, and the value of its stock was plummeting. There were rumors that the company was slated to be picked up by Sun Microsystems. Another story had that the computer maker would be taken over by an investor group headed by Larry Ellison, the founder of Oracle.

But now the Apple PC is finally out of the ICU, much to the joy of the estimated 600 million diehard Mac fans. The company announced profits of $309 million for 1998 – it's the company's first full year of profit since 1995, and a hopeful sign that Jobs knows what he's doing. The firm's shares have gained more than 70 percent since 1997, largely helped by the new product introductions. The company's stock is trading briskly, and the new iMacs have been selling faster than limited edition Nikes.

Credited with the turnaround is Steve Jobs, a co-founder of the company who was pushed out in 1985. He returned to Apple in 1996 as a "consultant," and was named interim chairman and CEO in September 1997. CEO Gil Amelio resigned in July 1997. A month later, the computer industry was stunned when Jobs announced that former nemesis Microsoft would become Apple's ally, cemented by the purchase by Gates & Co. of $150 million in Apple stock. That deal soon led to a 48 percent increase in Apple's share price.

Then Jobs began a major restructuring, and cut expenses by scaling back a number of departments and scrapping a number of research projects. He then rescinded agreements with manufacturers of Mac clones, which were draining core sales. In September 1997, Apple paid clone manufacturer Power Computing $100 million for its key assets, which resulted in the shutdown of its clone business by the end of the year. Jobs then reduced Apple's product line to only four models – one desktop and laptop each for business and home users. He also picked up on Michael Dell's direct sales and build-to-order strategies, a move that has proven extremely lucrative. Early in 1998, he scrapped production of the Newton, Apple's version of the Palm Pilot. (Jobs actually tried to buy Palm Pilot from 3Com in 1997, but the networking hardware company wisely refused to sell.)

Is Apple Back?, cont'd

Jobs then attacked the new Apple operating system. When he came back to the company, engineers were in the midst of developing a new operating system – one that would require software makers to completely rewrite existing applications. Jobs halted that project in favor of a less extreme route. The new OS will now run most programs, with some minor adjustments. Finally he pushed development of the new iMac, an alternative translucent construct which boasts a superfast G3 microprocessor and sells for only $1300, including monitor. It was released in August 1998, and consumers snapped up 278,000 units within six weeks.

In an interview with *Zdnet*, Jobs revealed that his strategy is to make Apple "the Sony of the computer business." He's certainly set his sights high. Though the computer maker is definitely doing well, it still has to compete with Wintel, and will be especially hard-pressed to convince the business world to switch over. Because it contains a Motorola PowerPC processor, the iMac handles many tasks at much faster speeds than the most powerful PCs now available running Intel Pentium II chips. However it doesn't have a disk drive – which is difficult to overlook. (Jobs contends that most data is now transmitted via e-mail or zip drives.) Also important is getting more software developers to write for the Macintosh platform – one of the reasons people have moved to PCs is the relative dearth of software available for Macs.

Semiconductors

What makes these computers work? Microprocessors (sometimes called chips) are made up of as many transistors as engineers can pile on an electronic circuit. The more transistors it has, the more functions a chip can handle, at greater speeds. And for these little miracles, you can thank the hardworking semiconductor industry – the folks who manufacture the necessary materials, and the companies that piece together these computing engines in superclean "fabs" – specialized microprocessor factories. And it's expensive to keep the research engines going – the average fab costs over $1 billion to develop.

Microprocessors – tiny integrated computers with their entire CPUs on one chip – are essential components of personal computers, servers and workstations today. At the high end of the market are the superfast chips that run workstations and servers; at the other end are cheaper chips you'll find in sub-$1000 PCs. High-performance chips, like Intel's Pentium II, earn their place in powerful personal computers, while even more powerful industrial-strength computer models rely on chips produced by Hewlett-Packard and Sun Microsystems. Historically, Intel has dominated the microprocessor industry, with over 80 percent of the market. But the Asian economic crisis and the deluge of sub-$1000 PCs have resulted in falling chip prices – so chip manufacturers now have to compete in terms of price as well as just power. Industry leader Intel introduced a lower-priced chip, called Celeron, which Dell uses in its version of the low-cost PC – the Dell Dimension. However, for the time being, consumers have shown a preference for higher-end computers – thus keeping chipmaker margins steady.

As more consumers get PCs and get online, the demand for computing power has increased in kind. The microprocessor industry has responded by finding ways to reduce chip size while significantly increasing performance. According to Moore's Law, named after Gordon Moore, the Silicon Valley guru who helped found the legendary Fairchild Semiconductor (which eventually spawned Intel), chip power doubles every 18 months. Intel's Andy Grove predicts the law will hold at least through the year 2013.

Major semiconductor firms

- Advanced Micro Devices
- Applied Materials
- Cyrix
- Intel
- LSI Logic

The Chipmaking Process

Chips are manufactured in "fabs," which are ultra-clean microprocessor factories. Chips are made from the essential material silicon, which arrives at fabs in footlong "ingots," which are then sliced into "wafers," disks roughly the size of pie tins. Several hundred semiconductor chips can be made on a single wafer. The wafers are then treated with aluminum and layered with a material called photoresist. The photoresist behaves like a film negative in ultraviolet light, allowing the chip pattern to be traced on the silicon dioxide underneath. The wafer is then "doped" with ions that permit current to flow along the etched pathways. After treatment, the wafer is sliced up into squares – the microprocessors, up to 1000 of them. Some companies, such as semiconductor equipment makers like Applied Materials, only manufacture the raw materials and the equipment necessary for fabs, while others, like Intel and AMD, complete the process and market the microprocessors. Fab!

WINTEL

Throughout the high-tech industry, the word is that the Microsoft/Intel relationship is on the rocks. It's no secret that Intel (though no angel itself) is sick and tired of Bill Gates' self-serving attitude and bullying tactics, and both companies have their hands full with their respective antitrust trials. Is there truth to the rumors? Or are the 'competitive' partnerships on the part of either company just smoke and mirrors? Only time will tell, or perhaps the judge on the case.

The Wintel (Windows/Intel) partnership was spawned in 1981 – IBM brought the two together when it put Intel's microprocessors and Microsoft's operating system on its PCs. IBM's role eventually decreased, but Microsoft and Intel continued to work in tandem as they grew into industry standards. They were also encouraged by corporate customers, who pushed them to coordinate their research efforts. Today they are among the most influential players in the high-tech world. But this relationship has significant ramifications on the entire industry. Because Windows is engineered to work perfectly with Intel chips, products with other (less expensive) chips are generally less compatible with Windows; and vice versa – other operating systems will not work as well as Microsoft does on an Intel processor. Every new version of Windows Microsoft's NT (now called Windows 2000) system happens to require a pretty potent chip – which just happens to be the latest version from Intel. Finally, Microsoft and Intel share marketing muscle – it's hard for smaller companies to compete when two of the richest are promoting each other.

On the outside, Wintel looks like a comfortable arrangement – but industry observers say things are not so cozy on the inside. The biggest reason – Microsoft is a bully, even to its pal Intel. It sees nothing wrong in cooperating with Intel's competitors, but it gets in a huff when Intel tries to play tit for tat. For example, Microsoft has been working closely with Advanced Micro Devices and Cyrix – companies that produce less expensive clones of Intel's microprocessors. And Microsoft's Windows CE (consumer electronics) software, used in mobile and handheld computers, uses chips from Hitachi and MIPS, not Intel. Meanwhile, Microsoft CEO Bill Gates has actively tried (and succeeded) to squelch any efforts by Intel that might hinder the Microsoft agenda. When the chipmaker has ignored direct threats, Gates has gone ahead and formed partnerships with Intel's competitors.

As the story goes, the first crack appeared in the spring of 1995 when Microsoft forced Intel to halt development of its NSP (native signal processing) multimedia software. Microsoft tried to characterize the NSP project as a misguided endeavor that would have hampered the release of new computers loaded with Windows 95. In reality, Microsoft

Wintel, cont'd

was just upset that Intel was moving in on its territory. Microsoft was equally peeved when it learned Intel planned to support and develop a software engine for Sun Microsystem's Java programming language. Such a chip would enable programs written in Java to run on any operating system, thus undermining Windows' hold on the market. Reportedly, Gates pressured Intel CEO Andy Grove to restrict the company's software development efforts.

Intel hasn't been utterly passive. Over the years, the company has forged a number of partnerships that ticked off old Bill. Intel was an early investor in Steve Jobs' NextStep, an operating system intended to compete with Microsoft [the NextStep workstation suffered poor sales]. It also supports two of Microsoft's biggest rivals in the Internet software business – Netscape and Sun Microsystems. Recently, the chipmaker has also been working with the makers of BeOS, an operating system that offers much more advanced graphics capabilities than Microsoft. And in October 1998, Intel invested in Red Hat, a company that distributes Linux software, which competes with Microsoft's Windows NT operating system.

Further driving Wintel's demise is the fact that both companies presently face government inquiries about their business practices. The Microsoft antitrust trial gets almost as much attention as the impeachment hearings these days, and the FTC filed an antitrust suit against Intel in June 1998. The latter is accused of withholding vital information about its chips from Compaq, Digital Equipment Corp., and Intergraph Corp. in an attempt to get them to turn over rights to patented technologies. The computer makers needed the information to develop their own products, which are based on Intel chips. Intel is also charged with stifling innovation in the microprocessor industry in order to enhance its dominant position. Insiders from both Microsoft and Intel have testified against the other company during the trials. During its antitrust trial, Microsoft turned on Intel, affirming that the latter did exploit its dominance in the chip market by leveraging information about its chips to force computer makers to give up patent rights.

Despite all the bickering and stress, the fact is Gates and Grove communicate regularly – quite often to discuss conflicts of interest – and they frequently invest in the same start-ups. In the works these days is the oft-delayed Windows 2000, intended to run on Intel's powerful Merced chip.

Software

If hardware represents the brawn of the computer, software is the brains. Software programs tell computers what to do, whether that be to create a spreadsheet or simulate the effect of car crashes to test automotive safety. Every computer needs an operating system – an overall control program that schedules tasks, manages storage, and handles communication with peripherals like modem and printers. The operating system presents a basic user interface, and all applications must communicate with the operating system. In 1997, companies spent an estimated $191 billion on custom software and support services. Businesses and consumers spend an additional $107 billion on packaged software. That doesn't include all the "free" software out there – either legitimately downloaded from the Internet, pirated or otherwise copied. Still, software is a lucrative business, enjoying fat profit margins. The cost of making one copy of a software program is as little as 10 percent of the retail price.

Many speculate that software has become the true engine of growth and innovation, now that hardware and chips are becoming commodity products. The classic example is Netscape, the scrappy web browser company that grew from a value of close to nothing (in 1994) to $6 billion after its IPO. In November 1998, AOL made what is by now old hat in the high-tech industry, entering the software industry by acquiring a software maker – in this case, Netscape. Other companies that have entered or expanded their interest in the software industry through acquisition include IBM and Sun Microsystems.

In the software business, the crucial feature is getting "mindshare," making sure that its software products are widely used. Economists have proposed a theory of "path dependence." Once a software product is installed, its user will tend to buy upgrades and compatible products from the same company. The most important thing is not necessarily to be best – just to be first and most prominent.

Right now, Microsoft holds a commanding share of the software market. Its operating systems are at work in over 80 percent of all PCs. Some believers think Java will challenge its industry dominance. Created by Sun Microsystems, the Java programming language can function with any operating system (or "platform"). Java breaks programs up into small "applets" that allow for swift Internet transmission. This enables centralized servers to do much of the work currently handled by installed software (a.k.a. Microsoft's bread and butter). Another challenge to Microsoft is Linux, a free operating system with open source coding similar to UNIX.

Leading Software Companies

- Computer Associates
- Lotus (unit of IBM)
- Microsoft
- Oracle

Vault Reports Profile

Nathan Myhrvold, CTO: Microsoft

Microsoft Chief Technology Officer Nathan Myhrvold is the creator of the famous "Interstellar Propeller Flying Beany Cap," hence the title "chief propeller head." He's Bill Gates' right hand man, and the co-author of the self-proclaimed visionary's bestseller *The Road Ahead*. He stands at the helm of Microsoft's advanced technology and research group, and has been a major force in the development of some of the world's most important products – including the ubiquitous Windows.

Bill's right hand

One of Myhrvold's biggest responsibilities is the care and feeding of the Microsoft Research center, a division he created in 1991. It is Myhrvold's dream to make Microsoft Research into an R&D 'institution,' along the lines of Bell Laboratories and Xerox's PARC. Since the center's founding, he has managed to attract 245 of the world's most innovative scientists – lured away from ivory towers (Carnegie Mellon) and competitors (Apple Computer and Silicon Graphics) with promises of intellectual freedom and irresistible compensation packages. It is speculated that the group boasts a combined IQ of about 40,000. Myhrvold's collection of brainiacs at "Bill Labs" focuses exclusively on software. They work on projects that are sometimes straightforward (simple improvements on existing tools), sometimes funky (a real-time chat program through which participants' animated aliases actually change their expressions according to what they type), and quite often seemingly impossible (a super-smart NT operating system that is as easy to interact with as another person).

Myhrvold's life story reads like that of a typical high-tech genius: After graduating from high school at 14, he went off to UCLA, where he earned a BA and a Master's degree in geophysics.

He received his PhD in theoretical and mathematical physics, plus master's degrees in space physics and mathematical economics from Princeton at the age of 23. Still feeling uninspired, Myhrvold began a fellowship at Cambridge, where he studied under renowned theoretical physicist Stephen Hawking. He then founded his own software company, called Dynamical Systems, which was acquired by Microsoft in 1986. He was brought on board at Microsoft as director of special projects, moved up to Group VP of Applications and Content, and then to Senior VP of the Advanced Technology Division. Microsoft Research was his brainchild, and continues to be one of his most important responsibilities.

Myhrvold thrives on research, though he'll admit that his area of interest is extremely specific. According to *Fortune* magazine, Myhrvold's interest lies in the understanding of "that portion of time when the universe was about $^{10}/33$ of a centimeter in diameter, up to when it was about the size of a grapefruit. After that," he claims, "it's all sort of history as far as I'm concerned." Myhrvold's recent activities do suggest, however, that he is broadening his scope to include a slightly more recent time in history (specifically, the Jurassic period). In 1998, he co-authored a paper with Philip Currie, one of Canada's most famous paleontologists. The two used computerized dinosaur simulations to prove that the giant creatures were able to create sonic booms by flicking their tails. Titled "Cyberpaleontology-Supersonic Sauropods," the paper was a finalist for the ComputerWorld Smithsonian Innovation Awards, and is included in the Smithsonian Institution's 1998 Innovation collection. With Myhrvold leading the way, both the road ahead and the road travelled long ago, it seems, are more interesting places.

Internet/New Media

Who would have thought that the Internet would make a dime? Created by the United States military in the 1960s, the Internet (then called Arpanet) was intended to be a communications system that would survive a nuclear attack. Today, it functions as a gateway for electronic mail between various networks and online services, and it is spawning new avenues for big business. Entrepreneurs and established businesses alike take advantage of the web's unlimited real estate, and almost instantaneous transmission.

E-commerce and online advertising are the biggest moneymakers on the Web, though both niches are constantly evolving. When e-commerce first arrived on the scene, people thought that e-malls were the way to go. But that soon faded. By 1997 the online retail market was saturated. Now it looks like the big winners will be the ones that either fill a niche or have serious marketing muscle: business-to-business commerce sites, extensions of existing businesses like Dell Computer, Ticketmaster or 1-800-Flowers, Internet pioneers like Amazon.com, and travel services like Expedia (owned by Microsoft). Consumer online services now represent a $3 billion business, one that is expected to grow to $15 billion by the year 2002. Internet commerce, the selling of products via the Internet, and Internet advertising, the plastering of advertisements and other enticements all over the Internet and Web, are also expected to boom in the next few years.

The Internet also helps companies do better business. Corporations can now expand their intranets into extranets that are accessible to business partners. By using open software standards, corporate customers can access business information and collaborate on projects. For example, Robert Mondavi Corp., a U.S. winemaker, buys satellite images from NASA to check for problems in its vineyards, and makes those images available on an Extranet so that independent growers that supply the company can troubleshoot weather in their own areas. And since 1995, General Electric has operated its Trading Process Network, an electronic marketplace where large buyers meet up with 2000 suppliers. GE used the service to purchase more than $1 billion worth of goods and services in 1997.

NET CONTENT

When the Internet first hit it big in 1996, it was commonly predicted that content would be the big attraction – the "killer ap," as it were, of the Internet. Microsoft invested in *Slate*, a high-flying online magazine with high-powered talent, such as Michael Kinsley, late of prestigious magazine *The New Republic*. The problem? No one could figure out how to make money on content – since there was so much free content already available on the Web. In 1998, however, the increased prominence of "community" web sites and increased traffic on the Internet may make targeted content sites, such as iVillage, a portal and community targeted at women, increasingly viable.

MAJOR INTERNET COMPANIES

- Amazon.com
- America Online
- Excite
- Infoseek
- Yahoo!

SILICON ALLEY

Of all the interactive ghettoes in the world, New York's Silicon Alley is undoubtedly the coolest. It's the artsy, "scene-y" version of the high-tech hotbed. And we're not just saying that because Vault Reports is based there.

Alley startups, launched and staffed by creative, twentysomethings, account for the most robust small-business growth in New York these days. They started springing up in downtown Manhattan in 1994, and focused on software, the Internet, web design, interactive marketing and electronic commerce. New York new media companies benefit from being at the hub of the publishing, advertising, journalism and financial industries – in fact many of their founders are formerly from these industries. (There are hardware and software companies in NY, but they are decidely in the minority. Industry insiders say there's little chance New York will become a hotbed of hardware tech like Silicon Valley.)

Silicon Alley begins in the Wall Street area, and depending on who you ask, stretches anywhere from the Flatiron building in the Chelsea district to 42nd Street and beyond. The New York Techno Hub lures startups and established companies with a variety of commercial and real estate tax incentives. New media businesses have also emerged in other parts of the state, as well as in Connecticut, and New Jersey. Many of the area's hottest companies got started at 55 Broad Street in Manhattan, which houses more than 70 technology companies. The building was the home of the Drexel Burnham Lambert brokerage firm until the stock market crash of 1987. Now called the New York Information Technology Center, 55 Broad is one of the few 'fully wired' buildings in the world – with satellite accessibility, video conferencing facilities, fiber optics, high-speed copper wire transmission and high bandwidth Internet access.

New York's high-tech workers are a more eclectic breed than the traditional monitor-tan sportin' West Coast variety. Not surprising, since they're more likely to be using the latest computers and software than making them. The Alley's artists and writers fill the all-important need for quality content on the Internet. They design corporate web sites, write for myriad online publications, and develop advertising and e-commerce strategies for retailers.

According to *Crain's New York Business*, however, after burning through a combined $1 billion, Silicon Alley startups have yet to generate any profits. More than 80 percent of the city's new media businesses have annual revenues of less than $1 million. The rate of growth in the industry has made the industry pretty volatile. More than a sixth of high-

Silicon Alley, cont'd

tech startups launched in New York over the past 18 months have already folded, and turnover among the young people who populate these companies is 70 percent over the first two years. However, there are still investors who see new media as a lucrative industry that will power New York in the future. Building an industry takes time, and according to a recent study by the New York New Media Association, Silicon Alley is well on its way. Combined, Silicon Alley companies generated $5.7 billion in revenues in 1997, and the industry has seen an increase in employment of close to 50 percent in the past two years.

Will it last? Only time will tell, but most think new media is here to stay. One study reveals that investment in New York new media by venture capital firms jumped from $167 million in 1996 to $240 million in 1997. In 1998, a survey by Price Waterhouse showed New York-area venture capital increased to $319 million. Among the lucky beneficiaries were online ad maven DoubleClick ($40 million), community site WebGenesis ($20 million), and e-commerce company InterWorld ($15). Every major investment bank and accounting firm is courting new media clients. Finally, City Hall, in collaboration with some private utility companies, has helped the industry by creating a venture capital fund aimed at investing in local high-tech businesses.

Hot Alley Companies

- DoubleClick
- Earthweb
- iVillage
- Jupiter Communications
- N2K
- NetGrocer
- TheGlobe.com

Vault Reports Profile

Jerry Yang, Co-Founder: Yahoo!

How much am I worth today?

If you search Yahoo! for Jerry Yang, you'll only come up with five hits, one of which is his home page – last updated in 1995 (though it does have a link to Yahoo). He's been kind of busy since then, working to build a multi-billion dollar business from what started out as a simple list of his favorite Internet sites. In 1997 he was already ranked No. 32 on *Forbes* technology's richest list.

In 1994, Yang was studying for his PhD in computer engineering at Stanford. In typical student fashion, Yang spent a great deal of time surfing the Net. Back then, the World Wide Web was a tangled mess, so Yang and his buddy, (fellow student) David Filo, took it upon themselves to bring some order to chaos. They created a list of links to their favorite sites, dubbed "Dave and Jerry's Guide to the World Wide Web." The pair e-mailed the URL to a few friends, and everything snowballed from there. People sent submissions for sites to add to the list, and pretty soon, Yang and Filo were spending 40 hours a week reviewing sites and adding links. The number of visitors to their site increased at an exponential rate. Soon, their PhD advisor advised them to move their little hobby off campus – it was clogging up the university's computer system. By February of 1995, after fielding offers from a number of venture capitalists, they finally accepted a $1 million offer from Sequoia Capital. Shortly thereafter, they abandoned their academic pursuits to concentrate on the business. The company went public in April 1996, and as of late 1998 was valued at more than $20 billion.

Yang and Filo hired Tim Koogle, another Stanford engineer, to be the company's CEO. Filo took the title of Chief Technical Officer, while Yang has focused on business strategy and public relations. As "Chief Yahoo," Jerry Yang is truly the public persona of the company.

One might say he's the ultimate procrastinator. He not only found a way to make a living fooling around the web, he now employs people to do it all day. At Yahoo headquarters, which looks more like a playground than the office of one of new media's most important businesses, he personally manages a staff of more than 100 "surfers" who review sites from all over the world and manually index each site under a complex framework categories and sub-categories. (Other search engines, like Lycos and Excite, use computer programs and algorithims to index every word on each web page.)

Yang says the company is modeled the company after Einstein's dictum to "make everything as simple as you can, but not simpler than that." And this no-frills approach to searching the Net seems to have a lot of appeal. Yahoo is the most popular search tool on the Web, with an index of about one million sites. Between 15 and 25 million visitors visit Yahoo each month.

Networking

How do computers zap information around your office and the world? Network systems. The firms that connect computers to each others make routers, which serve as central storage areas that accept different "packets" of information, and then send them to their correct destinations; switches, which set up high-capacity data channels between two points on a network (a switch is like an interstate highway; a router is like a cloverleaf); and for those at home, modems, which let computers talk via phone line, and hubs, connection points on a network that let several PCs talk to each other (sort of like a parking lot). Networking products can be LAN (local area networks, normally for computers within a discrete area, like a building, that are connected by cable) and WAN (wide area networks, connected via phone lines and satellite dishes). The modem modulates the digital data of computers into analog signals to send over the telephone lines, then demodulates back into digital signals to be read by the computer on the other end; thus the name "modem" – modulator/demodulator.

The next big networking thing is the cable modem, which carries information over cable television lines at a rate eight to 10 times faster than ISDN. Internet service provider (ISP), @Home, offers its services exclusively via cable modem and avoids bottlenecks on the web by maintaining a "parallel Internet," called @Network, that duplicates popular web sites.

Major Networking Companies

- 3Com
- Bay Networks
- Cisco Systems
- Lucent Technologies

On the Job

As the high-tech industry booms, jobs proliferate. The U.S. Commerce Department has conducted studies that reveal that between 1996 and 2006, over 1.3 million new systems analysts, engineers, programmers and computer scientists will be required by industry in the United States. That doesn't begin to cover the legions of content providers, marketers and new media industry analysts the booming high-tech industry will require. Here are some top jobs in high tech.

PROGRAMMER

Programmers write software "code" – instructions in one of many computer languages. Other programmers are debuggers, who check for tiny errors. A single misplaced comma in one line of code can sink an entire program. Debuggers often try to make programs fail in order to detect the source of their failings. Some programmers design programs to specifications, creating new features and updating old ones, but programming jobs are not always intellectually stimulating. Many openings are for "code machines" – people who write endless code to exacting specifications, as opposed to conceiving and designing products.

Programmers may advance to being project managers. Project managers write technical specifications for new products, then follow their concepts through design and testing to ensure all goes smoothly. Project managers work daily with technical and marketing teams and graphic artists to create fully integrated products ready for the market. In general, a computer science degree is necessary for programming positions, though self-taught geniuses are always welcome. (Some high-tech firms now recruit at high schools!) Programmers must keep current in their languages, and learn new ones.

Engineers

Hardware engineers assist in the design, development and ongoing support of new PC boards and systems products. Systems engineers also handle integration and compatibility issues, as well as the design of plastic and sheet metal components. Much time is spent in the lab ensuring that these components remain bug-free and compatible with other computer ingredients. Hardware engineers may specialize in circuit design, systems integration, firmware development, and peripheral design and evaluation.

Marketing

In the technology industry, it's not necessarily the best product that wins out, but the one most in use. (It has "mindshare.") And in such a quickly changing industry, firms must constantly search for new applications for their products, and create new ones that need to be sold. That means that the duties of marketers are more essential than ever. At some companies, marketers in the field (often called product managers) identify product openings and work with engineers in research and development to produce what consumers say they want. And marketing isn't directed just at the faithful anymore – that is, the high-tech savvy audience. Witness Intel, which was one of the first hardware computer firms to "brand" its image in the mass marketplace. You don't necessarily need a technical background to work as a marketer for a technology firm, but it's useful. An MBA is helpful for most marketing positions.

At high-tech firms, promoting a product is largely a public relations job. As one high-tech product manager puts it: "People don't write articles on Tide anymore, but there's whole industry of magazines dedicated to writing about software." That insider estimates that public relations can sometimes comprise half of a product manager's job.

THE PRODUCT CYCLE MODEL

While programmers and engineers may put together software products, it takes a skilled marketer to get those coded brainchildren on the market. Software projects are led by program managers – what in other industries might be called product managers. These managers are "idea people" – they are usually not programmers, but they have a good understanding of technology and what is feasible for a product. Program (or product) managers oversee the group that plans and develops new products.

The process by which new software is made is called the "product cycle." There are a variety of people involved, including program managers, developers, testers and designers and programmers. The first step is to develop a "business case." The development team comes up with a concept, determines how much it would cost to create and manufacture the software, and calculates potential revenues.

The next step in the process is to "size the opportunity." After the development team figures out the possibilities for the product, it writes a "vision document", which details the proposed project. The vision document identifies what's going on in the software market, the current competitors, core customer needs and the "preliminary specs" – the basic features of the new product. For example, take a hypothetical application, a product called "Start the Day" that helps users organize their lives. Start the Day wakes you up in the morning, prints out your calendar for the day, and uses push technology to give you the weather forecast, top news stories, and dinner recipes – complete with grocery lists, so you can shop on the way home from the office. Developers will prove the need for the product and explain how they're going to win over the market. Product managers then figure out a schedule for development and an overall budget.

The product manager presents the document to the product group management team in a "kick-off" meeting that details the schedule, budget, and overall product strategy. If Start the Day is approved at this point, managers continue with the development process, meeting often to mull over details.

After a period of time (typically three to six months) group managers meet again to explore more in-depth specifications – how Start the Day does what it does (how, for example, it receives news stories), and what the user interface is like (maybe it looks like a newspaper). Then programmers begin writing the code for the application. On average, program managers work 65 hours per week, and that can easily rise to 80 or more.

Product Cycle Model, cont'd

As the end of the development process nears, everyone puts in excruciatingly long hours. High-tech vets say there is never enough time to get everything into the product.

Once the programmers finish their work (or the deadline catches up with them), the alpha phase commences. In this phase, the testers come in, test the code, and see if the product matches up with the specs. A database is used to track all the "bugs" found – basically mistakes where the action of the software doesn't match what the spec calls for.

When the programmers are completely finished, comes the stage dubbed "Code Complete." Start the Day is submitted to a testing team, which then "starts pounding on your product – looking for bugs," says a source from the software industry. After that is "beta testing" – when an external group of users tests the product for more bugs. The development team fixes any problems, and sends it back until all of the kinks are knocked out. When this process is finally over, it's RTM time – Release to Manufacturing. The developers hand over the product for reproduction and release to the market. Break out the champagne!

Systems Consultant

Computer systems aren't necessarily intuitive. Many technology firms require systems consultants who determine client needs and expectations and write appropriate programs and networks to meet them. For example, a database consultant might consolidate several billing screens into one, thus preserving the original database while increasing the efficiency of firm operations. Technical consultants often employ cutting-edge technologies like data mining, multimedia, computer telephony and electronic commerce.

INTERVIEW PROCESS

Interviews at high-tech firms tend to come in clusters. While standards vary, at many you may have up to six or seven interviews in the same day. Stay upbeat, try to relax, allow yourself to be creative and never assume that your interviewers haven't discussed you between interviews. Note the company's dress code before you interview by reading the company profiles in this Industry Guide, so you don't end up wearing a suit to your Microsoft interview or sandals to Intel. And be prepared to exhibit the skills you say you have – if you advertise a knowledge of C++, be ready to write code on the spot.

Technical questions

Programmers and engineers should already have a sense of the sorts of technical questions they will face in high-tech interviews. Programmers, for example, will often be asked coding questions, while engineers may be shown diagrams or asked to sketch out schematics.

EXAMPLES OF TECHNICAL QUESTIONS

1. **Write a sorted linked list insertion routine.**

2. **Write a function that takes as its arguments: 3 integers > 0, if the product of the integers is even, return the one with the lowest value.**

3. **Using two cubes, make a desk calendar that can cover all days of the month (1-31).**

   Engineers better have a good spatial sense. The trick to this answer is to use 6 and 9 interchangeably.

4. **What is a cache? What is the difference between a write back and write through cache? What is the difference between a direct mapped and a fully associative cache? What are the pros and cons of each?**

Brainteasers

Companies in the technology sector are famous for their brainteasers. Anyone, after all, can come up with a canned answer to display their leadership and management skills – but fewer people can quickly come up with three solid reasons that a manhole cover is round. Whether you're applying for a technical, corporate finance or marketing position, expect to get a few of these beauties. Creativity and mental flexibility and speed are of paramount importance to high-tech firms, and one surefire way to test these qualities are through these slightly offbeat questions.

If you field one of these brainteasers, your interviewer may give you a time limit. Don't become flustered. Simply try to think through the question from every angle you can. Answering brainteasers requires either logic, the ever-popular "out-of-the-box" thinking, or both.

Sample brainteaser questions

1. **A company has ten machines that produce gold coins. One of the machines is producing coins that are a gram light. How do you tell which machine is making the defective coins with only one weighing?**

   Think this through – clearly, every machine will have to produce a sample coin or coins, and you must weigh all these coins together. How can you somehow indicate which coins came from which machine? The best way to do it is to have every machine crank out its number in coins, so that Machine 1 will make one coin, Machine 2 will make two coins, and so on. Take all the coins, weigh them together, and consider their weight against the total theoretical weight. If you're four grams short, for example, you'll know that Machine 4 is defective.

2. **Design the ideal alarm clock.**

   Let yourself go! And try to relate your answer to the position you're applying for. If you're up for a technical or engineering slot, talk about how you would design or program the clock. If you're in a marketing interview, talk about how you'd market its features. And anyone in finance should try to figure out what it might cost to produce this idealized clock. (Also, don't be afraid to be creative, or even a little bit silly – at the same time, don't imply that you have trouble getting to work in the morning!)

3. **Bono, the Edge, Larry Mullen, Jr. and Adam Clayton need to get across a narrow bridge to a show. They have only one flashlight, and only two people can go across at once. Adam takes only a minute to get across, Larry takes two minutes, the Edge takes five minutes, and slowpoke Bono takes 10 minutes. A pair can only go as fast as the slowest member. They have 17 minutes to get across. How should they do it?**

   The key to attacking this question is to understand that Bono and the Edge are major liabilities that must be cross together.

   Bono and the Edge must go across together, but they can not be the first pair (or one of them will have to transport the flashlight back), and they can not be the last pair (which would mean that one of them would have make three trips). Instead, send Larry and Adam over first, taking two minutes. Adam comes back, taking another minute, for a total of three minutes. Bono and the Edge then go over, taking ten minutes, and bringing the total to 13. Larry comes back, taking another two minutes, for a total of 15. Adam and Larry go back over, bringing us to 17 minutes. The show must go on!

4. **You're playing three card monte. Two cards are red, one is black. You try to pick the black card. After you pick, the dealer shows that one of the cards you have not chosen is red. You are given the chance to switch your selection. Should you?**

   The short answer is yes. A longer answer is yes, because you will win twice as often. An even longer answer is that by switching, you are betting that the card you initially chose was red. By not switching, you are betting that the card you initially chose was black. And because two out of three cards are red, of course, betting on red is the way to go.

5. **You have 12 balls. All of them are the same except for one. That one differs from the others in that it is either hollow, and weighs less, or it is solid, and weighs more (you don't know if the 11 normal balls are hollow or solid). You have a two-sided scale that can tell you only whether the items on the left side weigh more or less than the items**

on the right side. In three weighings, determine which ball is odd, and whether it is solid or hollow.

This is a bitch of a brainteaser that actually has several answers. Here is one that we think is most elegant.

You split the 12 balls into groups of four. Take two of the groups and weigh them against each other.

If the groups of four balance, you know that the odd ball is in the unweighed group of four. Take three of your unweighed balls, and substitute them into one of the groups of four on the scale.

If they balance again, you know that the unweighed ball is the odd one. Weigh it against any other ball and find out whether it is odd because it is hollow or because it is solid.

If, when switching the group of three, you tip the scales so that the group with the replaced balls are heavier, you know that the odd ball is one of the three, and is solid. If the scales tip the other way, you know the odd ball is light.

Take two of the group of three and weigh them against each other. If the scale is imbalanced, you take the one that matches your knowledge of whether the odd ball is hollow or solid, gained in the second weighing.

If it is balanced, the unweighed ball of the three is the odd ball, and is hollow or solid depending on the information you gained in the second weighing.

If the initial groups of four do not balance, you know that the odd ball is one of the eight weighed balls, but you do not know whether the ball is solid or hollow. Take three balls that you know to be normal (the unweighed balls) and use them to replace balls on the scale (let's say the right side). Take those balls that are being replaced, and move them to the other side (let's say left) of the scale, removing three balls from that side.

If the scale balances, you know that the one of the three balls removed from the left side is the odd one. And, depending on whether that side was heavier or lighter in the first weighing, you know whether the odd ball is hollow or solid.

Take two of the balls from the group of three and weigh them against each other.

If the scale stays the same, you know that one of two unmoved balls on the scale is the odd one (either the ball on the right side that was not one of the group of three moved to the left, or the ball on the left that was not one of the group of three removed from the scale altogether). Take either of the balls and weigh it against a ball you know to be normal. If they balance, than the unweighed ball is the odd one, and, knowing how the scale looked in the second weighing, either hollow or solid. If they do not balance, the

If the scale switches, you know that the group of three moved from one side to the other (in our case, from the right to the left) contains the odd ball. You also know whether the ball is solid or hollow, depending on which way the scale tips. Take two of the group of three and weight them against each other.

6. **You have one gallon of water and one gallon of wine. Take one cup from the water and put it in the wine, then take one cup from the wine-water mixture and put it in the water. Is there more water in the wine or wine in the water?**

   There is the same amount of water in the wine as wine in the water. You work out all the fractions.

Finance Questions

Working in finance at a high-tech firm involves some unique challenges. The industry is much more rolatile than other industries – the group an MBA is assigned to may not exist in its present form in a year. Finance types at high-tech firms must be able to handle questions like these:

1. **If an engineer were to present you with an innovative product that you cannot compare to previous products (because it is so innovative) how would you evaluate its potential?**

   If you have any technical knowledge, flaunt it here. As for the financial side of things, say that you would calculate the NPV (net present value) of the product. You would calculate the NPV by discounting the product cash flows from the project using an appropriate "risk-adjusted discount rate."

2. **Let's say the company makes a product (for example, a motherboard) on which it has negative margins. Would you discontinue it? Why or why not?**

   Tricky! Many companies make several products that are not profitable, but lead to the purchase of products which do make hefty sums for the company. (An example is Intel's motherboards, which Intel sells at or below cost, in order to encourage the sale of its compatible microprocessors.)

Marketing Questions

Make sure you have a sense of the industry in which you propose to work. While you may not need to be able to write fluid code or engineer a transistor in order to market high-tech products, most companies will want to ensure you at least know what your prospective employer does.

1. **What will be Microsoft's (or Intel's, or Cisco Systems') next major strategic move?**

2. **What is your hobby? What kind of product that does not exist today do you think would be most useful in your hobby?**

   Creativity and enthusiasm are highly prized at technology firms. Your ability to conceive of new products will be a bonus in an industry that is always producing new things.

3. **You are Amazon.com. You've seen tremendous growth in core business, and market value continues to rise, but you're facing heavy competition from Barnes and Noble and other superstores now moving into online selling. What are you going to do to continue to grow the business?**

   An example of a marketing case question. You should think aloud when answering these questions, allowing the interviewer to be suitably impressed with your analytical thinking process.

Our Survey Says

High tech, though its industry segments range from silicon wafer manufacturers to new media startups, is surprisingly uniform in many ways. We asked hundreds of high-tech insiders to tell us about their workplace pay, culture, dress and much more.

Culture: Teamwork, meritocracy and monitor tans

Teamwork is a hallmark of most high-tech firms, as products are normally developed together. One high-tech employee tells Vault Reports: "The atmosphere is very social – a good-sized chunk of my day is spent in other people's offices or with other people in my office discussing how to accomplish some task or the latest industry news." At another technology firm, an insider says: "We work together in small feature teams, so even though this is a large company, you always feel like an important part of a team."

Another buzzword at most high-tech companies is meritocracy. Companies "expect you to speak up and express your viewpoint," say insiders. "If you can get your point heard, you can make it happen." "There is no problem with stagnant old boy networks," adds one insider. "Those who do not produce do not stick around. Those who are valuable are rewarded."

When people think of high-tech employees, many instinctively summon up the image of a unwashed, bespectacled, skinny guy working around the clock at a computer, his acne-ridden face awash in the pale glow of the screen. Well, we'd hesitate to say that the undernourished, sleeps-under-his-desk programmer doesn't exist, but he's hardly the standard for the industry.

Yes, employees do joke about their "monitor tans" – that is, a ghostly pallor. Yes, there is a relative paucity of women – but this varies from company to company. At some companies, women may compose an appropriate 50 percent of all employees – and not just in such traditional estrogen ghettos such as human resources and secretarial services. At others, say insiders, the introduction of any female is an invitation for "lots of desperate moves by deprived geeks."

Staying up until all hours – that is true. "We tend to keep really weird hours," admitted one programmer. "For example, I am an insomniac, and I often come to work at around 2 or 3 in the morning, then work straight for 10 or 12 hours. I'm never alone when I come in, either."

High-tech firms are known for their collegiate, casual atmosphere. "The culture here is focused around doing great work, but it is also a very fun place to be," says an insider. Many firms also feature entertainment for overstressed employees. "We have a basketball hoop in our hall, which is a great way to blow off steam in the evenings or the middle of the day," says one insider. "Those technical people – I've heard stories about them," says one financial analyst. "I've heard they have espresso machines in their offices and futons on the floor in case they have to program late into the night. I see them playing frisbee in the halls and most wear jeans all day long."

But not all high-tech firms intersperse childlike cavorting with intense stretches of work. Many insiders say their workplaces are "a little more uptight." And don't expect a lot of hand-holding, either. Most high-tech firms have " no executive training programs where you are eased into the culture," says one insider, adding: "I love it, because you have the freedom to take risks and make things happen, but many people find it a difficult environment."

Pay: Glorious options, average pay

You may have heard tantalizing tales of the profit to be made at high-tech firms, and indeed, the computer industry has produced many a millionaire (and several billionaires). But know that most employees who do get rich aren't collecting million-dollar salaries. Instead, they're cashing in on stock options, a crucial benefit for most high-tech employees. With stock options, employees have a direct stake in the fate of their firms. One top high-tech firm "only pays the industry average salary. That's been a source of gripes for me and my colleagues – but over time, the complaints taper off as we watch the stock climb." Flush firms may also distribute bonuses to their hardworking masses. One firm gave out four bonuses in one year: "a $1000 thank-you bonus, one for 15 days of pay, one for 17 days of pay, the last for five percent of yearly salary." At many high-tech firms, "top-level engineers pull down over $100,000 a year plus two profit-sharing checks and other benefits." Some hot-shot startups "do pay more, but you trade that extra money for security." At other firms, compensation is "based on relative performance, experience, and how well you can negotiate with HR."

Perks: Cash and cafeterias

Perks, perks, perks. If you want them, you can find them in high tech. Generous stock options, holding forth the possibility of riches, are perhaps the greatest plus on offer at high-tech companies, but they are by no means the only perks available. Many high-tech firms use their local muscle to get discounts for their employees at local restaurants, gyms, golf courses and apartment complexes. One high-tech employee raves about his company's cafeteria, that has even "garnered some local press for its tasty fare. Each cafeteria has a theme, and if you're in the mood for Chinese, or Indian, or Californian, you always know where to go." Many employees also enjoy the unlimited liquids, including "free sodas, and juices too, all you can drink, right here!" Other perks – "flexible scheduling," "bringing my dog to the office," and "cross country skiing trips." An insider summed up the full benefit of his perks: "As for health and retirement benefits, there's the full medical coverage (free), dental plan (free), vision care (free), life insurance (free), the 401(k)." One employee thinks "the greatest, absolute best perk I get is stock options and the ESOP [employee stock option plan]. That means cash, and sending my kids to college."

Hours: "We work like maniacs"

But if you want all those juicy stock options, be prepared to work for them. "My firm works its employees HARD," one insider tells Vault Reports. "That's why they have a sabbatical, to give people a breather." At many firms, "you can just come in whenever it fits your schedule," but "you had better get your work done, and if that means working all night, you work all night." Many insiders say that hours depend on approaching deadlines. An employee reports that "about three times a year, I'll work 80- to 90-hour weeks for a month. Two years ago, I frequently worked 100 hours or more a week to get a product to ship, though I've lost some steam since then." But job satisfaction can actually make long work hours appealing, say many insiders. "Because they let us have so much fun, we work like maniacs," one employee says.

Some larger, more established high-tech firms have more traditional hours. A contact at one big firm reports: "Hours start early here. I worked somewhere else in Silicon Valley and used to get in at 10:30 or so. Here, you have to be in by 8. It's no longer compulsory, but everyone still does it so the peer pressure gets you." "Unofficially, 10- to 12-hours days are the norm," says another engineer. Don't try to slack off, either – "Deadlines are set in stone," say many insiders. "There are no extensions." In general, we hear that "work hours really depend on the type of job you are doing and the department you work for." Financial analysts and marketers "may work about 60 or 70 hours a week," but "they usually don't sleep on the futon in their office" like programmers often do.

Dress: What dress code?

Dress code? Ha! Most high-tech firms are known for their "non-existent" dress codes. "If you're a hot-shot programmer, no one's going to stop and say, 'But he's wearing sandals!' Get real," says one insider. Many executives at technology firms flaunt their lack of dress code by coming to work in shorts and sandals. In fact, some companies insist that it doesn't even matter what you wear to your job interview. Still, that doesn't mean that everyone working in for a high-tech firm can put their pinstripes through the shredder. "We wear whatever we want to work (shorts, jeans, T-shirts are most common), but this varies according to what job you do. Marketing people and lawyers dress up more than software developers," says one insider.

At another firm, says an insider: "Dress code here is casual. During the summer, when I don't have to make presentations, I wear shorts." A few firms ostensibly do have dress codes, though in practice they are fairly loose. "There are a few rules here, but nothing serious," say one insider. "In theory, skirts and shorts are supposed to be within two or three inches of the knees, but I have seen some pretty damn short skirts," says another contact at one big high-tech firm. Another insider comments: "I wear shorts almost all summer. But if you work in a lab, open-toed shoes are pretty much out, and in the superclean labs, you can't wear any makeup or hair spray. Other than that almost anything is OK."

Those T-shirt-clad programmers aside, at many firms the managerial staff "always has to dress professionally if we're seeing customers," but even this dress is usually "only at the khaki level

– every day is 'business casual Friday' for us." Those employees not "in the public eye" regularly "wear jeans with a blazer or sweater. It's a much more relaxed corporate environment in Silicon Valley than elsewhere." Some top executives are even seen "walking around barefoot on nice days."

Diversity: Everyone welcome, but who gets the corner offices?

High-tech firms will take talent wherever they can find it. Most high-tech firms have an international focus, recruiting employees "from all corners of the globe." The result: firms with minority employees, "on every level" from the board of directors downward. One insider says high-tech employees should "enjoy technology, be smart, willing to work long hours, and be aggressive." Beside that, it's an incredibly diverse group. "Maybe I'm naive," says one techie, "but in my opinion no one cares if you're black, white, male, female, anything." Many firms provide funding for administrative costs and space for meetings and displays for employee organizations, such as African-Americans, Asians, Native Americans and Latino constituency groups, technical women, gay and lesbian organizations, and Christian and Muslim organizations. One woman insider tells us : "I was actually surprised at first how many women there are in my company, at least 30 to 40 percent, which is rare in Silicon Valley. Not as many in hardware design, but the numbers are pretty fair in software." Others cite an unusual dearth of female employees at their companies. "There are many women in the administrative part of the organization, there are some (but less) in the artistic ranks, and almost none in the technical ranks," says one woman insider. "It's not unusual for there to be fewer women than men in technical environments, but my firm is really lopsided," adds another female insider. "I do wonder sometimes about the glass ceiling in Silicon Valley," says a woman at another firm, "since there still aren't very many senior executives here who are women."

Social Life: Happyhours.com

Most high-tech firms encourage a large amount of social interaction, to bond employees to their co-workers and the company. Many tech companies sponsor employee sports clubs, or at least "set up basketball courts so you can play." Team sports are popular in the high-tech world – many insiders say their companies sponsor "volleyball, basketball and softball games." "From music groups to jugglers to various team sports, there's always something going on in the buildings or around the campus," says one insider. "In many ways, it's much like being in college." Because high-tech firms typically hire "a lot of young people," working there "is almost like being at school, making friends your own age." Although there are "lots of group activities," the number of people who participate is "actually pretty small." Also not dissimilar to college life.

High-tech companies are willing to spend a little of their cash on festivities as well. One company has an "annual company picnic" and "two annual beer blasts," while some smaller firms have "happy hours at five o'clock on Friday," "free food," and "free drinks." At another firm, employees appreciate the "lavish Christmas parties" held each year, as well as the social events held on a regular basis in each individual department. "These are less elaborate, but they can be even more fun." In general, it seems that at many high-tech firms "there's plenty of money around for company-sponsored good times."

Offices: Slides to cubes

Employee offices at high tech firms show great variability, from "my basement," to a "lush, campus-like quad, real offices – not cubicles, and a great cafeteria that has everything from junk food to gourmet specialties." However, it seems the majority of high-tech companies feature cubicles – "rows and rows of them. I call it the Cube Farm," says one employee. Don't get too used to your cubicle, either, since they are "constantly changing." As one insider reports: "My office is growing so fast that our cubicle size was reduced, from 9 x 9 to 6 x 9. This is called compression. It's a little cramped." Some companies use cubicles as a sort of shorthand for an egalitarian ethos. At some companies, everyone, including the CEO, sits in a 9 x 9 cubicle. Most technology firms, however, realize the benefit of keeping their employees well-fed (and from wasting valuable coding time driving out to the nearest Taco Bell). Most offices have "cheap cafeterias with excellent food." That doesn't mean there's a typical office layout at any firm – one exploding Internet company has "a big red twisty slide that connects the floors."

RELATIONSHIPS WITH SUPERIORS: OPEN DOORS AND FLAT HIERARCHIES

Many insiders praise their firms for "maintaining a small company feel, with open communication between management and employees" amidst rampant success. One contact says of her firm: "If you have a problem, you can go all the way up to the VP level to discuss your issues and they will actually be addressed!" "This is a very flat firm," say those at a smaller high-tech company. "We do not have a lot of bureaucracy and we work closely with everybody. Some companies will assign each new employee a mentor of senior rank, and many companies have an "open door" policy, which "loosens things up quite a bit." One engineer says: "I feel I can talk to my superiors freely." Other high-tech firms "don't even have offices, or doors, for that matter, just cubicles, cubicles everywhere, so that even the CEO just has a cubicle. This evens things out." This doesn't mean that every high-tech firm is a model of equanimity. One famed high-tech firm has a review process where "employees are ranked up against each other by their managers. This means your immediate supervisor can basically ruin your life," laments one insider. Some high-tech firms practice something called "constructive confrontation," where you can be direct with anyone without being offensive. "But you had better know your supervisor well before you try something like that!"

SELECTIVITY: "MILLIONS OF APPLICANTS"

One very well-known high-tech firm processes 12,000 resumes each month. This means, as one helpful insider points out, "with 100 percent turnover every other month, we would still have too many applicants. So we can afford to pick and choose." At other high-tech firms, employees have an equally high view of the selectivity of their firm. "There are literally millions of people who'd like to work here. Few of them get to," brags one of our contacts. Those who receive the honor of a position are "really smart, dynamic people." Most employees – though probably no slouches themselves – say they are "impressed with the quality of people" they work with. One well-known firm is said to be "extremely selective. They make sure they get the brightest engineers, the ones with the 4.0, the overachievers. It is, in fact, a bit of a shock when they arrive here and find everyone is as smart as they are," says one engineer. Selectivity, of course, may vary by position. "Most openings are for entry-level computer science majors, so any other job is quite competitive," a marketing manager at another firm tells Vault Reports.

PRESTIGE: HIGH-TECH REP SOARS

Everyone's heard about the high-tech boom, and most companies in the industry enjoy a halo of prestige, not just the biggest firms. "My company is very sexy," says one financial analyst. Another employee tells Vault Reports: "I am always getting hit up by people who want to work for my company, and I can't blame them." Many insiders think their employer is at "the top of our industry." "Essentially, we rock," says one enthusiastic insider. Working for a high-tech firm is considered "a bonus," especially "knowing that your products are on the cutting edge of technology."

ASCII: American Standards Commission for Information Interchange. It is mostly associated with a plain text file format, which lacks any special formatting, such as bold or italic text.

Application: Software that enables one to perform particular tasks. Examples include word processors, and database managers.

Bandwidth: The amount of data a network connection can transfer at any given time. Bandwidth is important because data is transmitted faster during high bandwidth connections than lower bandwidth connections.

BBS: Bulletin Board System. A BBS is a computer system with one or more modems used as a public or private source of information to people. A BBS can also be used as a message center.

Beta: A term applied to software still in a testing stage.

BIOS: Basic Input/Output System. The BIOS controls all input/output devices, such as a keyboard, mouse, monitor, and disk drives.

Bit: Binary digIT. A bit is the basic unit of storage in a computer, with possible values of zero and one. A bit has many uses, including complex applications when eight combine to become a byte.

Browser: A program that permits users to read and navigate between hypertext documents on the Web.

Bus: The computer main vehicle for data exchange. Information between the hard drive, expansion cards and CPU is transferred through the bus.

Byte: A composition of eight bits.

CISC: Complex Instruction Set Computer. CISC microprocessors process the majority of complex operations and data, as opposed to RISC chips (see below).

Clock speed: The speed of a microprocessor, in megahertz.

Cookie: Information sent from your computer to a World Wide Web server. A cookie can be used to maintain contact between your computer and a Web server. (So, for example, you don't need to continually type in a password for web site access.)

CPU: The central processing unit. The CPU controls computer operations, performs calculations and carries out instructions. Usually contained on a microprocessor.

Crash: A sudden failure in functionality of a program or hard drive.

Data: A general term for an item of information.

Digitize: To convert anything into something a computer can understand. Digitizing can take many forms from scanning a picture to typing a handwritten document.

Disk: A device used to store computer information. There are two types of disks: floppy disks and hard disks.

Download: The process of receiving a file from a remote computer to your local computer.

DRAM: Dynamic Random-Access Memory.

Driver: A software program that controls a hardware device. Often called a device driver.

DVD-ROM: Digital Video Disk (or Digital Versatile Disk) Read Only Memory. Hardware similar to (and compatible with) CD-ROM technology which offers more storage capacity than CD-ROM, and can play high-resolution, full-length movies on computers.

EDO: Extended Data Output. EDO is an attribute of RAM that allows it to transfer data faster than traditional DRAM.

Ethernet: A networking system used to connect PCs to other devices.

Expansion card: A circuit board that extends the computer's functionality when plugged into an expansion slot. Examples of expansion cards include sound controllers, internal modems, display controllers and network interface controllers.

Expansion slot: A socket that allows the computer to increase in functionality through connection with an expansion card.

File: A collection of computer information under a single name.

Firewall: A security program that prevents users on one network from accessing the Internet and/or other networks.

GPF: General Protection Fault. A Windows memory handling error that made programs crash. Older versions of Windows were more prone to GPFs, though current users might disagree.

GB: Gigabyte. A gigabyte is 1 billion bytes, or more technically, 1024 megabytes.

GUI: Graphical User Interface. An operating system or program that allows the user to navigate it using icons.

Graphics Accelerator: A graphics adapter used specifically to enhance display performance

Graphics Adapter: A video circuit board that produces graphics as well as numbers and text.

Hardware: The physical components of a computer system. Examples of hardware include a keyboard, mouse, motherboard, monitor, etc.

HTML: Hypertext Markup Language. HTML is a programming language used to write web pages, with text formatting and hyperlinks.

HTTP: Hypertext Transfer Protocol. HTTP is the language that is used to transmit hypertext files between web servers and clients.

Hypertext: The basic element of the World Wide Web. Hypertext is text that contains links to other documents on the Web

Internet: A global collection of networks and computers linked together.

Intranet: A private network of computers. Utilized primarily among businesses.

ISDN: Converts a regular telephone line into two digital lines, each of which is able to carry data at 64 kilobits/second.

Java: A programming language that can be used to create animation and interactive features on web pages. Java programs are usually embedded inside HTML documents.

Linux: A UNIX-like operating system designed to provide personal computer users a free or very low-cost operating system, comparable to traditional and usually more expensive UNIX systems.

Microprocessor: A computer that has its entire CPU on one integrated circuit. First developed by Intel.

Motherboard: The main circuit board in a computer. The motherboard contains the CPU, memory sockets, expansion slots, and the bus.

RAM: Random-Access Memory. Temporary storage the computer uses for quick retrieval. There are many different types of RAM.

RISC: Reduced Instruction Set Computing. A microprocessor design that speeds up operations by performing simple operations, not complex ones, requiring some operations to be performed by the software before the data hits the CPU. This means the operating system and applications must be specifically tailored for RISC architecture. The PowerPC, the Power Mac and the IBM RISC System use RISC.

ROM: Read Only Memory. ROM is often used to store vital information about a computer, such as BIOS, that cannot be changed, or written to.

Router: A piece of hardware that sits on a network, picks up messages, and sends them to their destinations using the most efficient route.

Server: The computer in client/server architecture that supplies data, files or services. The computer making the request is called the client. Some networks have dedicated servers.

Spam: Unwanted or "junk" e-mail, sent to numerous accounts. The term spam is taken from a Monty Python sketch about a diner that serves Spam in every dish – implying that you'll always get spam, whether you want it or not.

UNIX: A multiuser operating system written in the C language. UNIX allows multiple programs to run simultaneously and more than one user to access a single computer, making it popular with businesses, universities and Internet operating systems.

Vaporware: Software advertised before it is completed. Often used to preempt development of similar software by a rival company.

Web TV: A method of accessing Internet and e-mail by using a regular TV set, a phone line, a set-top box and an online service. Owned by Microsoft.

Wintel: The mighty combination of Windows software and Intel microchips. Some posit a "Wintelco," to reflect the ascendancy of Cisco Systems, leading networking equipment maker.

WYSIWYG: "What you see is what you get." Software that allows you to see what your finished product will look like on the page.

Y2K: Shorthand for "Year 2000." Refers to the date change problems many computers may experience at the turn of the century. Shorth-sighted programmers created computers that can only recognize the last two digits of years – and read 00 as 1900. The flaw, if not corrected, may cause computer systems to shut down.

Leading High Tech Employers

@Home

425 Broadway Street
Redwood City, CA 94063
(650) 569-5000
Fax: (650) 569-5100
www.home.net

LOCATIONS

Redwood City, CA (HQ)

DEPARTMENTS

Administrative
Cable Systems
Content Programming
Customer Service
Field Sales
Finance
Human Resources
Information Systems
International
Networking
Quality Assurance
Sales & Marketing

THE STATS

Annual Revenues: $7.4 million (1997)
No. of Employees: 329
No. of Offices: 1 (U.S.)
Stock Symbol: ATHM (NASDAQ)
CEO: Tom Jermoluk
Year Founded: 1995

@HOME

UPPERS

- Free soda and snacks
- Growing company
- "Generous" stock options
- Diverse, progressive co-workers

DOWNERS

- Long hours
- Often overwhelming responsibility
- Uncertainty because of AT&T acquisition

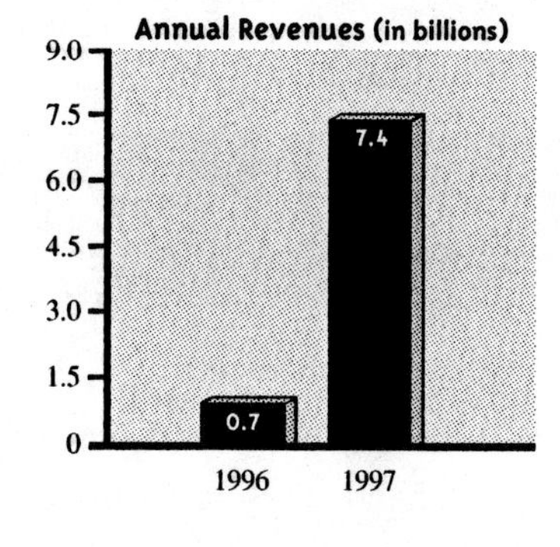

KEY COMPETITORS

- America Online
- Microsoft
- Time Warner
- US West

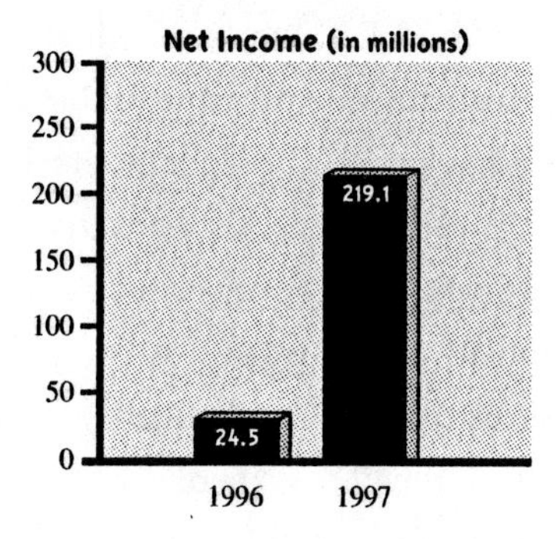

EMPLOYMENT CONTACT

Leilani Gayles
VP Human Resources
425 Broadway
Redwood City, CA 94063

Fax: (650) 482-4604
Or fill out the online application at:
http://www.home.net/corp/jobapp.html

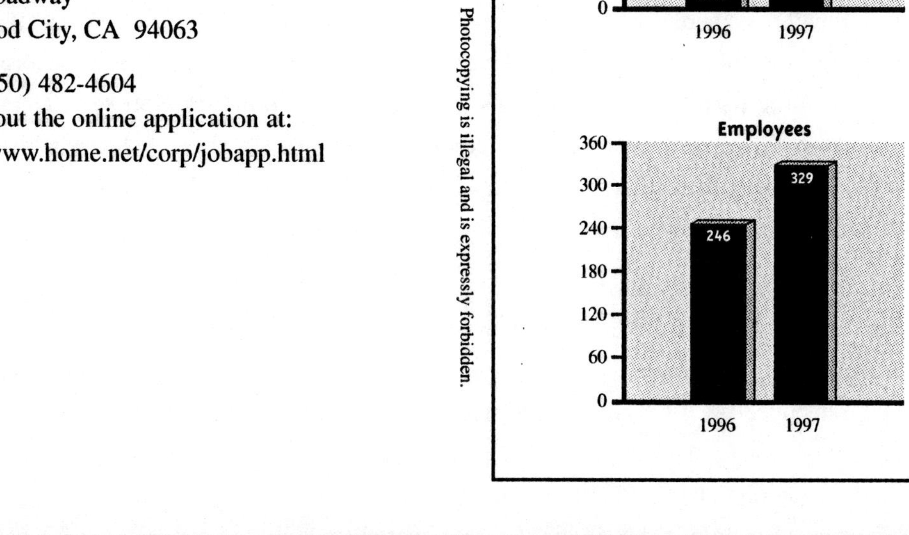

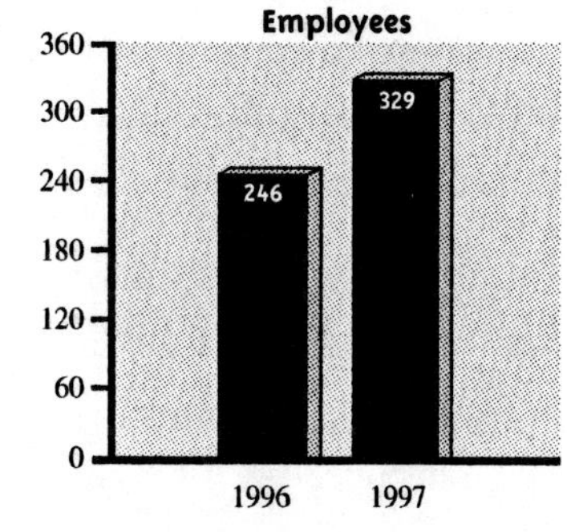

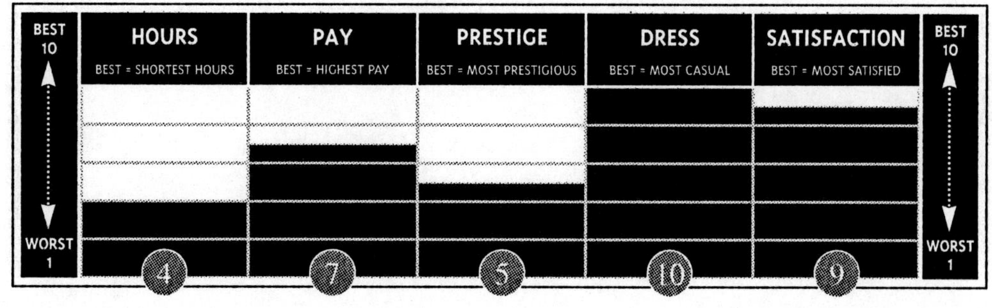

THE SCOOP

@Home is a member of the new breed of Internet service providers that wants to bring high-speed Internet users to the couch-potato level. Founded by cable television giant Tele-Communications, Inc. (TCI), @Home delivers Internet service via cable modems rather than traditional phone lines, which allows for speedier access (up to 300 times faster than dial-up connections) and improved overall service. Because it's faster, the Redwood City, CA-based firm can offer brighter colors, better graphics, and even short video clips on demand. @Home has also devised an innovative way to avoid bottlenecks on the Web. It maintains a 'parallel' Internet backbone, called @Network, that duplicates popular sites. When an @Home user clicks a link to the heavily trafficked New York Times Online site, for example, she actually sees the replicated page on the @Home backbone. Each parallel site is updated about every 20 minutes, so there's no worry about missing the latest updates. @Home also offers a multimedia component with multiplayer games, local weather and traffic information, and content from more than 60 big-name providers, including Bloomberg, CNN, Discovery Channel online and SportsLine. The company isn't just serving couch potatoes, either – its @Work service offers cable-based Internet access, telecommuting and LAN services for small- to medium-sized business customers.

The company was founded in 1995 by cable giant Tele-Communications Inc. (TCI), and Kleiner Perkins Caufield and Byers, a Silicon Valley venture capital firm. TCI owns a 42 percent share of @Home, which went public in May 1997. As part of AT&T's acquisition of TCI (announced in 1998) the long-distance carrier will receive the cable company's majority stake in @Home. This gives AT&T a head-start in the cable/Internet world, and sets up the communications giant as a major contender in the new media business. Though @Home's subscribers are only in the six digits (as compared to AOL's 14 million), AT&T is betting that this new alliance will put it @ the top of its burgeoning industry.

GETTING HIRED

@Home recruits talent on college campuses and at various job fairs around the country throughout the year. It also posts listings on its web site, at www.home.net. You can send resumes and cover letters via fax, e-mail and regular post, or apply using @Home's online form.

OUR SURVEY SAYS

Insiders describe @Home's corporate culture as "sort of undefined," "still in development," and "all over the map. "It's about as un-Corporate America as you can imagine," says one insider. You'll find everything from "alternative youth" to "suits" in this "progressive corporation," where "there is an inherent emphasis on diversity." "Yes, there is blue and orange hair in parts of the company," reports one insider. Our sources say representation of women and minorities "has improved vastly" over the past few years, and the company "is getting more diverse all the time." One insider reports that "there is a lot more support for gay/lesbian/bisexual employees here than at other corporations I've been with."

Our sources describe co-workers as "a great bunch" of "talented," "motivated" people who "treat each other with respect and professionalism." Dress is casual, "mostly jeans, but you will see some ties and jackets" among people in sales, management and marketing departments. Engineers report donning "shorts, T-shirts and sandals every day."

"@Home is an exciting place to work right out of college," say insiders. Schedules are flexible, "but the work is quite demanding." "The company is growing very fast so don't expect to be coddled. Expect huge amounts of responsibility." "People work hard," explains one source, "but that is not to say they work extremely long hours." Others say they try to "work smart, not hard." One source adds that "it really depends on how engaged you are in the kind of work you do." Another points out that the length of your day will also "vary depending on job type and susceptibility to workaholism." In general, "we are all gung-ho people and put in extra time for the sake of the company." Says one contact: "People work here all hours of the day, all days of the week." Engineers "come in between 10 a.m. and noon and work until 8 or 10 at night. In addition, many telecommute one or two days a week." In fact, "some of us work '24-7' on-call duty." Now th@'s devotion. After work, however, be assured that @Home employees "know how to party." "We have a pretty loose attitude," says one source.

"Salaries are competitive, plus you get stock options and an employee stock purchase plan," employees say. One source adds: "@Home is more generous with stock options than any other company I have ever worked for."

@Home encourages its employees to keep in shape by providing outdoor basketball courts, an indoor slide, pool table, and Ping-Pong at the company's headquarters. The company also provides other "typical engineer necessities" like "all the free soda and arcade games you want." Other perks include dry-cleaning service, shuttle transportation to San Francisco, a lunch bar in the building, and free snacks. As if that wasn't enough, "Excite [the World Wide Web search engine company] is next door, and we play roller hockey every day." @Homers might play a pick-up game with their industry comrades while you waiting for @Home's on-site mechanics to change the oil in their car – another perk the company offers.

"Everyone receives stock options as part of their compensation package, and that's what most people are counting on, more than salary."

– Internet insider

3Com

P.O. Box 58145
5400 Bayfront Plaza
Santa Clara, CA 95052
(408) 764-5000
Fax: (408) 764-6966
www.3com.com

LOCATIONS

Santa Clara, CA (HQ)
As well as 165 other offices located in 42 countries around the world

DEPARTMENTS

Advanced Technology
Corporate Services
Customer Service
Diagnostics
Finance
Hardware Engineering
Manufacturing
Marketing
Software Engineering

THE STATS

Annual Revenues: $5.4 billion (1998)
No. of Employees: 12,500 (worldwide)
No. of Offices: 165 (worldwide)
Stock Symbol: COMS (NASDAQ)
CEO: Eric A. Benhamou
President and COO: Bruce Claflin

UPPERS

- Paid sabbaticals
- Flexible scheduling options
- Nifty Palm Pilots
- Generous vacation time

DOWNERS

- Long workdays
- Lack of career support
- Merger induces lumpiness and bureaucracy
- Renaming Candlestick unpopular notion

KEY COMPETITORS

- Acer
- Ascend Communications
- Bay Networks
- Cisco Systems
- Lucent Technologies
- NEC

EMPLOYMENT CONTACT

Grace Soriano-Abad
Regional Staffing Manager
P.O. Box 58145
5400 Bayfront Plaza

(408) 326-1381
Fax: (408) 326-5959
college@3mail.3com.com

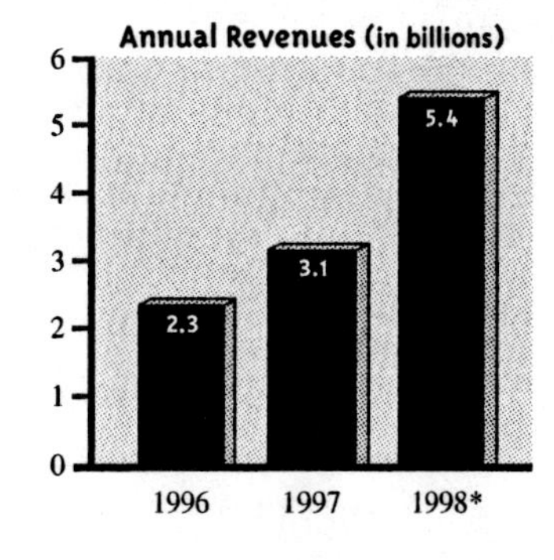

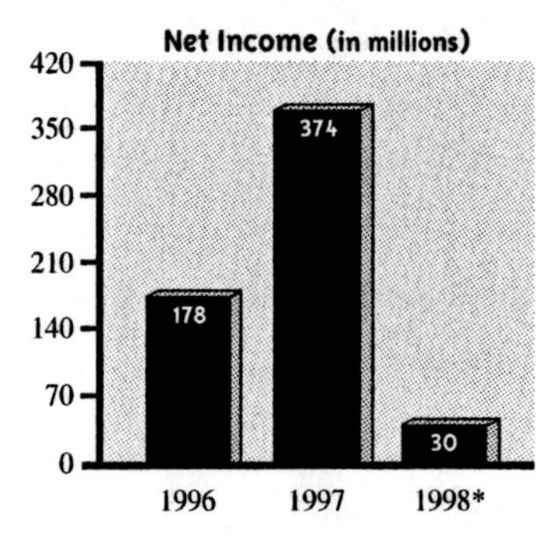

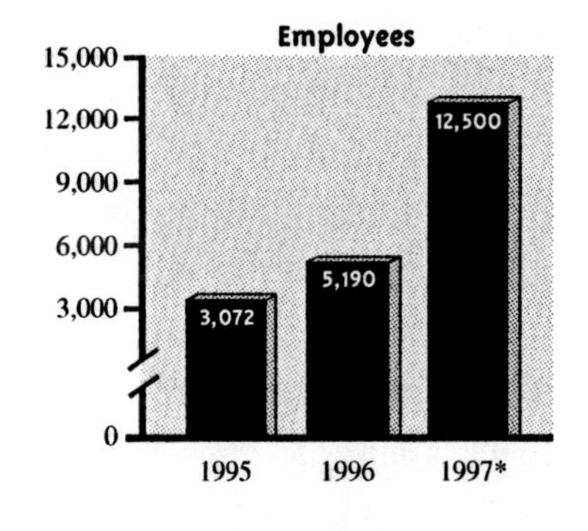

*Reflects merger with U.S. Robotics

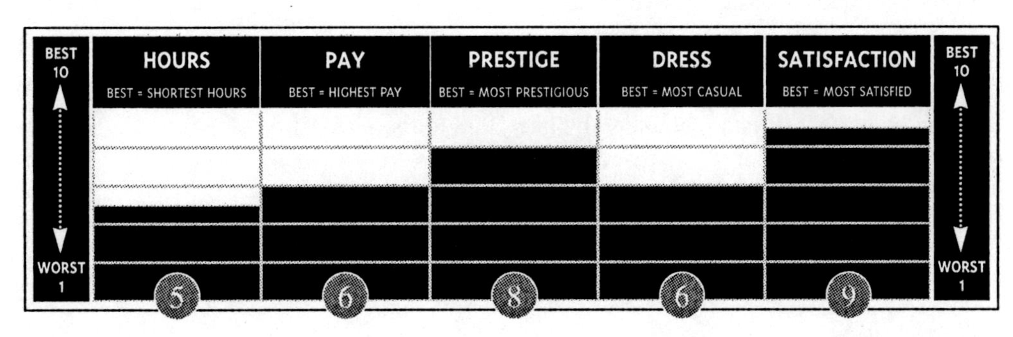

THE SCOOP

When Robert Metcalfe invented Ethernet in the famed Xerox research lab in 1979, he was a little ahead of his time. Metcalfe had developed a powerful networking tool that would, in 15 years, become the industry standard for connecting personal computers. But in 1979, few people had heard of PCs and the market for PC networking was miniscule. But when IBM came out with its first widely popular PC in 1982, the company saw sales take off. 3Com went public in 1984. Three years later, it acquired Bridge Communications, one of the first companies to sell networking equipment for PCs. The company's $8.5 billion purchase of modem maker U.S. Robotics in 1997 puts the company on the same financial scale as rival networker Cisco, though Cisco's lead in the networking business is almost as formidable as the leads Intel and Microsoft have in chip making and software.

3Com's name is simple and appropriate; it derives from three "coms" (computer, communications and compatibility) and describes what the company does. In a nutshell, 3Com helps computers communicate with each other. (The company's original line of business was making network cards – small circuit boards that connect computers to network wires coming out of walls.) Currently, 3Com is pushing a new product, called a "telecommunications access device," which is able to separate different types of data (for example, high-density data that require advanced transmission techniques versus data that can easily go through phone lines). One of the company's subsidiaries, Palm Computing, manufactures equipment for networking of a different kind: the wildly popular PalmPilot allows techno-enthusiasts to keep track of all contacts and appointments with a computerized gizmo that fits in one's hand.

When 3Com announced its acquisition of U.S. Robotics in 1997 in a $8.5 billion deal, analysts everywhere described the move as "bold," "dramatic" and "daring." They did not, however, agree the merger was necessarily "smart." After the merger was announced, a series of complicated negotiations and inventory problems dropped 3Com's stock price from nearly 40 to the high-20s. (It had rebounded by late 1998 to the mid-30s.)

3Com sees small- and mid-sized businesses and individuals as a huge market for networking products in the future. The familiar U.S. Robotics moniker, reasons the company, will be a name these new clients automatically turn to. To help build its own brand name, 3Com even rented a baseball stadium – or at least the naming rights to the stadium: Candlestick Park in San Francisco has been 3Com Park since 1996.

GETTING HIRED

New hires at 3Com are usually placed in the San Francisco Bay area, home to almost half of its workforce. The majority of entry-level positions at 3Com are in technical fields, but the company also hires recent graduates with backgrounds in finance, sales and marketing. The interview process with 3Com usually involves several rounds of meetings with both managers and potential colleagues. Although one insider tells Vault Reports that "some managers like to use brainteasers or riddles," this practice seems uncommon. Be ready for technical questions, however, even if you're not an engineer. "As a technology company, we are very concerned with having people who are technology literate even in non-technology positions," says one manager. "That does not mean that an individual needs to have a strong technical background, just technology and possibly industry awareness." "3Com really likes engineers," advises one former marketing MBA intern. "Know your technical stuff before interviewing." One product manager, suggests candidates "learn about [3Com projects] before you interview. A little knowledge goes a long way."

OUR SURVEY SAYS

Our sources note that 3Com's company culture is difficult to define in large part because it has grown through acquisitions, not organically. "The acquired company is turned into a division and tends to keep their pre-acquisition culture," explains one employee. One thing seems consistent, however – 3Com's work environment is marked by communication and teamwork. "3Com has tried to minimize the class divisions of more traditional companies. Everyone in the company, including the VPs and the CEO, have cubicles," says one employee. "They want to reinforce an open-door policy for all employees. Understandably, the executives' cubicles are larger, but still without a door." Says another, echoing the company's name origins: "Open communication is a very important part of our culture." "The culture is very open and honest, and exhibits great team effort," says yet another insider.

The absence of hierarchy, which one employee describes as "fraternalistic rather than paternalistic," does have its drawbacks: "You need to fend for yourself, but you will be supported in what you are trying to do, rather than having your manager responsible for steering your career," says one contact. "It is a hard-driving place, so anyone coming here should be ready to hit the ground running, and running hard and fast," confides another insider. Former U.S. Robotics employees describe much the same situation: "U.S. Robotics became big in a short time. Hence, the environment here is one of go-getters. There is less bureaucracy but is also a bit less organized."

Still, employees seem very satisfied. Says one employee, who has stayed with 3Com for more than a decade: "This [length of employment] is somewhat unheard of for high tech but 3Com is a company that works hard to retain its employees." Part of the reason employees may stay is that they get to spend a lot of time away from the company. Every four years, 3Com employees get an extra one-month paid "sabbatical." Workaholics despair – we hear the sabbatical is "mandated." This is in addition to an already whopping 28 days paid vacation for starting employees. Also, there is a "company-wide shutdown between Christmas and New Year's." To encourage employees to take time off, 3Com "doesn't let employees accrue more than 120 hours [of vacation]. If you earn more you lose it, because you haven't taken what you've got." Although many employees report working 9- and 10-hour days, and some complain about long hours, other insiders remark that "unlike the unwritten rule at many high-tech companies who really expect 50 to 60 hours each week or more, you work hard while you're here and then go home or go play. You're expected to work 40 hours a week."

And it's not as if life at the Santa Clara 3Com campus, located next to the San Francisco Bay is particularly austere. "We have a 24-hour, 7 day-a-week, 365-days-a-year on-site gym with aerobics classes, stationary and free weights, treadmills, bikes, stairmasters, and locker rooms for $20 a month," reports one health enthusiast. The gym features also include a sauna and a massage service, insiders say. "There is also a café, a cafeteria, movie rentals, dry cleaners service, a Starbucks coffee shop, commuter shuttles, and a car wash service," according to another employee. The loyal troops of 3Com receive "at least one free night at 3Com Park for a (San Francisco) Giants game." Insiders also "get to purchase personal use 3Com products for a discounted price, including modems, Palm Pilots, and various plug-in cards for PCs." Monetary perks include a "3reward" system that offers cash bonuses for outstanding performance, a stock purchase plan that lets employees buy company shares at a 15 percent discount (10 percent of gross pay can be set aside for this plan). And, all employees are offered stock options upon their hiring.

While 3Com may have the amenities of other Silicon Valley companies, it hasn't gone quite as far toward casual dress days. "The dress code is generally dress casual – by this I mean nice clothes, Dockers and maybe a button-down shirt for men," reports one employee. However, "on Fridays, you can wear just about anything you want." Engineers report casual dress; sales and marketing employees report wearing suits like the rest of the corporate world.

Adaptec

691 S. Milpitas Blvd.
Milpitas, CA 95035
(408) 945-8600
Fax: 408-262-2533
www.adaptec.com

ADAPTEC

LOCATIONS

Irvine, CA • Milipitas, CA • Longmont, CO • Miami, FL • Apple Valley, MN • Austin, TX • Grapevine, TX

DEPARTMENTS

Administration
Applications Engineering
Design Engineering
Facilities
Finance/Accounting
Human Resources
Information Services
Marketing
Operations
Quality
Sales & Support
Software Engineering
Technical Services
Test Engineering

THE STATS

Annual Revenues: $1.0 billion (1998)
No. of Employees: 2,250
No. of Offices: 7
Stock Symbol: ADPT (NASDAQ)
CEO: Larry Boucher
Year Founded: 1981

UPPERS

- Unusually diverse employee mix
- Telecommuting

DOWNERS

- Recent downsizing
- Falling revenue and morale

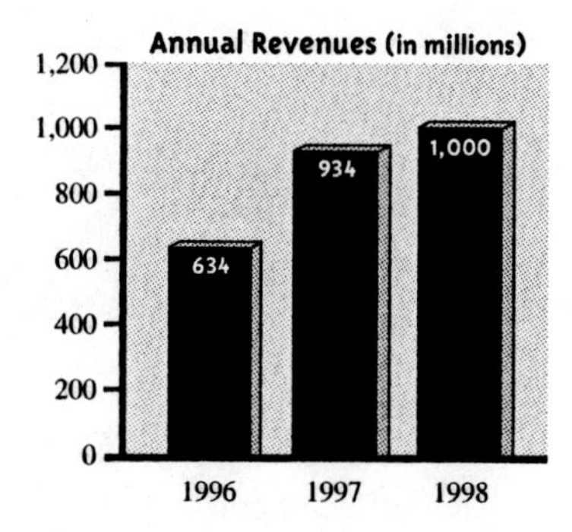

KEY COMPETITORS

- 3Com
- Asante
- Cabletron
- Cirrus Logic
- CMD Technology
- Digi International
- FORE Systems
- Fujitsu
- IBM
- LSI Logic
- Microware Systems
- Olicom
- Q Logic

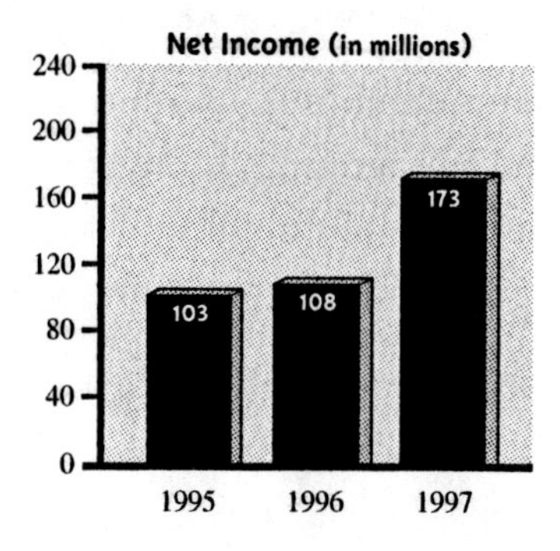

EMPLOYMENT CONTACT

E.J. Tim Harris
Adaptec Professional Staffing
Mail Stop 15A
Department WWW
691 South Milpitas Boulevard
Milpitas, CA 95035

Fax: (408) 957-7810

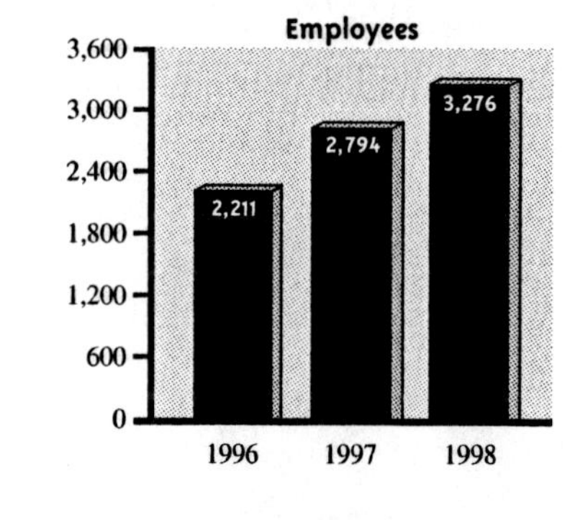

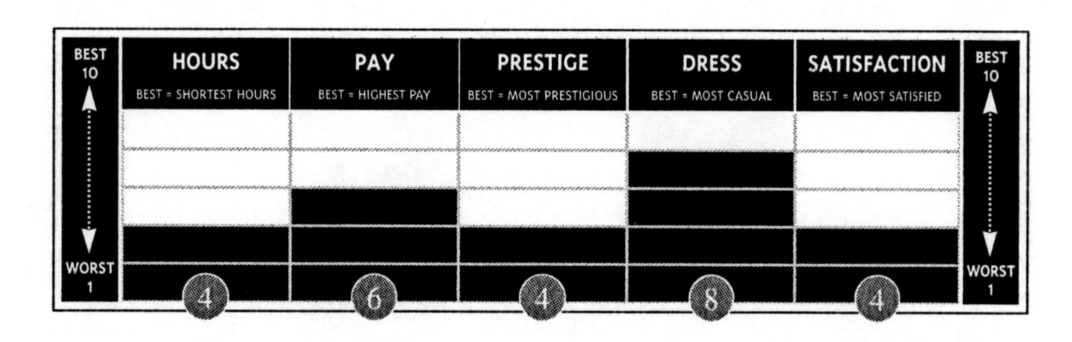

THE SCOOP

Based in Milipitas, California, Adaptec makes hardware and software that expedites data transfer between computers, networks and peripherals. Adaptec is the market leader in SCSI technology. Pronounced "scuzzy," this technology connects peripheral devices like printers or scanners to computers and servers. The company was formed in 1981, and sold its first SCSI products in 1983. Among the keys to its SCSI technology (which stands for small computer system interface) is the company's bandwidth management solutions, which increases the speed of data transfer. The company's input/output products are compatible with all of the leading processor platforms and operating systems.

Since the late 1980s, Adaptec's stock price and earnings had grown steadily, but the company hit a wall in 1998 and was forced to lay off 250 U.S. employees in April. A month later, the company announced that net income fell 100 percent to $23.4 million from $46.8 million for the previous year. Revenue dropped 23 percent to $204 million from $265 million for the corresponding quarter in 1997. The decline was attributed to excess inventory and price instability. Two months later, Adaptec eliminated 350 jobs, or 10 percent of the company's overall workforce, at its Singapore manufacturing facility. The company's founder, Larry Boucher, returned in August 1998 as interim CEO to lead a reorganization that has included the divestiture of unprofitable divisions. The company is in the process of building a new manufacturing facility in Singapore.

Along with data transfer products, Adaptec has in recent years worked on building other businesses, largely through acquisitions. The company makes the award-winning CD-recording software, Easy CD Creator Deluxe – which allows the home-PC user to record music, images and videos onto CDs. The company also offers a similar product for Mac users, called Jam for Macintosh; and a DVD software application called DVD Toast.

GETTING HIRED

"Hiring is mostly done by referrals," sources say – "the company offers cash incentives." These cash incentives are reportedly between $1000 and $2000 if a referral is successfully hired. One insider estimates that the company conducts "more than 50 percent" of its employees through referrals. In addition, "we go to job fairs and to college campuses, but we are not as strong in college recruiting as other companies." The company also posts job listings on its web site at www.adaptec.com. Applicants can submit their resumes with an online form on the site.

The interview process "is fairly relaxed," insiders report. In the first round, candidates "deal with four to six people, one at a time, including HR, senior members of the specific department, and then the supervisor." In the second round, candidates meet again with the supervisor, then with the manager. Expect "both technical questions and queries about other aspects about work life." If you're going for an engineering position, your interviewers might pose a design problem, but those "should be rather ambiguous and open-ended – to get you talking about a technical issue." One source says that applicants may expect to receive "some hypothetical situations involving the technology." Say insiders: "What really matters is the ability to step through a problem in a logical manner." Also important: "the ability to work with others, fit in, and augment the culture we have established."

OUR SURVEY SAYS

Most Adaptec employees say they are happy with the company – though a little distressed about the company's recent troubles. Says one insider: "I usually tell people I work for the best company in Silicon Valley. I've toned down a bit since the stock prices have tumbled so dramatically and since Adaptec suffered the first reduction in force in more than a decade." "The old Adaptec is dying fast," laments another source. "In fact, it has died. I would like to give you the glowing speech I used to give everyone, but it's just not like that anymore." Employees describe themselves as "bright and positive-thinking" and say "we enjoy one another's company," but add that "the company is having trouble hiring and keeping good people." Some complain that the corporation "no longer has a commitment to its employees," noting that "when they say 'We are primarily responsible to our shareholders,' they mean 'We are exclusively responsible to our shareholders.'"

Others are quite satisfied, however, noting that "managers take very good care of their engineers, sending them to conferences and classes, buying the best workstations, and allowing them to work from home one or two days per week." And in contrast to other Silicon Valley companies where burnout is a severe problem, one Adaptec insider reports: "I do think the company as a whole, and the HR group specifically – who are strongly connected through the company – actively stay involved with everyone to help manage work overload." Still, this doesn't mean Adaptec employees have it easy. While work hours are "flexible" – the most important thing is getting your work done. "If that means working long hours, so be it." says one insider.

The dress code in California is "no shorts and no ties." In Colorado, however, "it's more relaxed. They even officially allow shorts on Friday – though people wear them all the time anyway." Sources in California remark that "the racial mix is so diverse that it would be impossible to have any racial bias." As one insider elaborates: "We're always far too desperate to find good people to give any thought to race." And Adaptec reportedly has a much better distribution of men and women than other female-starved high-tech firms. "In engineering, we have about a 50/50 mix of men and women. Managment is also about 50/50. The executive staff isn't so even, but it's getting there."

Adaptec offers two weeks vacation, plus "the entire time between Christmas and New Years'." They also have a "very good" stock purchase plan, and employees receive "stock options for good performance." The stock purchase program allows employees to buy company stock at 85 percent of the market price at either the beginning or end of the fiscal quarter – whichever price is lower. Employees say "they could be worth a lot in five to 10 years." Techies also rave about the SCSI grant program – "If you think of a cool SCSI application for your home PC, they will pay $500 towards implementing it."

"You need to fend for yourself, but you will be supported in what you are trying to do, rather than having your manager responsible for steering your career."

– Hardware insider

Adobe Systems

345 Park Ave.
San Jose, CA 95110
(408) 536-6000
Fax: (408) 537-6000
www.adobe.com

LOCATIONS

San Jose, CA (HQ)
Seattle, WA • Various other U.S. cities

Edinburgh, Scotland (Europe HQ)
Argentina • Australia • Brazil • France • Germany • Hong Kong • India • Italy • Japan • The Netherlands • Singapore • Spain • Sweden • Switzerland

DEPARTMENTS

Administration
Engineering
Finance
Human Resources
Information Services
Marketing
Operations
Sales & Support
Software Q&A
Tech Support & Customer Service

THE STATS

Annual Revenues: $895 million (1997)
No. of Employees: 2,702
Stock Symbol: ABDE (NASDAQ)
CEO: John Warnock

UPPERS

- Exposure to cutting-edge computers and software
- Free drinks and candy
- 401(k) matched up to 25%
- Company gym
- Great stock options

DOWNERS

- High pressure environment
- Extremely competitive – difficult to get in

KEY COMPETITORS

- Corel
- Macromedia
- MetaCreations
- Micrografx
- Microsoft
- Quark

EMPLOYMENT CONTACT

Professional Staffing
Adobe Systems Incorporated
345 Park Avenue
San Jose, CA 95110

Fax: (408) 536-6818
jobs@adobe.com

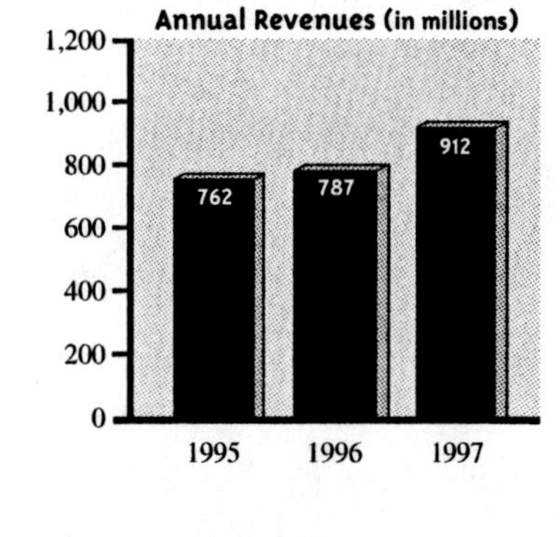

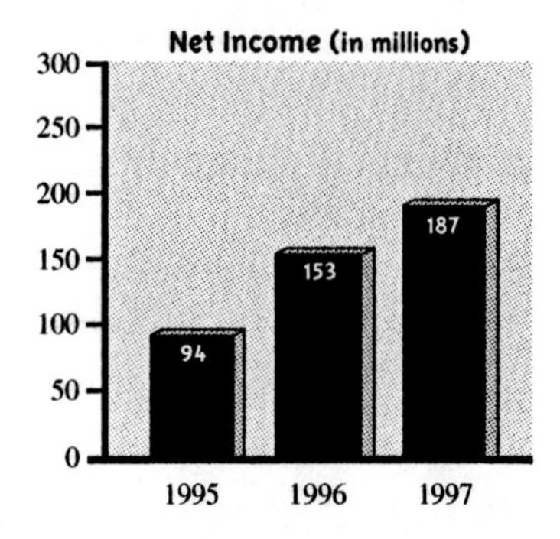

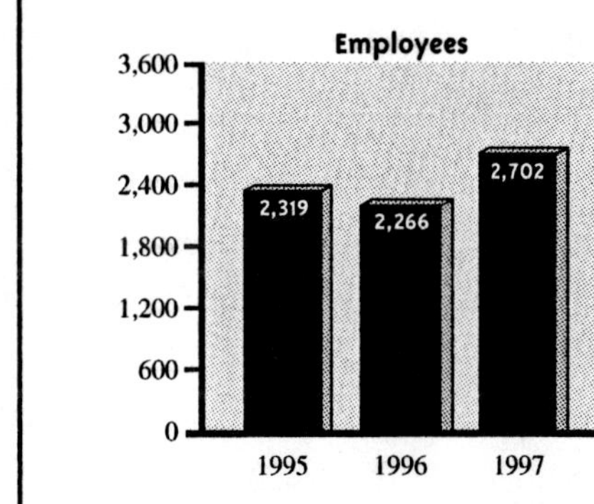

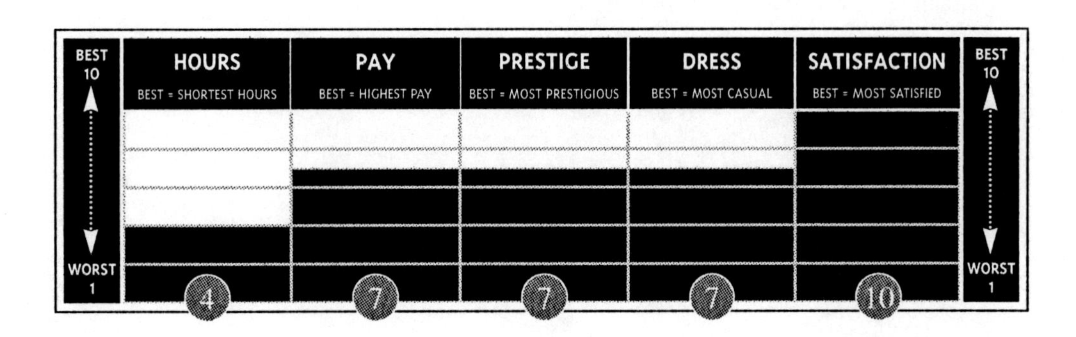

THE SCOOP

Chances are, Adobe is part of your everyday life and you don't even realize it. Many of the graphics you see in today's newspapers and magazines are created by Adobe programs like Illustrator and Photoshop. The opening scene of the blockbuster movie *Men In Black* was created using Adobe's After Effects application. When you surf the Net, many of the web sites you see are designed using Adobe Acrobat. And the PostScript inside your laser printer was also probably created by Adobe.

Adobe CEO John Warnock and President Chuck Geschke first met at Xerox's Palo Alto Research Center (PARC). Together they developed PostScript, a computer language that translates code into printable pages of type. After failing to convince Xerox to market the application, the two left to start their own company, called Adobe. But Canon came out with a cheap laser printer before they could, so they were forced to alter their path. Luckily, Steve Jobs approached them and convinced them to develop the PostScript technology for the Apple's Macintosh. The result was the Apple LaserWriter – and so began the story of desktop publishing.

PostScript quickly became the industry standard for desktop publishing, and Adobe later developed the software that would become Internet standards – including Adobe Acrobat, PageMill, and its WebType Fonts. Scan the employment ads of any newspaper, and the name Adobe will come up on just about every listing for a desktop publishing or web design job. Another of the company's major products, Photoshop, allows users to create designs and manipulate digitized photographs. The company also makes Adobe Premiere, a computer application that is used for video editing. Adobe offers applications that are completely compatible across platforms, allowing files to be moved easily between systems.

Adobe's products weren't always so compatible. In the 1980s and early 1990s Adobe established a strong relationship with Apple, but basically ignored the PC market. The company changed its tune after an $11.8 million loss in 1995, caused largely by the problematic acquisition of software maker Frame Technology Group. By 1996, with Apple foundering, the pressure was on for Adobe to tap other markets improve its software's compatibility with other systems, particularly Microsoft Windows. Adobe expanded its software line with in-house development and aquisitions of other companies. It also made successful efforts to promote its brand in Windows-dominated corporate and government arenas. Today, 56 percent of Adobe revenues come from sales of Windows applications. Though Adobe has concentrated efforts on wooing the Windows user, the company insists that its ties to Apple remain solid.

With a firm foothold among graphics professionals and in the corporate world, Adobe's latest target is the home PC consumer – a larger, more lucrative market. The company has developed user-friendly imaging software like PhotoDeluxe, and bundling its products at reduced prices to entice buyers. PhotoDeluxe retails at $49, versus $695 for Photoshop. Adobe plans to release more "consumer versions" of its professional software in the near future. The company is also looking to increase revenues by building more relationships with computer resellers in Asia and Latin America.

Two of the biggest problems Adobe has had to face are software piracy and copyright infringement. It won a landmark case (for typeface designers, at least) against Southern Software in 1997. The judgement means fonts are now classified as intellectual property, and subject to the same copyright laws.

After sustaining losses early in 1998, the company expects revenues to increase with the upcoming introduction of several upgrades and new products. (Part of Adobe's trouble is that it is facing a mature market in the U.S. – and weakening markets in Asia.) Adobe recently completed construction of its corporate headquarters and new $50 million facility in Fremont, WA. Finally, a rumor has been circulating that Adobe is planning to release two major new products in 1999 – a page-layout technology (code name: K2), and a workflow system (code name: Stilton). Sources say K2 has been labeled the "Quark Killer" among Adobe insiders, because its functions and features far exceed the capabilities of Adobe's formidable rival in desktop publishing software, Quark Xpress. Quark actually made an offer to buy Adobe in Novenber 1998, only to withdraw it days later.

GETTING HIRED

Adobe recruits on college campuses, and posts job listings on its web site. When applicants submit resumes online, the company sends back applications to fill out. Upon receipt, the resumes are scanned into a database that hiring managers can search when they're looking for new talent. Employees say the company also hires many of its interns as full-time employees, so an internship is a good foot in the door.

Interviews usually consist of two rounds, more if you're applying for an engineering position. Sometimes a hiring manager will call for a "phone screening," to determine whether they should bring a candidate in, insiders say. Insiders say interviews are usually not very stressful, "though some managers are more prone to put you on the spot technically than others."

OUR SURVEY SAYS

In general, Adobe is an "innovative," "employee-friendly" place, with "excellent managers." Ranked in the past as one of *Fortune's* "Top 100 Companies to Work For," Adobe does its best to accommodate workers, offering flexible schedules, telecommuting, and "a great many training classes to get up to speed with the products." Employees also enjoy "a great cafeteria," access to a company gym ("they even have aerobics classes"), and free fresh squeezed orange juice, soda, Starbucks coffee and "various other goodies" every day.

If you work best under pressure, Adobe is the company for you – "most of the time is pressure time," says one stressed Adobe inmate. "Our hours fluctuate between 40 and six million a week," reports one engineer, "but the marketing, sales and administrative people have different work patterns." Though they work a lot, engineers and other technical specialists "pretty much come in when they want if they're working on a long-term project," and get to "dress how they want." Plus, their salaries are "typical of the geek scale" (that means high). Tech professionals can start "anywhere in the high 30s," and "if you know C++, you can make up to $50 an hour!" The corporate employees who "get away with working 40 hours a week" also enjoy good salaries, but "they have to come in by 9, and dress professionally." Benefits are "excellent," including "very generous" stock purchase and profit sharing options, quarterly bonuses, full insurance, and a 401(k) matched 25 percent. Adobe insiders also get up to five weeks of vacation, eight paid holidays, and paid sick/personal days. Insiders also feel fortunate to have "access to cutting edge computers and software."

Finally, insiders say Adobe is a "progressive," "laid-back company" that "knows its people are its strongest asset." "The level of intelligence of the people here" is also a big draw. "We have some incredibly artistic people working here," says one programmer, "and just being able to talk to some of the engineers who wrote PhotoShop is a perk to me." Working at Adobe is also extremely challenging: "It's a good place to work, and work you will," says one insider. "We run a lean company – every single one of us has the job of at least two – but it keeps us on our toes."

Advanced Micro Devices

One AMD Place
P.O. Box 3453
Sunnyvale, CA 94088
(800) 538-8450
www.amd.com

AMD

LOCATIONS

Sunnyvale, CA (HQ)
Austin, TX • Singapore • Bangkok • Thailand • Dresden • Germany • Sales offices worldwide.

DEPARTMENTS

Computations Products Group
Communications Group
Memory Group
Vantis

THE STATS

Annual Revenues: $2.4 billion (1997)
No. of Employees: 12,800 (worldwide)
No. of Offices: 40+ (worldwide)
Stock Symbol: AMD (NYSE)
CEO: W. Jeremiah Sanders, III

UPPERS

- Tuition reimbursement
- "Feelin' Good!" Health program
- Stock purchase plan
- Paid sabbaticals

DOWNERS

- Long workdays
- Intel market dominance

KEY COMPETITORS

- Atmel
- Cypress Semiconductor
- Cyrix
- Intel
- LSI Logic
- Lucent
- National Semiconductor
- Rise
- VLSI Technology

EMPLOYMENT CONTACT

Employment or University Relations
Advanced Micro Devices
Dept. WWW
One AMD Place
P.O. Box 3453
Sunnyvale, CA 94088

Fax: (408) 774-7024
jobs@amd.com

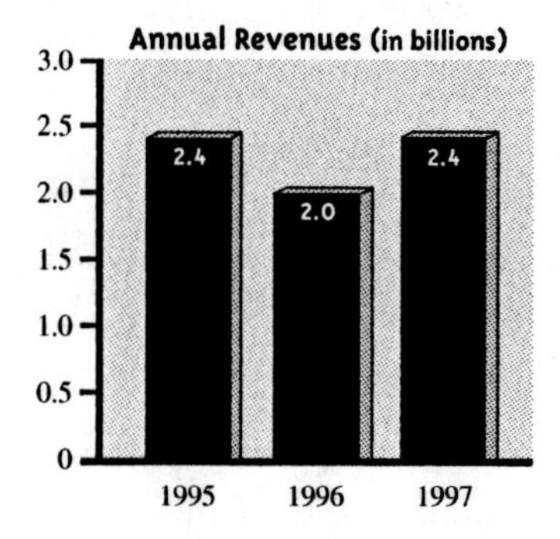

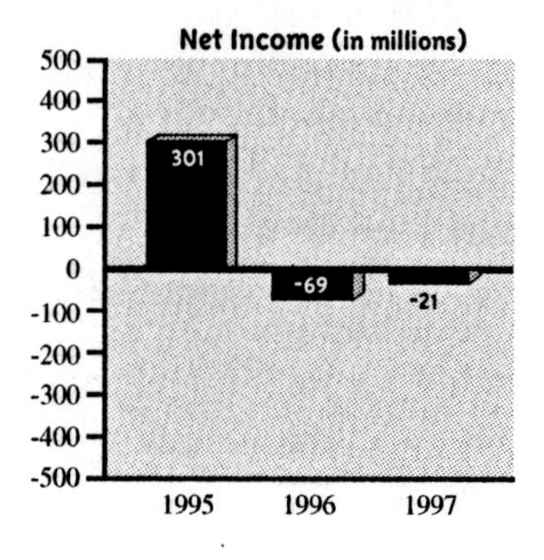

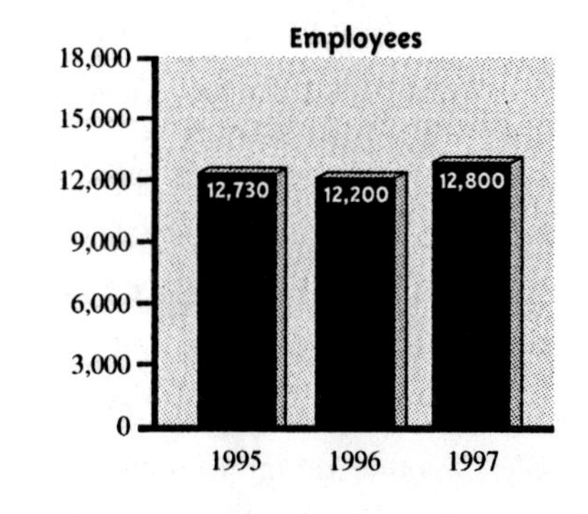

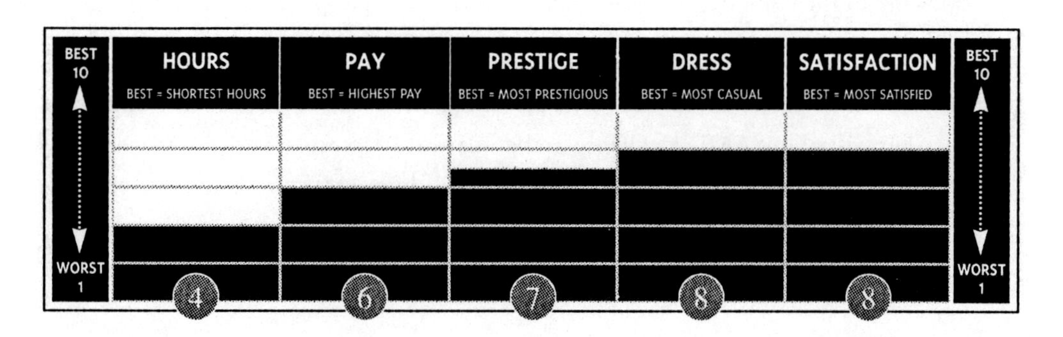

THE SCOOP

With a 9 percent share of the semiconductor market, AMD is a David of the industry looking to knock off the Goliath of semiconductors, Intel. Advanced Micro Devices (AMD) is the world's largest supplier of integrated circuits (ICs) for personal and networked computers, and public communications infrastructure. AMD products include processors, flash memories, programmable logic devices, and networking and communications products. The company began in 1969 when marketing whiz Jerry Sanders was fired from Fairchild Camera & Instrument. By putting his marketing know-how to work, Sanders was able to overcome a capital shortage and a lack of management experience to launch his company and go public in 1972.

After the company began developing its own chips, problems arose with competitor Intel, embroiling both companies in legal battles throughout the early 1990s. Sanders and company tried to make their point by sending prospective Intel buyers Monopoly board games, but scored an even greater victory in 1994 when a federal jury ruled in AMD's favor over Intel, which had tried to sue AMD for copyright infringement.

Lately, AMD has been winning a few battles with Goliath Intel. AMD's recent K6 microprocessor has tested as nearly as fast as Intel's Pentium chip, but sells for 25 percent less. Digital Equipment (DEC) was the first top-ten PC-producer to use the new chip. Although they have not yet moved entirely to the K6, Compaq (which bought Digital) and Hewlett-Packard are both using AMD chips in their increasingly popular sub-$1000 models. In 1998, IBM announced that it too would include the K6 in some of its systems. Despite its mammoth 89 percent market share, Intel has reacted furiously to any and all challenges from its small competitor: In the summer of 1997, Intel drastically cut the prices on its microprocessors in order to keep pace. AMD retaliated with price cuts of its own.

AMD has also forged an alliance with the ailing Motorola to exchange technology. The partnership will allow AMD to license and co-develop Motorola's copper technology (faster than presently-used aluminum), while Motorola will join AMD's research into flash memory – the technology that allows devices to keep data while switched off.

GETTING HIRED

AMD accepts resumes through regular mail, fax, and e-mail, and also through an online form available at its web site. AMD hires at two major U.S. sites – Sunnyvale and Austin, Texas – both of which have their own mailing address, fax number, and e-mail address. This information, as well as current job postings and descriptions, can be obtained at the company's web site (www.amd.com). In addition to a variety of positions, AMD offers different "rotational" training programs that draw from both undergraduate and graduate programs. These include the Wafer Fab training program, which offers a career in manufacturing engineering; the Technical Sales Engineer training program, aimed at producing a top-notch sales force; and the Technology Development Group training program, which sponsors engineers in an individualized program with a sponsor and four mentors.

OUR SURVEY SAYS

Although AMD is frequently cited in the book series the *100 Best Companies to Work For In America*, Vault Reports hears mixed reviews. Some employees at AMD say that it "is one of the best companies that [they] have ever worked for," citing the "great benefits" and the "chance to work with cutting-edge technology" as sources for such enthusiasm. However, another says: "There have been quite a few people dissatisfied with their jobs in the last year and have left our group." "One [employee] left after failing to get along with a certain individual, who also left," according to that insider.

Part of the reason for the varied response is that AMD employees are subject to the vicissitudes of their work. "It's a bit of a roller-coaster financially, but that's the nature of the industry," according to one. "When times are good, the perks are good." This up-and-down nature is especially true when it comes to work hours and intensity. "I've put in 17-hour days and I've put in five-hour days," says one employee. Another says he spent most of a year working 9 to

5, but as his group approaches a target date "sometimes I am up till 1 or 2." However, most employees say their bosses are flexible with their hours, as long as they get their work done.

Although a few contacts describe AMD as "a little frantic," employees generally give good reviews of their work atmosphere. "AMD is less stressful than many other high-tech companies I am familiar with," says one. Another says employees are "not workaholics." One describes a "tight community that's very friendly." Another source has joined the company's basketball league and a bowling league. As far as their bosses go, one worker says he's "proud" to report that "upper management really cares what its employees think." "We are viewed as individuals, rather than cogs in a machine," says another. And dress at AMD is business casual to casual. Many employees report wearing jeans and T-shirts. "We dress very casual," says one. "Finance is the only department I know of who wear ties everyday."

The vast majority of employees surveyed say AMD is a diverse company when it comes to women and ethnic minorities, and that these employees are treated well. Any variations on this theme are chalked up to the ways of the world itself, not AMD. "I have noticed that most of the senior executives – as with many Silicon Valley companies – are white males," says one. "As far as treatment of women and minorities, they have taken huge strides to address this issue. As with all companies, however, all is not perfect," according to one woman insider. "I've known some folks who felt discriminated against, but rarely if ever have I seen this feeling justified," says one manager.

Employees report that AMD pays salaries standard in the industry. One engineer remarks: "I guess they could pay me more, but then I guess I could stand to win the lottery too." But, AMD has "a good profit sharing program" and "1½ percent match on our 401(k)." An employee stock purchase plan may turn out to be a special plus from a company "on the horizon of greatness." Other perks include "discounts for amusement parks," "free workout programs," an on-site daycare center, computer purchase and tuition-reimbursement plans, and, in recent years, "free Rolling Stones tickets." The corporate headquarters in Sunnyvale "are beautiful" and include a fitness center with two basketball courts. But the perk most-commonly cited by employees is the company's 'sabbatical' system, which offers employees two extra months of paid vacation after they complete seven years.

Amazon.com

1516 Second Avenue
Seattle, WA 98101
(206) 622-2335
Fax: (206) 622-2405
www.amazon.com

amazon.com

LOCATIONS

Seattle, WA (HQ)
New Castle, DE

DEPARTMENTS

Business Development
Customer Service
Editorial and Production
Finance
Human Resources
Marketing
Merchandising
Operations
Software Development
Systems and Network Operations Center

THE STATS

Annual Revenues: $147.8 million (1997)
No. of Employees: 1,130
No. of Offices: 2 (U.S.)
Stock Symbol: AMZN (NASDAQ)
CEO: Jeffrey P. Bezos

UPPERS

- Casual atmosphere
- Stock options
- Non-bureaucratic environment

DOWNERS

- Low pay
- Long hours
- High stress

KEY COMPETITORS

- Barnes & Noble
- Blockbuster
- Books-A-Million
- Borders
- Cendant
- Crown Books
- Viacom

EMPLOYMENT CONTACT

Amazon.com
Strategic Growth
P.O. Box 80387
Seattle, WA 98108

techjobs@amazon.com

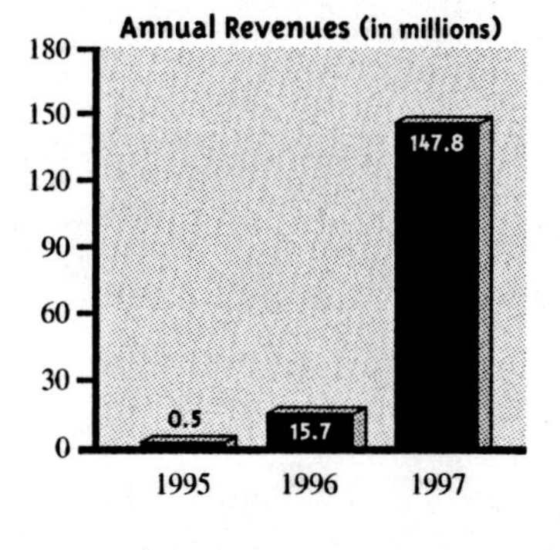

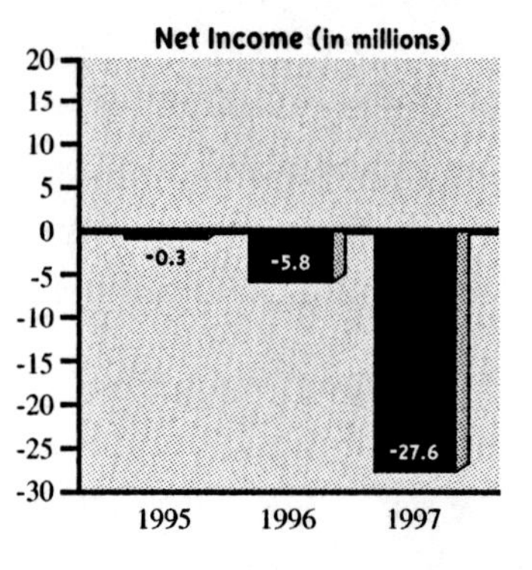

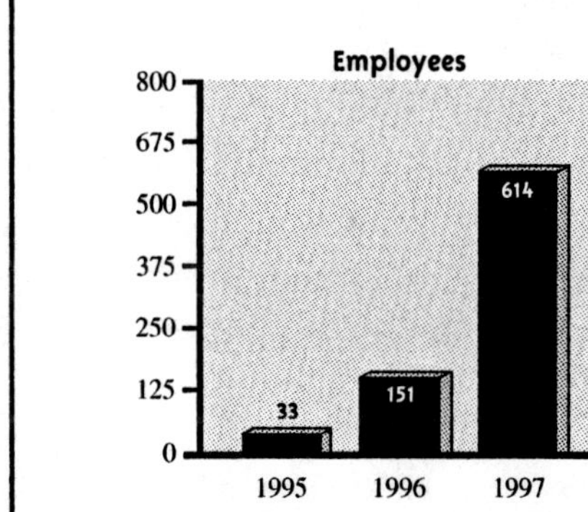
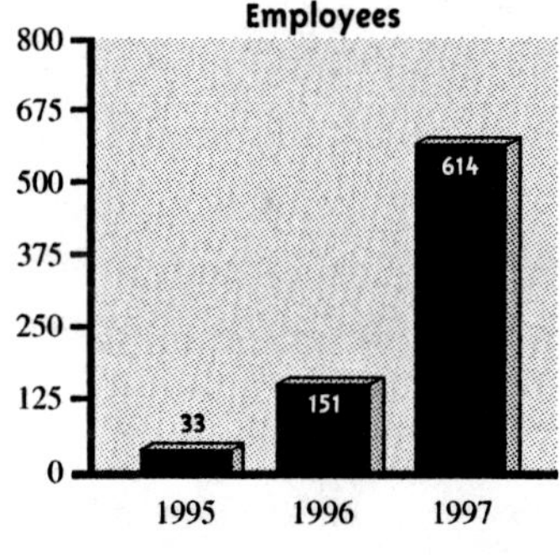

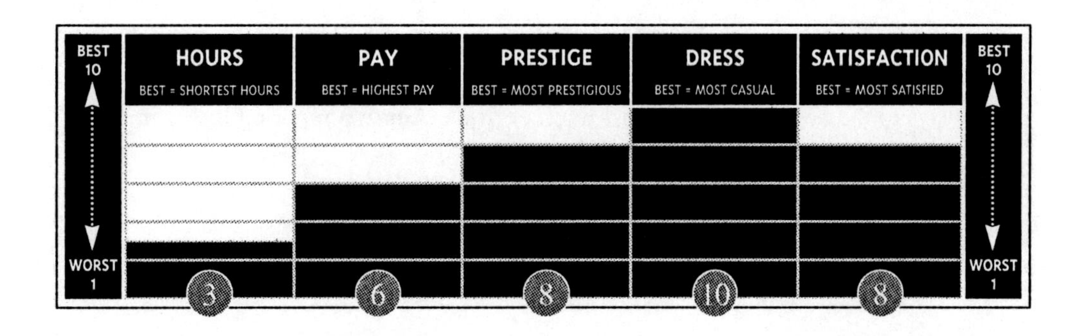

THE SCOOP

One bookseller has been riding the surging tide of e-commerce, today a $3 billion industry. Billed as the "earth's biggest book and music store," Amazon.com has proved to be one of the early bright spots in the nascent world of Internet commerce. Since its 1995 founding, Amazon.com has become the poster child of Internet retail; its fortunes have been watched closely as an indicator of the progress of electronic commerce. Amazon.com reports some of the highest visitor numbers of all commerce sites. Based in Seattle, the company sells 3 million titles through its web site. The site also features CDs, videos, and audiotapes, and Amazon.com has plans to further increase its range of product offerings in the future.

The company was the brainchild of computer guru Jeff Bezos. Bezos had already helped build one of the most successful hedge funds on Wall Street by the 1990s when he turned his eye to the burgeoning field of Internet retail. He founded Amazon.com to fill the gap between the broad field of book publishers titles, and bookstores that couldn't begin to carry all those titles. The company opened in July of 1995, in Bezos's suburban Seattle garage. The site boasted 1.1 million book titles, dwarfing the number carried by the average bookstore (130,000), or even a book superstore (175,000). Amazon's online catalogue allowed visitors to search for books by author, title, subject, or keyword. Dealing directly with distributors and publishers, Amazon.com offered discounts of up to 40 percent. Within a month of its opening, book lovers from every state and 66 countries had made purchases from the site.

The company quickly drew the attention of bookstore powerhouse Barnes & Noble. In May 1997, the successful retailer sued Amazon.com over its claims to have a "far greater" selection of books than Barnes & Noble. The suit was settled in October, with neither party claiming liability nor paying damages. Both companies said they would rather compete in the marketplace than in court. Amazon.com has risen to the challenge. The company established exclusive deals with America Online, Excite and Yahoo!, essentially locking up three major gateways to the Web. But, like most Internet firms, Amazon reported losses in 1997, despite climbing sales and a growing customer base.

In 1998, Amazon expanded its virtual aisles to offer CDs and DVDs online. The music branch of the site offers a bevy of audio clips, reviews, detailed genre synopses, and a personalization option that distinguishes Amazon from older online music stores. In 1998, Amazon also bought UK-based Bookpages and German Telebooks, both Internet bookselling competitors (now transformed into amazon.uk and amazon.de), and the Internet Movie Database, an extensive movie information site that will undoubtedly play a role in launching Amazon's upcoming video branch.

Amazon's potential has certainly caught the imagination of investors. At one point in 1998, its market value was twice that of Barnes & Noble, even though the company has yet to post an annual profit. Amazon will probably continue its conflicted existence until it achieves global brand recognition and automates its ordering process. This central goal to "get big fast" includes plans to step up advertising, publicity, and alliances with Internet portals. In 1998, Amazon's revenues hit $1 billion.

GETTING HIRED

Amazon.com looks for smart people with experience and drive. A growing company, Amazon may have a job for you, even if you aren't a "techie." One contact notes, "Unix and PC skills are a big asset here," but "the key to working here is flexibility." The same contact reports, "It's fairly easy to move within jobs or departments, depending on your skills and interests. I have a BA in Romance Languages, an MA in Italian, and an MLS – the combination of humanities with 'techiness' is what has allowed me to move around a lot. It's cool to work somewhere that's both techie and book-centered at the same time." With Amazon expanding abroad, another insider notes, "If you have any foreign language skills, they would be highly valued." Depending on what job you apply for, one insider says, "They may want writing samples, SAT and GRE scores (if you have them), and several references." One insider reports that "the No.1 factor for being hired [at Amazon] is intelligence, and standardized test scores are one way of measuring that." Still, being technically proficient certainly doesn't hurt. One insider reports: "You should get some experience under your belt, like knowing HTML and perhaps some creative programs like PhotoShop."

Be prepared for a straightforward but rigorous interview process, which can take several months to complete. Candidates typically go through several rounds. Though questions vary by position, and the interviewer's personal style, most "can expect some to be highly technical." Whatever you do, be original, and don't fudge on your references. "When interviewing," one insider notes, "you should know that you can challenge conventional wisdom, find creative solutions, think analytically and quantitatively, and stay focused on your goals." In addition, one insider warns, "Reference checking is extremely thorough." Visit www.amazon.com for a complete listing of openings in the company's Seattle headquarters, and newer Delaware location. Resumes can be e-mailed to jobs@amazon.com, or sent by mail to the company's Strategic Growth address, listed on Amazon's web site.

OUR SURVEY SAYS

Amazon offers the long hours and casual atmosphere typical to the high-tech field. One insider sums up Amazon as "dynamic, hectic, casual, professional, and everything in between all at once." Most insiders praise Amazon's creative and friendly culture, and say that the long hours and hard work are worth it. "It's hard work, and folks don't blink at a 60-hour workweek," one insider reports, "but everyone is here because they love it." With an "extremely casual" corporate culture, one book-lover notes: "Even the president, Jeff Bezos keeps dress shirts in his office because he's often in jeans." The same insider credits Microsoft for much of Amazon's atmosphere, "This being Seattle, there are lots of Microsoft refugees, and they've imported the relaxed yet high intensity culture."

In fact, "high intensity" may be an understatement. "It is extremely demanding," one contact warns, "and if you do come to work for Amazon, you will work." Depending on your position, the long hours may not be reflected in your salary. One insider reports: "The pay is fair, although probably not the highest. It depends on what you're after. If you're a programmer (especially C++ & Oracle), your pay will likely be at the high end of the scale, but if you're an editor, the pay may be slightly lower than at other companies." However, the same contact notes: "The upside is that everyone receives stock options as part of their compensation package, and that's what most people are counting on, more than salary."

The people, many report, may be the best part of working for Amazon. "The people are great, very open-minded and casual," one Amazon insider reports. "In fact, my co-workers are one of the best reasons to work at Amazon.com. Everyone who works here is an overachiever in one way or another, so there are plenty of intelligent and supportive people to work with." "It is as close to a meritocracy as any organization I know," another insider reports. "There are women and minorities at every level of the company." Another contact notes: "The treatment here of everyone is wonderful, and the female energy is very strong." "I see no barriers nor favoritism for women or minorities." Although Amazon has only two female vice presidents, "the majority of the next-level managers are women."

Amdahl

1250 E. Arques Ave.
Sunnyvale, CA 94088
(408) 746-6000
Fax: (408) 773-0833
www.amdahl.com

amdahl

Chairman: John C. Lewis
President & CEO: David B. Wright

LOCATIONS

Sunnyvale, CA (HQ)
Bellevue, WA • Dallas, TX • Indianapolis, IN • Jersey City, NJ • New York, NY • Phoenix, AZ • San Ramon, CA • Santa Clara, CA • Tampa, FL • Calgary, Canada • Halifax, Nova Scotia • Brussels, Belgium • Paris, France • Munich, Germany • Sinapore • Syndney, Australia • As well as numerous other offices around the world.

DEPARTMENTS

Compatible Systems
Customer Services Organization
DMR Consulting
Open Systems Group
Software Business Group

THE STATS

Annual Revenues: $2.2 billion (1997)
No. of Employees: 12,000 (worldwide)
No. of Offices: 151 (worldwide); 75 (U.S)
Subsidiary of Fujitsu Ltd.

UPPERS

- 401(k) plan
- Growing business
- Frequent dress-down days
- On-site gym

DOWNERS

- Potentially limited advancement for minorities
- Political hijinx at higher levels

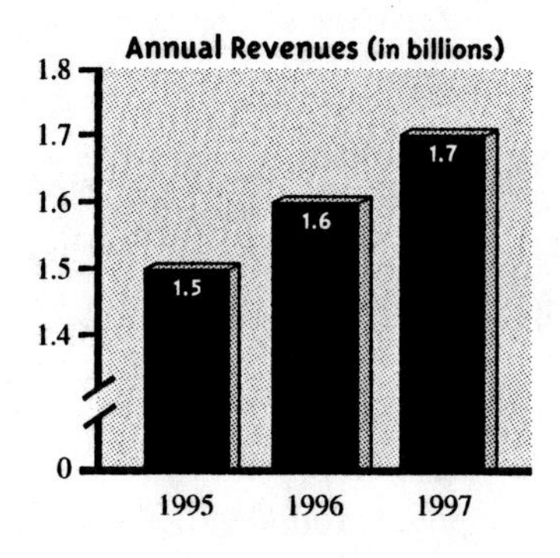

KEY COMPETITORS

- Andersen Consulting
- Cap Gemini
- Digital Equipment
- Hewlett-Packard
- Hitachi
- IBM
- NCR
- Unisys
- Wang

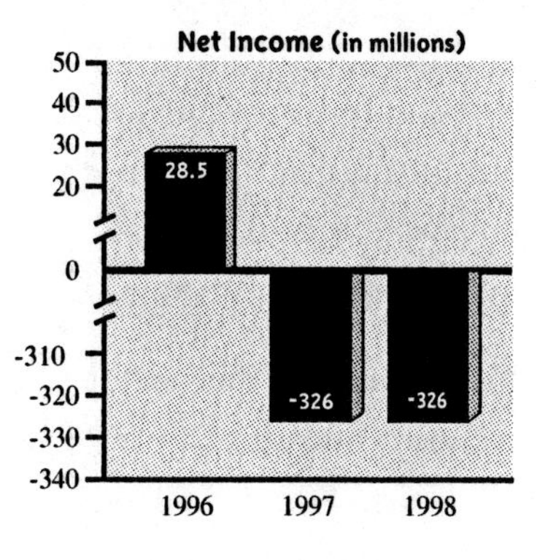

EMPLOYMENT CONTACT

Gene Plonka
Director of Staffing Programs and EEO
Amdahl
1250 E. Arques Ave.
Sunnyvale, CA 94088

(408) 746-6065
Fax: (408) 992-2389

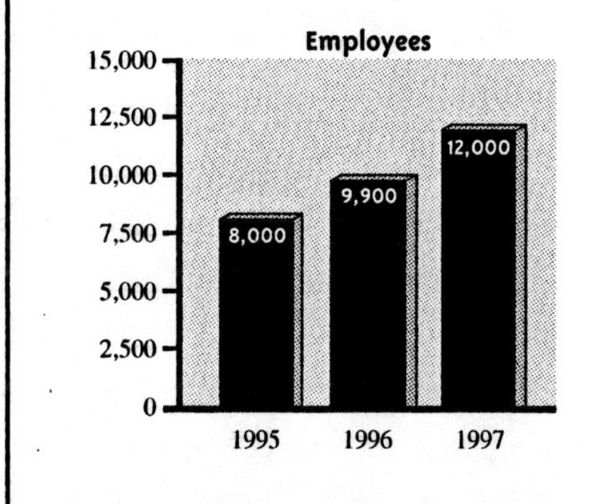

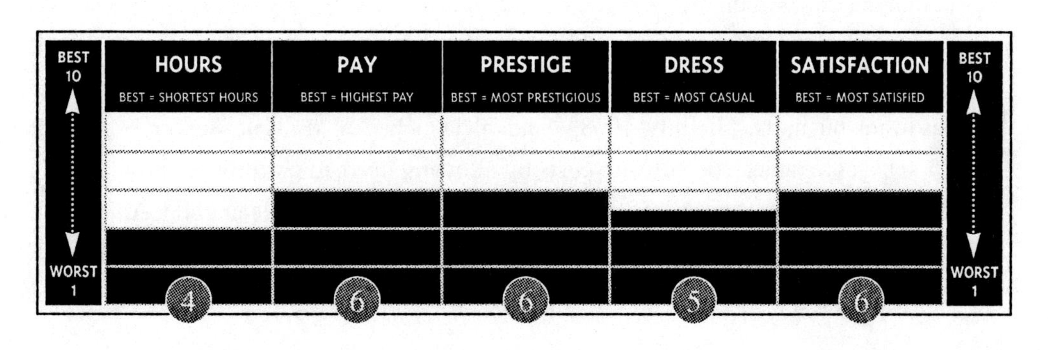

THE SCOOP

Gene Amdahl, the principal designer of IBM's popular System/360 family of mainframe computers, founded his own eponymous company in 1970 after Big Blue rejected his idea for a more advanced computer. Amdahl released his IBM clone in 1975 that ran faster and cheaper than IBM machines, while boasting compatibility with IBM software and peripherals. Initial success convinced him to take his company public in 1976. IBM retaliated by slashing prices and announced the impending release of its own improved product, heralding the arrival of competition in the large-scale computing business. Amdahl himself left in 1980 to found Trilogy, an ill-fated mainframe and computer chip maker.

Under Eugene White and Jack Lewis, Amdahl's successors, Amdahl grew through acquisitions and diversification. In 1995 and 1996, Amdahl acquired two information-technology service providers, TRECOM Business Systems, and DMR Group, a Canadian company. In 1997, Fujitsu took over Amdahl (it already controlled a 42-percent stake), but has allowed the company to run as a separate subsidiary.

Amdahl's present strategy is to exploit its more lucrative support services arm and to transform from a simple hardware/software company to a full-service technical "solutions-driven" enterprise. Amdahl's strategy consists of supplying a corporation's hardware needs and then offering related operational and consulting services. Thus far, Amdahl has benefited from the corporate rush to prepare for the Year 2000, or Y2K, situation. It has also developed specialized groups to serve different areas of business. In 1996, Amdahl formed a SmartCard group to produce processing systems to support worldwide electronic cash networks. And in 1997, Amdahl formed a Telecom Business Group to market its services to the telecommunications industry. Today, 60 percent of the company's business comes from its support services. The company has business partnerships with many other high-tech companies, including Microsoft. But growth in services doesn't mean that Amdahl's neglected its hardware business. In July 1998, Amdahl launched a Multiple Server Facility (MSF), which helps companies cut software costs by allowing users to partition their mainframes into multiple servers, then run separate applications on each section. Meanwhile, Amdahl has new 1000 MIPS Millenium mainframes, which will hit the market in early 1999, will rival both

IBM's S/390 and Hitachi's Skyline systems in terms of pure processing power and capacity. And since most mainframe users currently need only about 400 million instructions per second in processing power and won't immediately require Amdahl's full 1000 (MIPS) processing power, the company has made its system scalable to about 85 MIPS.

GETTING HIRED

Like any high-tech company worth its salt, Amdahl posts its job listings on its web site. Although the company recruits at 35 campuses, employees reveal that "internal referrals generally hold more weight than someone coming in off the street." So if you know someone, use him or her – especially if you don't have much experience. "The key for recent college graduates is to have a contact who can be your 'champion.' Otherwise you are just one of many applicants," says one insider. Amdahl also recruits on college campuses – check out the company web page for a list of the schools the company visits every year.

In most cases, interviews consist of several "relaxed" rounds. Employees say they are "usually an all-day event for software/hardware people," though sometimes the interviews take place over two days. Luckily, "Amdahl doesn't take the hard approach – there are questions, of course, but not too many tests."

OUR SURVEY SAYS

Insiders say "there isn't really a distinct corporate culture" at Amdahl. "It's very political at the higher levels," but at the departmental level, "it's very fragmented," and "each department has a small shop feel." This "family feeling makes for lots of fun," say employees. Insiders also seem content with Fujitsu's recent acquisition of the company. "We work very closely with Fujitsu developing products; we are, however, a U.S. company," says one employee. In addition to increased funds for R&D, sources report that "it's a fantastic time to be here as we change from a hardware/software provider to a solutions-driven company."

In the company's headquarters and other corporate offices, say insiders, "there is definitely a dress code – business casual from Monday to Thursday, and dress-down Fridays (meaning jeans and T-shirts)." In creative services and engineering departments, casual day is every day, and "jeans, T-shirts, shorts, etc." show up regularly.

"The company is very good to women and minorities at the lower levels," reports one insider, "and we have an outreach program here specifically to draw minority employees and interns." "This program, called InRoads, relies on relationships between Amdahl and 12 professional assocations for minorities. However, warns one employee, all but one VP and "other high-level people here are all white males." And minorities interested in climbing the corporate ladder be warned: some sources say "this isn't the place to be." It's also worth noting that "the average age of employees is higher than most of Silicon Valley." Salaries are standard for the industry, and the benefits package includes a 401(k) plan, childcare resource and referral service, healthcare (including coverage for domestic partners), on-site fitness facilities and health services, and tuition reimbursement.

"We're always far too desperate to find good people to give any thought to race."

– ***Hardware insider***

America Online

22000 AOL Way
Dulles, VA 20166
(703) 448-8700
www.aol.com

LOCATIONS

Dulles, VA (HQ)
Major locations in CA • NY
Call centers in AZ • FL • NM • OH • OK • UT

DEPARTMENTS

Corporate Communications
Corporate Development
Customer Service
Finance
Human Resources
Interactive Services
International Division
Legal
New Market Development
Sales & Marketing
Technology Development

THE STATS

Annual Revenues: $2.6 billion (1998)
No. of Employees: 8,500 (worldwide)
Stock Symbol: AOL (NYSE)
CEO: Stephen M. Case

UPPERS

- Company gym
- Casual dress
- Stock options
- Free AOL account for you and a friend

DOWNERS

- Frantic environment
- Lots of smileys :)

KEY COMPETITORS

- AT&T
- CNET
- Excite
- Infoseek
- Lycos
- MCI Worldcom
- Microsoft
- Yahoo!

EMPLOYMENT CONTACT

Mark Stavish
Senior Vice President for HR and Facilities
America Online
22000 AOL Way
Dulles, VA 20166

(703) 365-2666

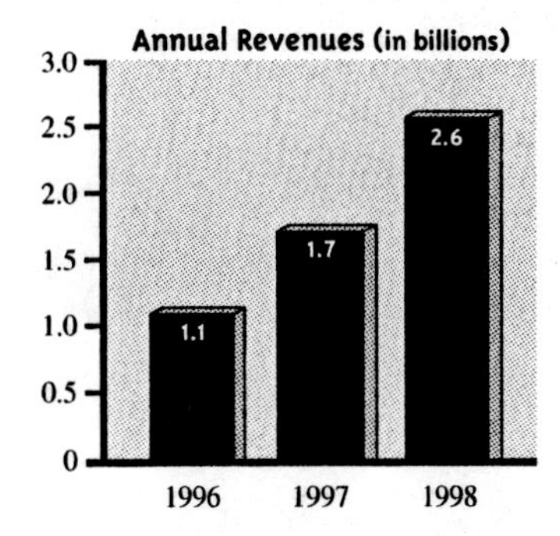

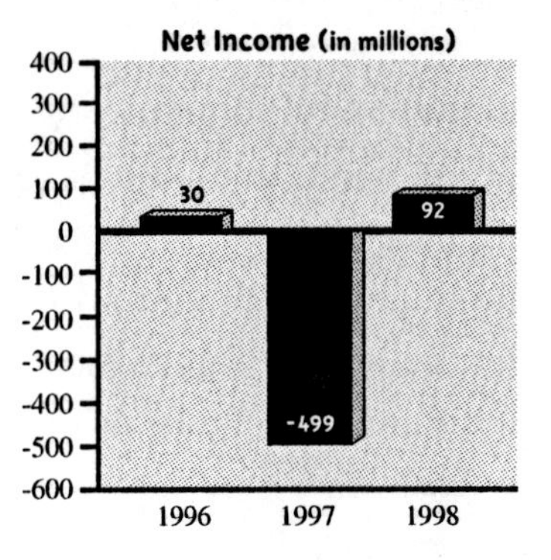

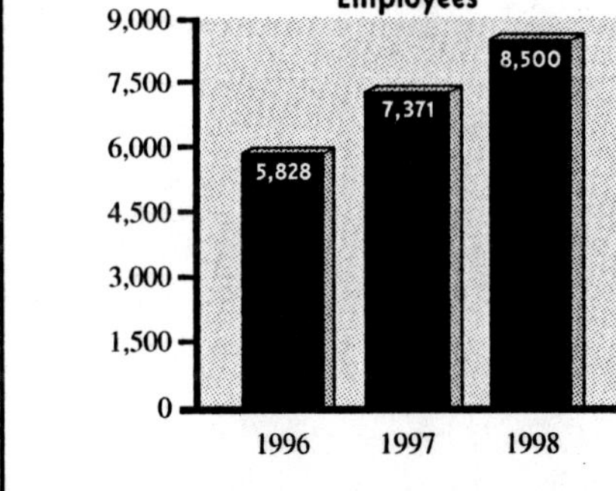
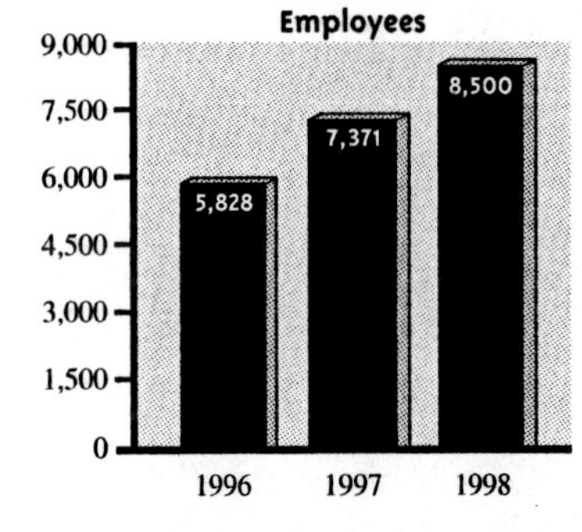

HOURS	PAY	PRESTIGE	DRESS	SATISFACTION
BEST = SHORTEST HOURS	BEST = HIGHEST PAY	BEST = MOST PRESTIGIOUS	BEST = MOST CASUAL	BEST = MOST SATISFIED
4	5	9	10	9

BEST 10 — WORST 1

THE SCOOP

America Online (AOL) is the world's most popular online service provider and the first billion-dollar company in the Internet industry. With over 14 million subscribers, successful marketing techniques, and improved modem access, AOL has assured its primacy in the market. Founded by current CEO Steve Case in 1989, AOL was the first online service to provide World Wide Web access in addition to Internet services and e-mail. After going public in 1992, AOL expanded its services in a successful effort to surpass its competitors. Since then, the company has improved its position by adding the Netscape and Microsoft browsers to its service. In 1996, the company grabbed headlines when it began offering unlimited Internet access for a low monthly fee. The plan was so successful that AOL's equipment could not handle the demand at first, leading to widespread complaints and service delays. The company addressed the problem by increasing its modem capability. AOL acquired arch-rival CompuServe the following year, and maintains its lead through alliances with such companies as retail and travel conglomerate CUC International, online bookseller Amazon.com, and online supermarket NetGrocer.

The company's spreading global reach had led industry analysts to speculate about its potential takeover by a media or telecommunications giant, but AOL stunned the Internet world when it announced its own takeover in November 1998 – of rival Netscape for $4.2 billion. With the acquisition, America Online is expected to aggressively pursue the electronic commerce support business. The company that has become the player in the Internet industry on the back of chat rooms, stock quotes, and other consumer-oriented services now hopes to draw more of its revenue from companies. With the deal, AOL, Netscape and Sun are betting that AOL's media presence (its 14 million subscribers), Netscape's software (its browser and e-commerce software) and Sun's hardware, Java programming language, and sales and support force, can defeat titans like IBM and Microsoft. The partners (Netscape will continue to be run as an independent business) envision a world in consumers log on to the Internet using AOL through not only computers, but – because of Java – through appliances like pagers or cell phones as well. In this world, companies would pay AOL and Netscape for software to help run businesses utilizing e-commerce (this software would be sold and implemented in part by Sun sales and support staff). And in this world, the partners would be able to use "online real estate," that is, advertising backed up by AOL's enormous traffic, as an enticement to work with the companies. Ambitious plans for a company only a decade old, yes, but nobody's ready to scoff at 14 million subscribers – power in the new media biz truly belongs to "eyeballs."

GETTING HIRED

To find out about current job opportunities consult the "Careers" section of AOL's web site, located at www.aol.com. In addition to sending a resume to Human Resources, applicants should write directly to the department in which they are interested. AOL insiders advise candidates to use e-mail aggressively in making company contacts. The company hires people of all professional backgrounds. Opportunities run the gamut from programming and software development to editing and writing. Whether you're interested in marketing or accounting, there may be a place for you at America Online. Technophobes need not apply. With employee growth over 25 percent for 1997, AOL is a fast-growing enterprise interested in hiring young people. Finding a job with AOL takes initiative, to make things easier, the company has begun on-campus recruiting.

AOL employees report that the company offers many entry-level opportunities but that more advanced positions require "dogged persistence." They advise new employees to "be aggressive" and "make sure you have the projects that you want." For business school grads, the "easy access to VPs and sometimes even the CEO" can be very rewarding, although there is no set path of entry for MBAs because it is difficult to predict the number of positions available for business school grads.

OUR SURVEY SAYS

The culture at AOL is described as "pretty laid-back yet intense": "It's hip high tech – a mix of young people in a fascinating new industry." Insiders describe co-workers as "young, energetic, and motivated." One employee says walking through the halls in AOL's Virginia offices, he sees "people hanging out in small groups chatting about projects they're working on – always seeming so casual, yet there's a sort of burn in their eyes. It's hard to describe."

There's nothing stuffy about the dress code at AOL. "Everyone here wears shorts, jeans, or any particular clothing that covers the erogenous parts of the body," says one insider. Loincloths aside, the quirky computer-rebel image sticks: "Most of the employees here are the 15-earrings type. There are no drug tests here, so who knows what kind of people we have." More commonly, employees wear "jeans, T-shirts and sneakers everyday." However, on the corporate side, employees are expected to dress in a more businesslike manner. Still, for techies and others with a lax dress code who don't have enough of the standard company uniform, AOL is happy to oblige: "I have two drawers full of AOL shirts." This isn't just one T-shirt hoarder talking – a programmer reports that "there's lots of free T-shirts." One employee at the Florida call center cavils that free T-shirts constitute a main AOL benefit, wryly commenting that "AOL isn't just a job, it's a wardrobe."

Where hours are concerned, many insiders report that they have flexible hours: "I'm not a morning person, so I usually do noon to 9 p.m. or so." "Personally, I set my own hours," brags another contact. Flexible, however, doesn't mean short: "It's not uncommon for people to put in 60 hours a week," says one insider. Reports another: "The hours can be long, and can take a toll on your personal life, especially if you are on a track for increased responsibility." And: "The company only requires eight-hour workdays officially, but realistically, many people work longer hours than that." Insiders attribute the long workdays largely to a culture of constant innovation: "It can be difficult for those who don't like to work hard, since we're always pushing ourselves to do things that have never been done before."

Marketing and accounting representatives describe their atmosphere as "informal and less stuffy than most places." This is partly attributed to the relative youth of AOLers (one insider estimates the average to be "in the mid- to late-20s," though the company says the average is around 32). Others say the age of the employees also pays dividends when it comes to an equal opportunity environment. "Maybe it's because of the age of the company or the relative youth of its employees, but the concept of an 'Old Boys Network' or cabal of one group or another does not seem to exist here." Says another: "It's completely safe to be an out gay, bisexual, or lesbian here." "Black, white, brown, women, men – they will hire anyone that knows jack about computers," says one techie. Fair enough.

AOL employees are not as effusive when it comes to pay. "It certainly pays the bill and leaves me some left over to play," reports one source. "It's probably not the highest you'll find, but competitive," comments another insider. However, insiders say that AOL has a 401(k) plan that matches 50 cents on the dollar up to 6 percent of an employee's salary. Other perks include both annual stock grants (through which employees are given AOL shares) and an employee stock purchase plan (through which employees can buy stock at a discount), a company gym, a "decent health plan," and a "free AOL account for you and one other person of your choosing.

Most important, however, is that most company insiders report an extraordinary level of job satisfaction. "I still wake up in the morning looking forward to coming into the office," says one employee who has been with the company for four years. "Out of almost six years," says another, "there's probably been less than 10 days that I haven't looked forward to coming into work." Getting repetitive? "Overall, I've never regretted it," says yet another. "I can't imagine working anywhere else."

Apple Computer

1 Infinite Loop, MS 75-2CE
Cupertino, CA 95014
(408) 974-7411
Fax: (408) 974-5691
www.apple.com

LOCATIONS

Cupertino, CA (HQ)
Atlanta, GA
Austin, TX
Chicago, IL

DEPARTMENTS

Finance
Hardware Engineering
Interactive Media
Marketing/Product Marketing
Operations
Service and Support
Software Engineering

THE STATS

Annual Revenues: $5.9 billion (1998)
No. of Employees: 9,700
Stock Symbol: AAPL (NASDAQ)
Interim CEO: Steven P. Jobs

UPPERS

- Profit sharing
- Stock discounts
- Product discounts
- Substantial vacation time
- Fee-based health club
- Training courses
- Brown bag lunches with execs
- Casual dress

DOWNERS

- Diminished reputation of company
- Constant re-organization
- Long workdays
- Steve Jobs isn't going to be there forever

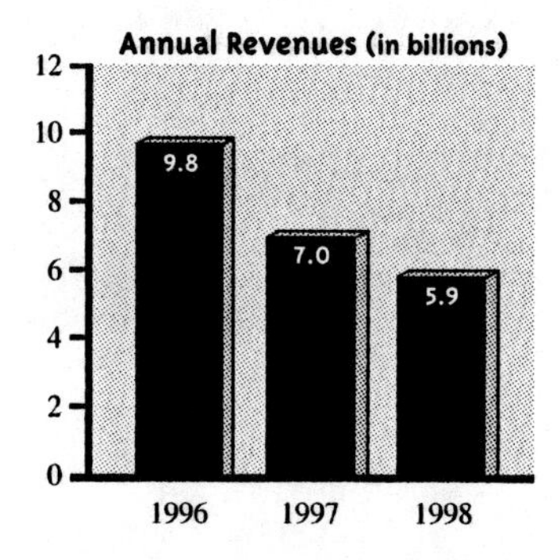

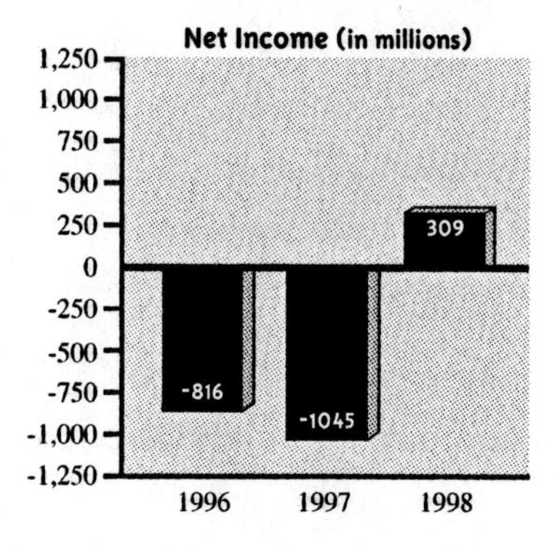

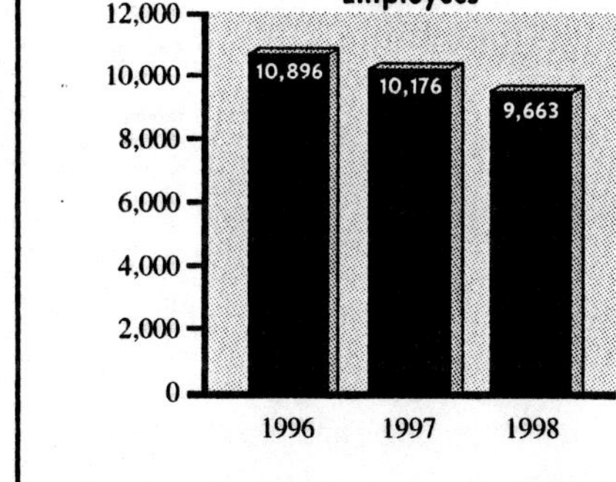
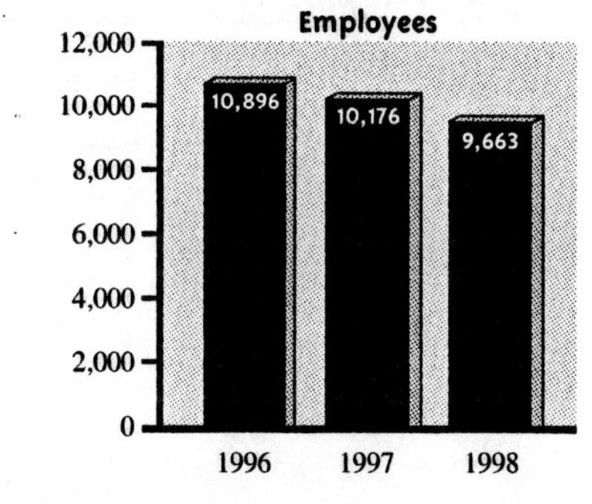

KEY COMPETITORS

- Canon
- Compaq
- Dell Computer
- Gateway
- Hewlett-Packard
- IBM
- Intel
- Microsoft
- Novell
- Sony
- Sun Microsystems

EMPLOYMENT CONTACT

Human Resources
Apple Computer
1 Infinite Loop, MS 38-3CE
Cupertino, CA 95014

applejobs@apple.com

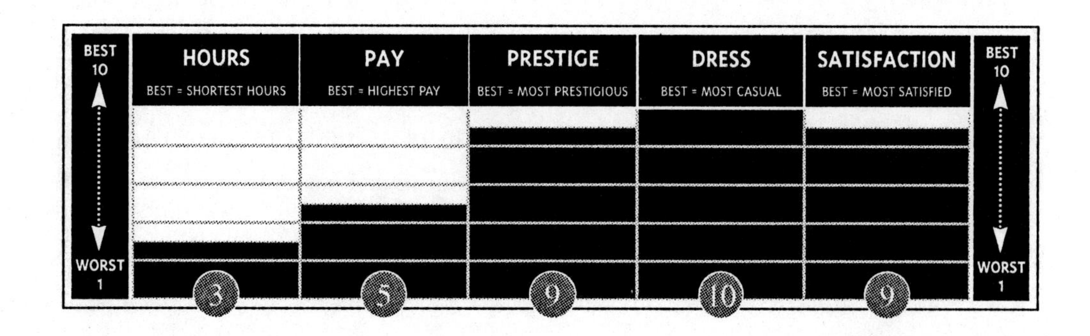

THE SCOOP

No "two guys in a garage" have managed to capture the nation's hearts and minds like Steven Jobs and Steven Wozniak did over two decades ago. In 1976 these two Silicon Valley dreamers emerged from their makeshift laboratory with a PC that would change the face of computing. Apple's Macintosh offered consumers the most advanced, user-friendly interface consumers would see for the next dozen years, until Microsoft finally managed to ape much of Apple's finest attributes for the Microsoft Windows 95 system. Nonetheless, since the release of Macintosh, Apple's home PC, Apple's string of innovative and inspired products have won them an almost cult-like loyalty from their users.

But the finest operating systems, regrettably for Apple, don't always translate into marketplace success. On the open market, the Macintosh has been sorely beaten by Microsoft's MS-DOS system. Apple, outflanked by Microsoft, collapsed. Its market share flumped from 18 percent in 1994 to a pitiful 6 percent in 1996, prompting some industry insiders to predict the company's imminent demise. Nevertheless, Apple's line of PowerPCs, Performas, and PowerBooks remained popular with students and graphic designers. In an effort to go back to the source, then CEO Gil Ameilo acquired NEXT, Steve Jobs' second enterprise, because of the new operating system it was developing. The move brought Jobs back into the Apple arena, exactly where stockholders wanted him.

When CEO Amelio left Apple, Steve Jobs proved an inspiration to Apple once again. Agreeing to take control of Apple as interim CEO in September 1997, Jobs immediately embarked on a series of major changes designed to return the computer maker to stability. Within the first weeks of his tenure Jobs reached an agreement with longtime corporate enemy Bill Gates, CEO of Microsoft, whose support, Jobs knew, was essential to the survival of underdog Apple. Gates agreed to $150 million in software investment for MacOS programs. Jobs also killed the production of Mac clones, which were draining the already meager sales from Apple's own line.

Jobs pared down Apple, cutting jobs and reducing its PC line from 15 computers to a smaller product line aimed at Apple's core of customers: educational institutions, desktop publishers, designers and home users.

On May 11, 1998, standing before a room of Macintosh programmers, Jobs unveiled what may be the last great hope for Apple's future success: the G3 computer line. Touted as running "up to two times as fast as Pentium processors," the G3 arrived in three incarnations: the sleek black Power Book G3, the Power Mac G3 desktop, and the startling iMac one-piece consumer desktop.

The iMac, with its bulbous, colorful, translucent shell, is undoubtedly attractive and innovative. But is it the lifeline Apple needs? Apple's designers have made the unusual decision not to include a floppy disk (3.5) drive. Jobs holds the opinion that most information nowadays is transferred via the Internet, CD-ROMs and Zip Iomega drives. (*Fortune* scoffs that this might be true "in about five years.") Apple's lack of a floppy drive may also prevent its educational customers – which support many computer labs – from upgrading to the iMac the move may prove to be detrimental. Apple currently has 44 percent of the educational market, though it is weakening in colleges and universities.

Regardless, early response to the computer has been favorable. Christmas sales in 1998 were brisk – same-month percentage of the PC market rose to 11.6 percent in October, versus 7.6 percent in October 1997. More important to the future of Apple, nearly a third of iMac buyers have never owned a home PC before, and 12 percent were formerly Windows-PC users. Just as promisingly, Jobs – the so-called "interim" CEO – looks like he's staying put.

Apple has further changes up its silicon sleeve for 1999. At the Macworld expo in January 1999, Steve Jobs – or, as he jokingly described himself, the "iCEO," introduced the new series of iMacs, which not only retail for $100 less than the original iMac, but come in an assortment of five fruity hues (tangerine, blueberry, strawberry, lime and grape). (Some retailers fear that, much like the new Volkswagon Beetle, one shade of iMac will become wildly popular, causing supply delays.) The less superficial should note that Apple has also introduced a new Apple attribute with "NetBoot" capability that allows system software to run directly from a server.

GETTING HIRED

A bachelor's or master's degree in computer science, electrical engineering, or mechanical engineering is usually required for most entry-level positions at Apple. Interested parties should either mail, fax, or e-mail their resumes. (E-mailed resumes should be pasted into the message box, not attached as separate document.)

Apple's rocky financial history makes it difficult to join the team; applicants must not only be of the highest quality, but also be specifically enthusiastic about joining the Apple family. (Quick: what kind of computer do *you* have at home?) Apple's looking for people who know SAP, OOA/OOD, EDI/AS400, WebObjects, EOF, Rhapsody, NeXTStep/OpenStep, UNIX, C, Object C or NT. To search current job listings in a variety of divisions, check Apple's site http://www.apple.com/employment/jobs.html, where "you can submit a plain text resume via e-mail."

OUR SURVEY SAYS

Talk to some Apple employees, and you might come away with the impression that there's no better place to work than Apple. "Funny, brilliant, relaxed" co-workers and "modern, spacious, beautiful" offices filled with comfortable couches and huge picture windows make work time a pleasure. While most admit "Apple has changed over the past 10 years," insiders stay put, comments like "absolutely no downers," "everyone is an essential part of the team," and "Apple remains my dream company," could make anyone want to pack their bags for Cupertino.

Then again, some insiders still have worries about the "depressing" negative publicity surrounding Apple; concerns about "a high turnover rate" and a "sometimes gloomy mood" bother some staff members. (The recent popularity of the iMac has certainly made Apple coverage more favorable, and Wall Street has been mollified by the surge of Apple stock prices, up to around 40 from a low of 12 3/4.) "If you put your heart into the 'Mac vs. Windows' jihad, it's very frustrating to see an inferior OS getting all the attention," says one programmer. The downside of the unusually creative atmosphere is that sometimes trying to enforce a corporate strategy is like "herding cats." The pay is "not bad, but not the best in the Bay Area," though some contacts add "the people make up for the pay."

Most of our contacts say they have no plans to leave Apple anytime soon. A dress code that is casual "with a capital C" and a uniquely Californian "work hard – play hard" culture make up for the hours toiling at the office. "Working at Apple is a wonderful experience," employee says, noting that the environment is "intellectually challenging, inspiring and rewarding." "The culture is the best," another insider gushes.

The main R&D campus in "Cupertino is pretty posh, with a huge lawn in the middle, a very nice cafeteria, and a big indoor atrium with a café." Perks at the liberal icon include "same-sex domestic partner medical benefits." And while the company is mostly white and male, "Apple does look for women and minorities." Insiders say "it's common knowledge that Apple's progressive nature is a large part of its image."

Applied Materials

3050 Bowers Avenue
Santa Clara, CA 95054
(408) 727-5555
Fax: (408) 727-9943
www.appliedmaterials.com

APPLIED MATERIALS®

LOCATIONS

Santa Clara, CA (HQ)

Research, Development and Manufacturing
Austin, TX
Tel Aviv, Israel
Chunan, Korea
Horsham, CA
Narita, Japan
Hsinchu, Taiwan

Sales and Service offices
Sales and Services offices are located in the United States, Europe, Israel, Japan, South Korea, Taiwan, Singapore, and China

DEPARTMENTS

Administration
Business Operations
Human Resources
Manufacturing Operations
Marketing
Product Operations
Quality Assurance
Research & Development

THE STATS

Annual Revenues: $4.04 billion (1998)
No. of Employees: 12,000 (worldwide)
No. of Offices: The company has more than 95 locations in 14 countries
Stock Symbol: AMAT (NASDAQ)
CEO: James C. Morgan

UPPERS

- Stock options
- Profit sharing
- Tuition subsidies
- Wellness programs and company gyms

DOWNERS

- High pressure work environment
- Irregular work schedules

KEY COMPETITORS

- Advantest
- Canon
- KLA-Tencor
- Lam Research
- Nikon Corporation
- Novellus Systems
- Teradyne
- Tokyo Electron

EMPLOYMENT CONTACT

College Programs
Applied Materials
3050 Bowers Avenue, MS 1826,
Dept. TCE0996PW
Santa Clara, CA 95054

staffing@amal.com

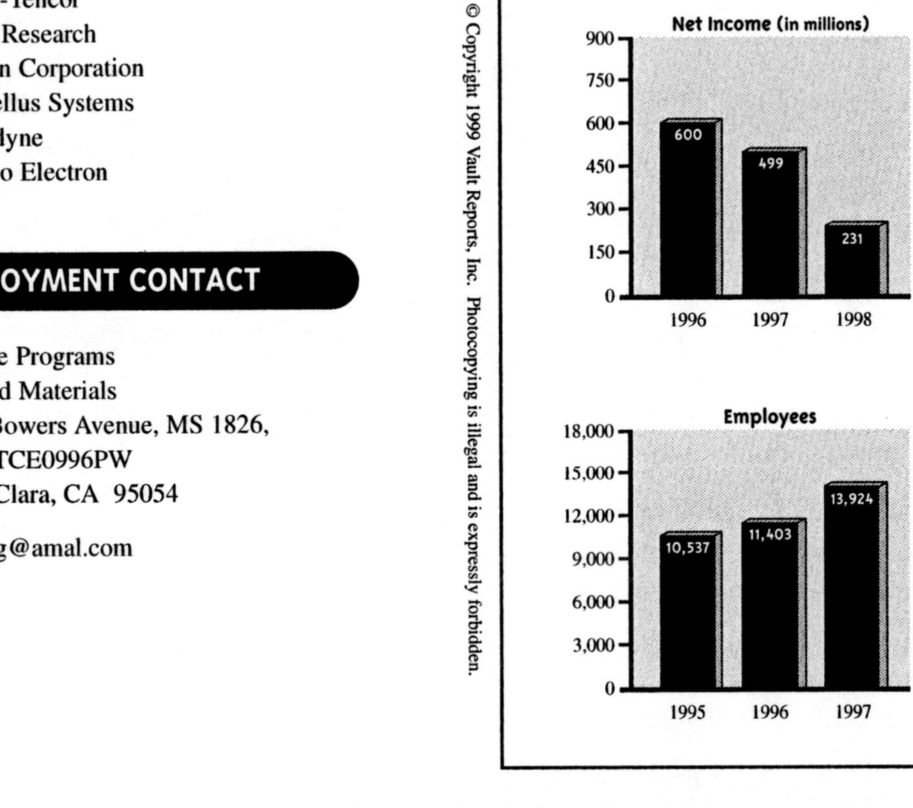

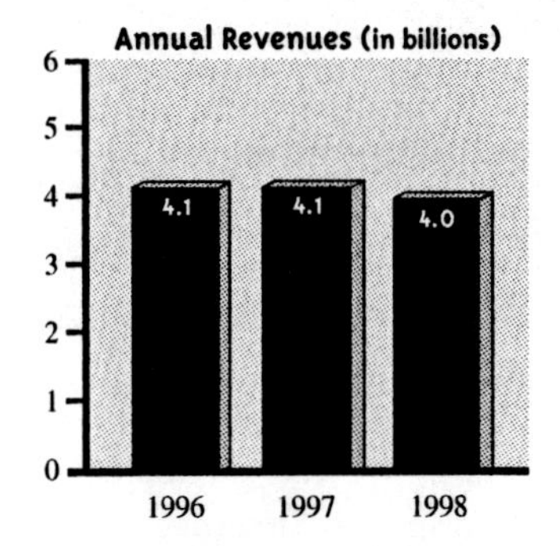

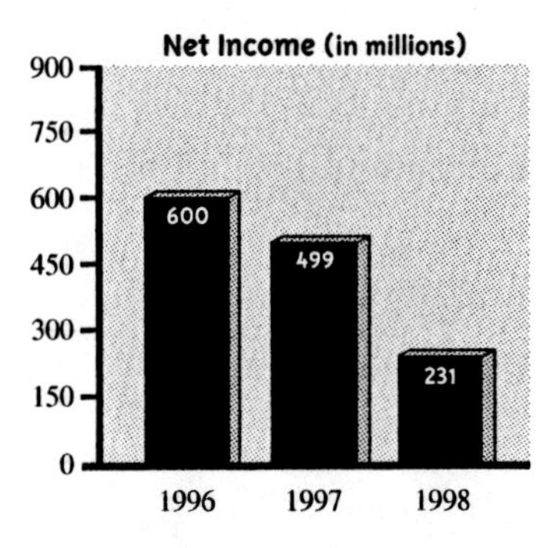

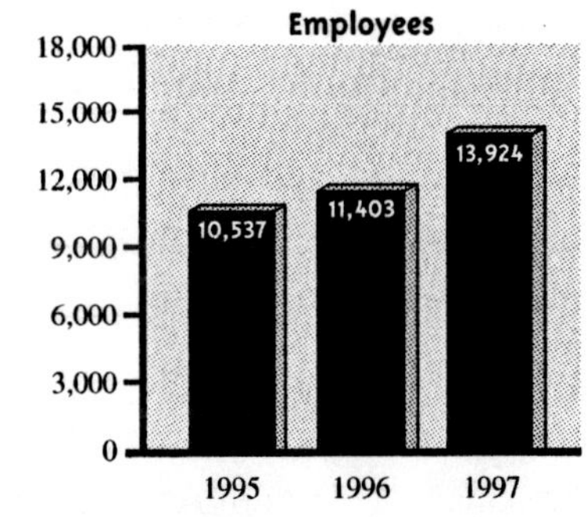

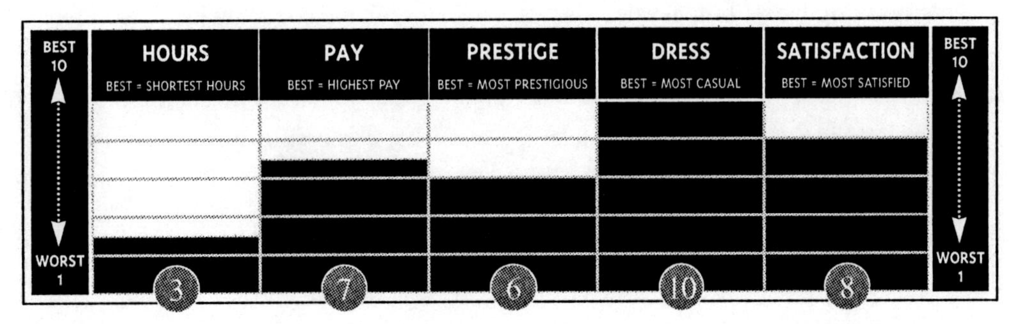

THE SCOOP

High-visibility companies like Microsoft and Intel may be leading the parade onto the Information Superhighway, but Applied Materials helped pave the silicon road. As the world's largest producer of silicon wafer fabrication equipment, Applied Materials' machines execute the precise, multi-stage processes of making silicon chips for today's computers. Applied Materials competes in various segments of the chip-making industry, such as deposition, etching, and ion implantation. As a leader in a field where knowledge and technology translate directly to billions of dollars, Applied Materials also spends its energy defending its patents. In 1997, rival company Novellus paid Applied Materials more than $80 million to settle a patent infringement lawsuit.

The company hasn't simply been fighting competitors in court – it's also bought a few. At the end of 1996, the company purchased Opal, Inc. and Orbot Instruments, Ltd., for which Applied Materials paid $175 million and $110 million respectively, both in cash. Opal supplies CD-SEM systems (which measure dimensions of circuits during the manufacturing process); Orbot builds systems that inspect wafers during the process.

Globalization is another important part of Applied Materials' strategy. The company's sales outside of North America are more than double those on the continent. Applied Materials has done business in Asia successfully since it set up its first joint venture, Applied Materials Japan, in 1979. Applied CEO James Morgan even wrote a book based on his experiences working with the Japanese, called *Cracking the Japanese Market.* The company continues to prepare for growth in Asia's emerging markets.

If the past is any indication, Applied Materials should have no problem building on its success. After breaking the $1 billion sales mark in 1993, it took the company just three more years to quadruple that figure. In March 1997, these "wafer warlords" demonstrated the first system ever capable of producing 300mm wafers, pitching the company once more into the forefront of a new technology. Four months later the insatiable corporation announced it was throwing another $430 million into expansion of its home-base facilities in Santa Clara, California. Applied has experienced some growing pains, however. A persistent slump in the global semiconductor industry in recent years has compelled Applied Materials to restructure. In

June 1998, Applied Materials offered some employees a voluntary separation package in an effort to pare down its workforce by 1000 at its Austin and Santa Clara locations.

GETTING HIRED

Applied Materials provides extensive hiring information at its web site, located at www.appliedmaterials.com. In addition to job listings, the page provides links to resume submission and college recruitment pages. College students with backgrounds in materials science, computer science, physics, chemistry, and mechanical, electrical, chemical, or manufacturing engineering information systems should consider Applied Materials' New College Graduate Training Program. Applied Materials is especially interested in students who speak Mandarin Chinese, Korean, Japanese, French, German, Italian, or who have experience with semiconductor wafer processing. For more information, write to College Programs, 3050 Bowers Avenue, MS 1826, Dept. TCE0996PW, Santa Clara, CA 95054; the fax number is (408) 986-7940.

OUR SURVEY SAYS

Employees at Applied Materials are "tough and competitive" people, who are "proud" of their equally tough and competitive company. As one technical supporter engineer puts it: "The general mindset is that it is our right to be No. 1 in market share with every one of our products." Applied Materials employees "don't settle for mediocrity" and "do whatever it takes to get the job done." Working for a high-tech firm is considered "a bonus," especially "knowing that your products are on the cutting edge." The result of this sort of attitude is a "high pressure environment" where employees "commonly work 50 to 60 hours per week."

Employees describe themselves as "quite well compensated." Although "the pay is far from astronomical," one new employee is satisfied that "my stock options have already doubled in value." There is also a profit sharing program with a maximum payment of 15 percent of salaries. Still, many complain about the difficulty of finding "affordable" housing near the company's Santa Clara, California headquarters despite the "impressive" compensation packages that they receive. Nonetheless, insiders "appreciate" the "excellent medical and dental benefits." And for those who want the chance to advance within the company or elsewhere, the "superb tuition reimbursement program" gives them the chance.

As a Northern California high-tech company, the "Silicon Valley dress code" is the standard. Employees say that the "comfort of sneakers and jeans" helps to foster an "informal," "social" atmosphere. For minorities, employees say there is "no glass ceiling." Because of the company's longstanding ties in Japan, "much of the upper management" is Asian or Asian-American. But it is not just Asians who are well-represented at the company; Applied Materials' international focus ensures that it recruits employees "from all corners of the globe." The result is a firm with 35 percent minority employees, "on every level" from the board of directors downward.

"It's a bit of a roller-coaster financially, but that's the nature of the industry. When times are good, the perks are good."

– ***Semiconductor insider***

Ascend Communications

One Ascend Plaza
1701 Harbor Bay Pkwy
Alameda, CA 94502
(510) 769-6001
Fax: (510) 747-2300
www.ascend.com

LOCATIONS

Alameda, CA (HQ)
Numerous locations throughout the world

DEPARTMENTS

Engineering
Marketing
Materials

THE STATS

Annual Revenues: $1.2 billion (1997)
No. of Employees: 2,500 (worldwide)
No. of Offices: 11 (worldwide)
Stock Symbol: ASND (NASDAQ)
CEO: Mory Ejabat

UPPERS

- Good pay
- Gangbusters growth
- Beneficial alliances

DOWNERS

- Doubt raised by Cascade merger
- Complaints of rigidity

KEY COMPETITORS

- 3Com
- Bay Networks
- Cabletron
- Cisco Systems
- Digital Equipment
- FORE Systems
- General DataComm
- IBM
- Lucent Technologies
- Motorola
- NEC

EMPLOYMENT CONTACT

Human Resources
Ascend Communications
One Ascend Plaza
1701 Harbor Bay Pkwy
Alameda, CA 94502

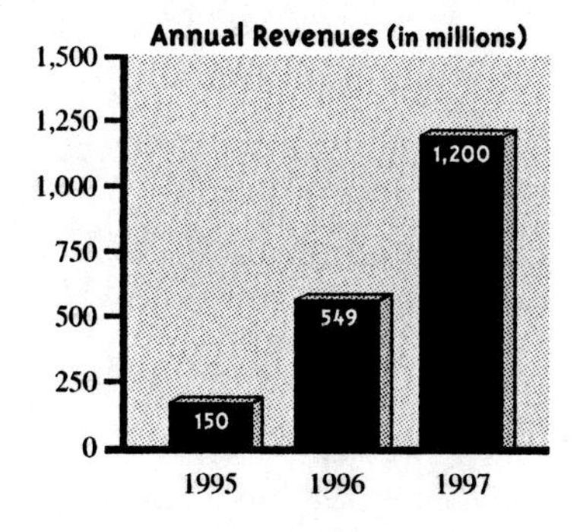

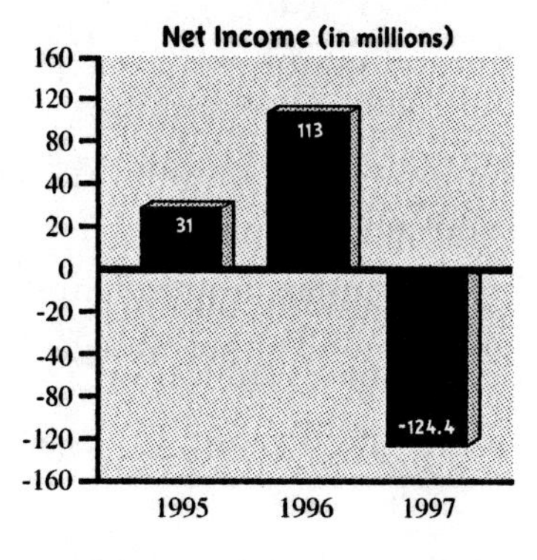

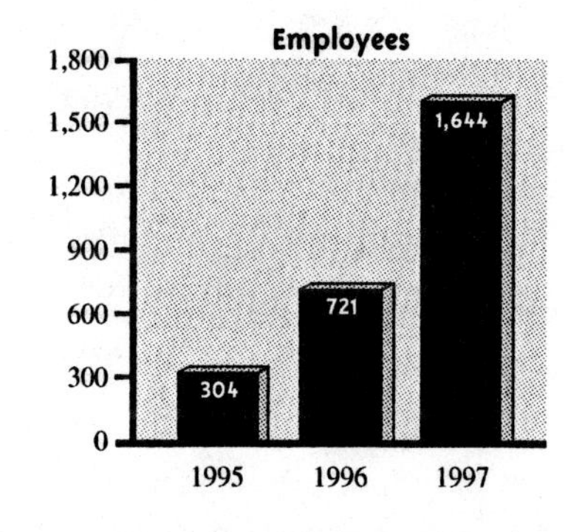

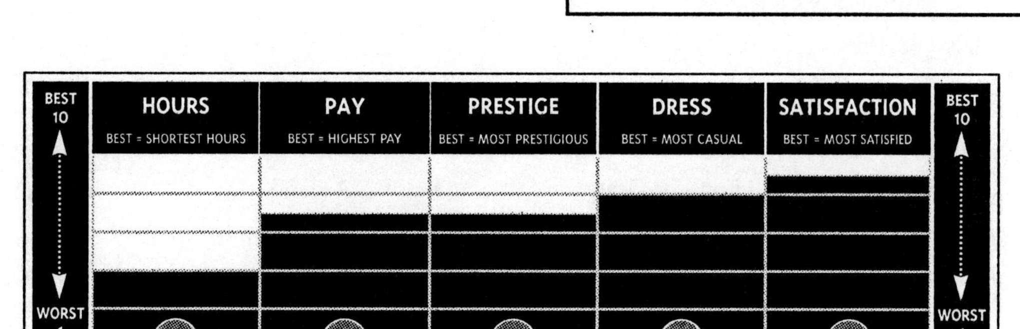

THE SCOOP

The surging popularity of the Internet has Ascend Communications seeing dollar signs. Founded in 1989, Ascend makes remote networking access products that allow users to dial up a central computer from any location. Its MAX (multiband access) WAN (wide area network) access switches account for more than half of the total WAN access concentrator ports worldwide (54 percent, to be exact). Ascend has also sold over a million ISDN ports, and over 500,000 digital modems. Ascend has become a market standard as well – 28 of the 30 largest Internet service providers in North America have standardized on Ascend solutions. All these products mean healthy profits for Ascend. The market for remote accessing products, currently $1.2 billion, is expected to grow to $6 billion by the year 2000.

In 1996, the company expanded its product line and market share through several acquisitions, including NetStar, a maker of high-speed computer switches, and Morning Star Technologies, a provider of Internet security. The same year, Ascend teamed up with Microsoft to introduce technology for secure VPNs (virtual private networks) for international data communications. Ascend recently allied with rival Cisco Systems to develop technology that eases the flow of Internet traffic.

The company's $3.7 billion acquisition of Cascade Communications in 1997 made Ascend a $890 million, full-service WAN systems provider. With the purchase of Cascade, Ascend is now able to offer all three major technologies that Internet service providers need: fast switching technology, high-capacity servers, and routers. Ascend now poses a real threat to Cisco Systems, the leader in networking services. However, some analysts are skeptical about the expanded Ascend, commenting that neither Ascend nor Cascade are experienced with large-scale mergers.

Ascend doesn't ignore its customers once they've purchased its products – the company offers "Ascend Advantage Services," which supplies a range of services from installation to maintenance work. For corporate customers, there's Ascend Channel Services, an out-sourcing support service program. That program is run in conjunction with IBM Global Services in order to provide the services in a wide geographic range. Ascend products are currently used in more than 50 countries worldwide.

GETTING HIRED

Visit the "Career Opportunities" section of Ascend's web site for job openings and descriptions. Qualifications vary by position. Send, fax, or e-mail resumes to the addresses listed on the web site. Most positions, even those not in engineering, require some background in WAN or ISDN.

OUR SURVEY SAYS

"Team spirit" and "high morale" make Ascend "an excellent place to work." "We are a little company gone big, and we still have an enviable corporate personality," opines one engineer. Others agree: "I think this is the coolest place in the world," says a sales manager. According to one engineer at Ascend, "employees work hard and are treated well." Not only that, they are "incredibly nice." Employees praise competitive salaries and an excellent benefits package, which includes stock options and 401(k) plans. Ascend employees are kept well informed on company moves, and enjoy good relations with upper level co-workers. Employees are also well-nourished, at least on Fridays, when the company offers free food and drinks after 5 p.m. Fridays also offer the chance to sport casual dress (Mondays through Thursdays are business dress). Ascend insiders are proud of their co-workers. "I am surrounded by professionals," says one employee. "Out of the 1600-plus employees here I don't know of a single loafer." These Ascend professionals are a diverse lot. "We probably have as many "minorities" as "majorities." I work with guys and gals from all kinds of different ethnic backgrounds." This equity is somewhat reflected in top management; one woman currently serves on the company's six-member board of directors.

Ascend prides itself on being a choosy firm. An employee with an extremely high opinion of Ascend asks: "Do you consider yourself the best at what you do? Do you rank in the top 5 to 10 percent of your peer group? If not, you need to go work for Cisco, 3Com or Bay Networks.

We feel at Ascend that we have hired the best-in-breed employees and we are helping to shape the next generation global public data network." Some employees think Ascend is a bit bureacratic – an employee at recently acquired Cascade Electronics (now the Core Switching Division of Ascend) thinks Ascend allows "little individual effort and autonomy to get stuff done." Despite this complaint, other insiders say Ascend seems to go out of its way to please its workers. "Teleworking is a priority at Ascend," say insiders; one employee reports working "two hours south of corporate headquarters."

"Out of the 1600-plus employees here I don't know of a single loafer."

- Hardware insider

Atmel

3525 Orchard Parkway
San Jose, CA 95131
(408) 441-0311
Fax: (408) 436-4200
www.atmel.com

LOCATIONS

San Jose, CA (HQ)
Colorado Springs, CO
Rousset, France

DEPARTMENTS

Engineering
Human Resources
MIS
Multimedia & Communications Design
Sales
Technical

THE STATS

Annual Revenues: $958.3 million (1997)
No. of Employees: 4,589 (worldwide)
No. of Offices: 3
Stock Symbol: ATML (NASDAQ)
CEO: George Perlegos
Year Founded: 1984

ATMEL

UPPERS

- Flexible work hours
- "Small company" feel
- Relaxed dress code
- Heterogeneous workforce

DOWNERS

- Pay is low for the industry
- Recent downsizing

KEY COMPETITORS

- Actel
- Altera
- AMD
- Catalyst Semiconductor
- Cirrus Logic
- Integrated Device Technology
- Intel
- LSI Logic
- Microchip Technology
- National Semiconductor
- NEC
- Oki Electronics
- Xicor
- Xilinx

EMPLOYMENT CONTACT

Attn: Human Resources Department
Atmel Corporation
2325 Orchard Parkway
San Jose, CA 95131

Fax (408) 451-4828
resume@atmel.com

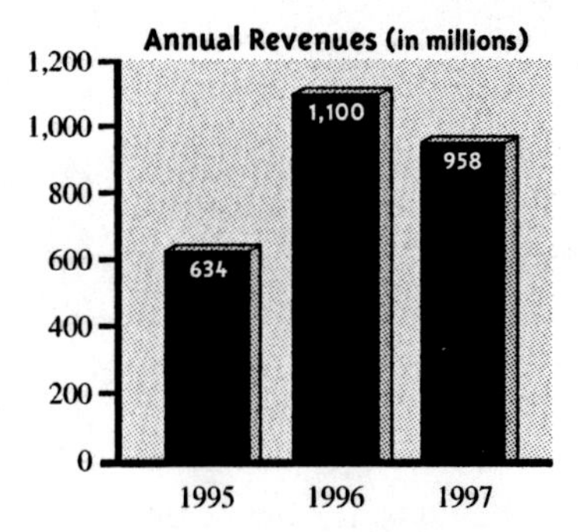

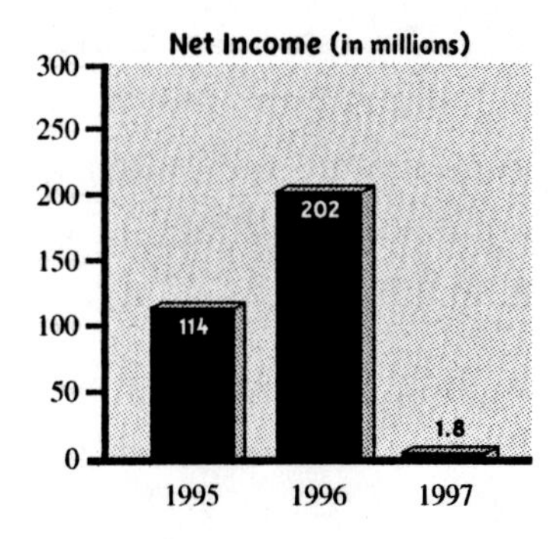

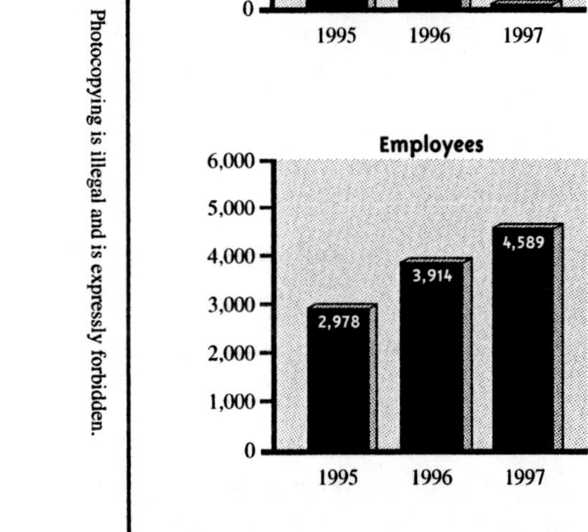

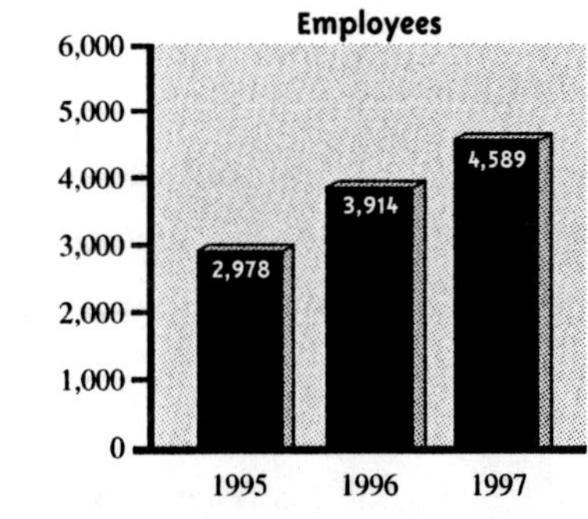

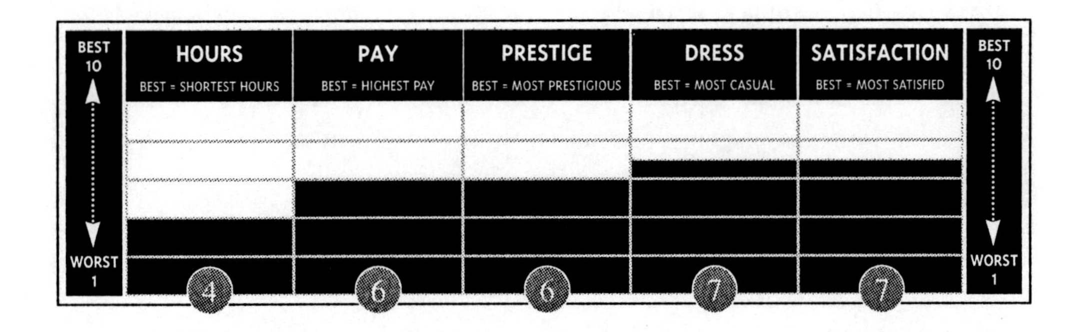

THE SCOOP

Atmel (Advanced Technology for Memory and Logic) was founded in 1984 by CEO George Perlegos, a former design engineer for Intel. Atmel manufactures logic chips (which process information) and nonvolatile memory chips (which retain memory when a device is turned off). These products are used in portable electronic devices like pagers, telephones, computers and smart cards. Atmel went public in 1991, and has grown into the world's largest manufacturer of Parallel and Serial EPROMs (erasable programmable read-only memory chips) and EEPROMs (electrically erasable programmable read-only memory chips). Both EPROMS and EEPROMs are types of nonvolatile memory chips.

By 1996, nonvolatile memory was in high demand, so profits at Atmel were soaring. But in 1997, the Asian financial crisis, compounded by falling prices for its core technologies, led to serious losses – earnings plummeted 99 percent per share, and revenues fell $112 million from the previous year. The company was forced to implement a restructuring program, including a 10 percent workforce reduction, decreased production, and a focus shift to system-level integration products and more advanced technologies.

Fortunately, Atmel has made some moves that it expects will return it to greater profitability. In March 1998, the company purchased Temic Semiconductor, a European chip maker previously owned by auto giant Daimler Benz. Operating as an independent entity, the acquisition has been a source of increased revenues since its acquisition. Company execs say Temic will also be instrumental in the development of new technologies – a combination of both companies' products. They intend to develop new chips for use in products like multi-band digital cell phones, airbag systems for automobiles, and remote keyless entry systems.

A particularly interesting new Atmel product is AT24RF08, the high-tech industry's first "asset identification" chip. Introduced in April, this security device identifies computers based on the configuration of components. Similar to the anti-theft devices used in clothing stores, equipment with this chip prevents unauthorized movement of computer equipment and prevents unauthorized memory access.

GETTING HIRED

Atmel actively recruits on college campuses and accepts resumes and cover letters at its HR departments in each outpost. Look for job listings on the company's web site, located at www.atmel.com.

In general, interviews are "relaxed and informal," and consist of one-on-one interviews 30 minutes to one hour in length," insiders tell us. If you are applying for a technical position, "you are sure to encounter a set of technical questions," which will vary according to the level of the job. "Expect to meet with the hiring manager, three or four engineers, and an HR representative." Insiders advise aspirants to come in with "a basic understanding of what the company does." In addition, "our major concern when interviewing people is how well we think they will fit in to the group."

OUR SURVEY SAYS

Despite the fact that it's a billion-dollar company with more than 5000 employees, sources say working at Atmel "feels like working in a small company." Insiders love working in this "very informal atmosphere" where "there is opportunity for advancement;" and where the "high quality" employees are "relaxed and encouraging." As one San Jose engineer puts it, "we work hard, have fun and make money." "I have had numerous other opportunities around the country," notes one employee, "but I'm still here after four years, and I have no intention of leaving. I wouldn't be here if the experience was not enjoyable." This sentiment seems to be widespread: one source reports that the average member of his team has been with the company for more than eight years. This makes for "a very close team," he adds.

"The dress code is generally casual," says one insider, "depending on the amount of outside contact you have." "The culture places more emphasis on performance than appearance," explains one source. People in the "fab," or manufacturing plants, must wear all-encompassing, sterile, static-free "bunny suits" every day, and "must adhere to strict work and break times." Those in management and marketing wear business attire when they travel or have meetings. Pay is "not as high as in some other companies," laments one source, but the work is "challenging and rewarding," and "the hours are flexible." The official workday is 8 to 5, "but if you are an exempt employee, there is no fixed time to come and go," one source reveals, "as long as you are reasonable and get your boss'. approval, you can set your own hours." Other than the "reasonably standard" benefits, employees say "there are very few 'special perks' at Atmel." "I've been to companies in the Valley that go overboard to accommodate their employees with fancy cafeterias and frequent parties, but we don't do that here." Sources note that this style is typical of the company's "straightforward" founder, George Perlegos. "The only difference between his cubicle and mine is size," one employee reports. "He doesn't care to see anybody trying to get too fancy, especially at his company's expense." The company's 'basic' benefits include 401(k), medical benefits and corporate stock purchase plan.

Another insider says "the management organization is relatively flat, which tends to reduce political game playing." So no need to worry about co-workers "jockeying for positions up the corporate ladder." The company as a whole "is fairly average in the way it treats its employees overall," says one source, who adds that "there can be vast differences within the company" as you go from one department to another. One employee describes the company as "a stew of many nationalities." An insider from the Colorado Springs office agrees. "A good portion of our technical staff are women and of various ethnic origins."

"The general mindset is that it is our right to be No. 1 in market share with every one of our products."

– ***Semiconductor insider***

Bay Networks

4401 Great America Parkway
Santa Clara, CA 95054
(408) 988-2400
Fax: (408) 495-5525
www.baynetworks.com

LOCATIONS

Santa Clara, CA (HQ)
Billerica, MA
Gaithersburg, MD
Andover, MA
26 international offices

DEPARTMENTS

Finance
Hardware Engineering
Sales and Marketing
Software Engineering

THE STATS

Annual Revenues: $2.1 billion (1997)
No. of Employees: 5,960 (worldwide)
No. of Offices: 2 (worldwide), 2 (U.S.)
A Subsidiary of Nortel
CEO: David L. House

BAY NETWORKS

UPPERS

- Discounted movie, theme park, and ski lift tickets
- Company sports teams
- On-site fitness facilities
- Community volunteer program
- On-site dry cleaning and photo developing pick-up

DOWNERS

- Inter-office clashes
- Uncertainty because of acquisition by Nortel

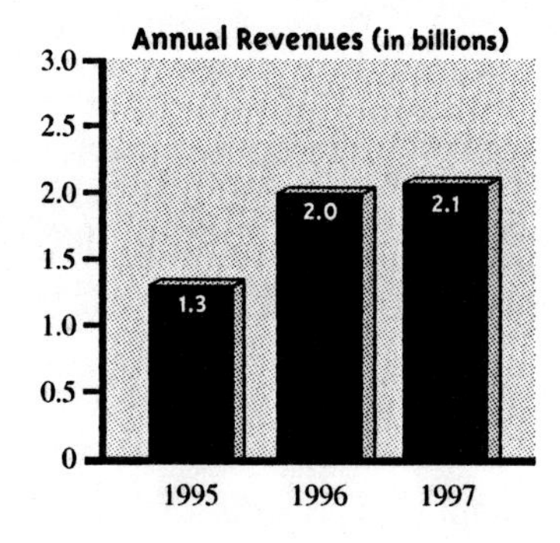

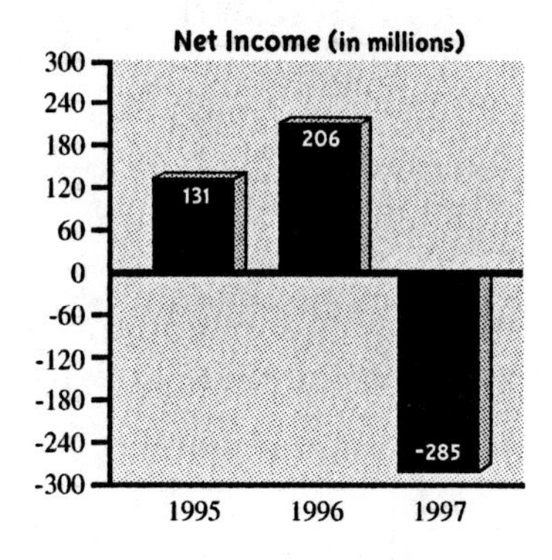

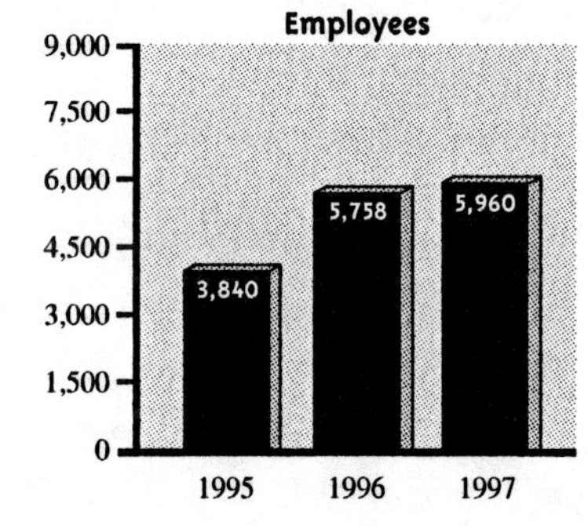

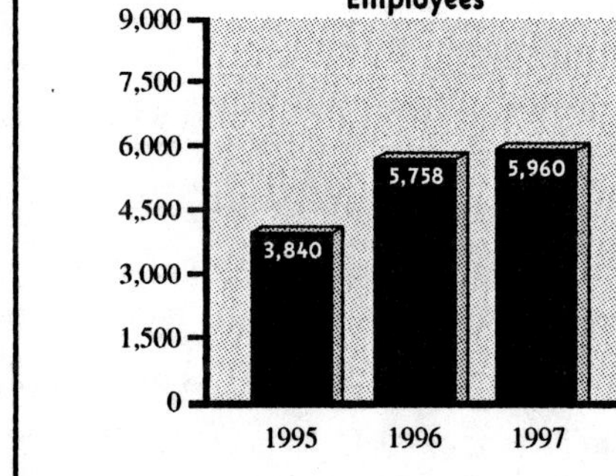

KEY COMPETITORS

- 3Com
- Ascend Communications
- Cabletron
- Cisco Systems
- Digi International
- FORE Systems
- Hewlett-Packard
- Lucent Technologies

EMPLOYMENT CONTACT

Staffing
Bay Networks
4401 Great America Parkway
Santa Clara, CA 95052

Fax: (408) 495-1898
baynetworks@isearch.com

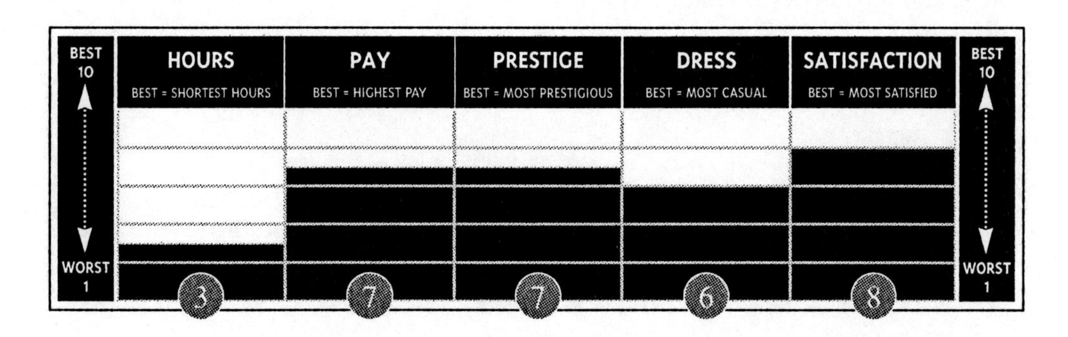

THE SCOOP

Bay Networks makes networking products that control the flow of information within the computer networks of small and large businesses. Founded through the 1994 merger of California-based Synoptics and Massachusetts-based Wellfleet, Bay Networks holds the No. 3 position in the industry, behind Cisco Systems and 3Com. The company's current product line includes hubs, routers, switches, network management software, and other LAN and WAN products that allow computers to communicate with one another over the Internet.

In 1994, Bay Networks received a $20 million contract from Chase Manhattan to network the banking giant's computers. The next year Bay introduced new technology that has improved the way that ATMs share information. However, Bay Networks was soon plagued by product delays, bi-coastal division, poor marketing, and the flight of many of its prominent engineers. To combat these problems, Bay Networks hired Intel executive David House in 1996 to become CEO. That year that the company purchased Penril DataComm Networks' modem business. Two other companies that Bay Networks has acquired – Centillion Networks (which manufactures token ring networks) and Xylogics (a leading router manufacturer) – continue to function as independent subsidiaries.

In June 1998 Northern Telecom (Nortel), the Ontario-based telephone equipment manufacturer, acquired Bay Networks for $9.1 billion. As a result, Bay Networks became an independent subsidiary of Nortel, but secured the corporate channel marketing of a larger company in the process. Bay Networks CEO David House is to serve as Nortel's president under Nortel CEO John Roth. Whether the merger will help revenues remains to be seen. A month after the deal, Bay Networks bought the German cable-telephone networker, NetServe, in an effort to lure AT&T into its client fold.

GETTING HIRED

Bay Networks most frequently hires hardware and software engineers. The company accepts e-mailed resumes sent in ASCII text format. Bay Networks' web site, located at www.baynetworks.com, lists current openings. Both the company's California headquarters and its Massachusetts office accepts resumes submitted by regular mail or fax. Bay Networks has positions open for technical support engineers, product management, professional services, and sales. Bay also recruits through classified ads and job fairs.

One Bay insider recommends that candiates "look smart, relax and be confident. You should be aware that a manager might ask you to try to solve a complex problem to see how strong your troubleshooting skills are. It isn't important to get the answer right because there might not even be a right answer." That employee continues: "They will also ask you questions to see how well you work with others. For example, 'What would you do if you were working with a team member and they decided to take all the credit for a particular accomplishment you both did?' There isn't a right or wrong answer. They just want to see how you attack the problem. Also, they will try to figure out whether you like repetition or learning new things, if you are a people person or a loner, if you like to travel, if you like to work or surf the Net all day. They just want to make sure they put you in the right kind of position."

OUR SURVEY SAYS

Although they agree that the California and Massachusetts cultures first clashed after the merger that created Bay Networks, insiders say that the company's new CEO, David House, has improved the "inter-office communication" and morale "dramatically." While these areas "still require attention," employees are "optimistic" about Bay Networks' "impressive turnaround." One insider says that "cultures vary from coast to coast. The atmosphere on the

West Coast is a bit more formal, though things have loosened up considerably since the merger between Wellfleet and Synoptics. The East Coast, on the other hand, I would characterize as more relaxed." An insider willing to generalize says that "both coasts are very intense. Deadlines are extremely tough and pressures can be relatively high. However, when the crunch is done and we make our deadlines, things lighten up." We hear that "offices are pretty decent. They are primarily cubes, decent size, some larger than others. Lots of windows, a nice campus in which to work. Most offices are reserved for management, personnel and employees with seniority. We have dining facilities that are pretty decent, I guess, for cafeteria food." Bay Networks' "flexibility" helps it meet employee needs, and on-site perks such as fitness facilities and dry cleaning pick-up helps them cope with "demanding schedules that can become more grueling as project deadlines near."

Should you consider working at Bay Networks? You should know that our contacts seem to enjoy the culture at Bay. According to one insider: "The culture here is very open, honest and healthy. Our mission is always to get the job done. We are not a company filled with busybodies trying to stab each other in the back." One employee says "I think opportunities for recent college grads are on the rise. But [new employees] are required to continue learning as they grow their career. It's a fast-paced, highly technical environment." Required courses for employees include "time-management" and "presentation skills," as well as "updates on the latest technology." Naturally, "Bay pays for the courses, and you may take them off-site or take company seminars."

In return, Bay treats its employees well. An insider reports that "I have honestly never been with a company that rewards its people as well as Bay Networks. There are award programs, promotion opportunities, and of course, monetary compensation." This pay "consists of base, bonus (a combination of individual and corporate achievement goals) and stock options." As far as diversity, "Bay scores big," says one employee. "I don't know as many other major computer companies that are as fair and as open to women and minorities as Bay. As a minority, I don't have to prove my existence every day. I can come here, do a good job, and be treated with fairness like everyone else. No joke." A veteran employee says: "Those of us who joined early are getting used to the big company atmosphere. But I would say that Bay

Networks is an excellent place for a career. Bay Networks is changing. I would say that its influence on the industry is on the upswing. You can really make an impact on the industry here."

Brøderbund Software

500 Redwood Boulevard
Novato, CA 94948
(415) 382-4400
www.Broderbund .com

Brøderbund

Subsidiary of The Learning Company
CEO: Joseph P. Durrett

LOCATIONS

Novato, CA (HQ)
Boston, MA
Chicago, IL
Cleveland, OH
Columbia, MD
Dallas, TX
Fremont, CA
Los Angeles, CA
Mountain View, CA
Philadelphia, PA
London, England

DEPARTMENTS

Customer Support
Engineering; Marketing
Management Information Systems
Quality Assurance

THE STATS

Annual Revenues: $190.8 million (1997)
No. of Employees: 1,129 (U.S.)
No. of Offices: 11 (U.S)

UPPERS

- Company gym
- Tuition reimbursement
- Free software
- Symphony and sports tickets
- Birthday celebrations

DOWNERS

- High turnover rate
- Low pay relative to software industry

KEY COMPETITORS

- Adobe
- Educational Insights
- GoldenBooks
- Intuit
- Microsoft
- Scholastic
- SEGA
- Viacom
- Walt Disney

EMPLOYMENT CONTACT

Human Resources
Broderbund Software
500 Redwood Boulevard
Novato, CA 94948

Job hotline: (415) 382-4404
resumes@broder.com

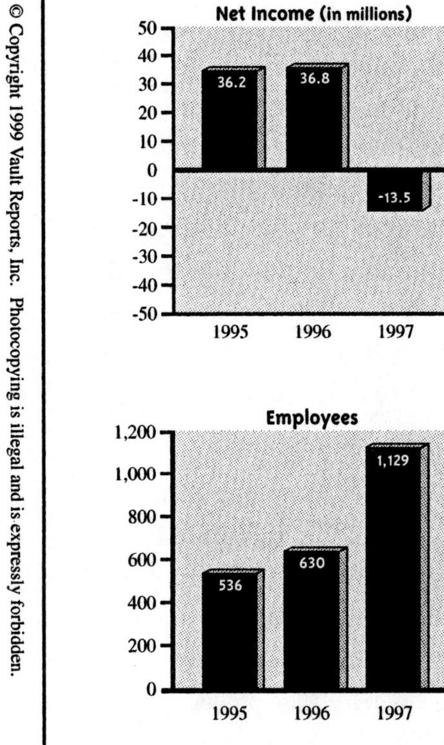

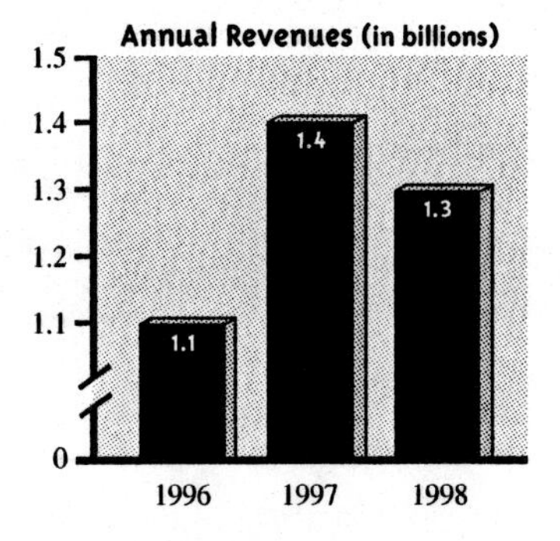

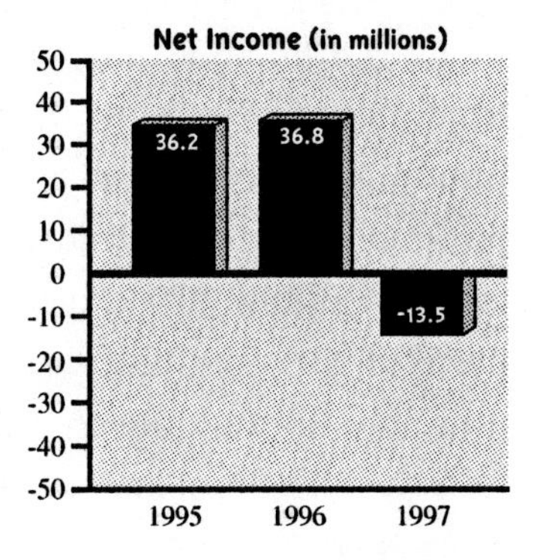

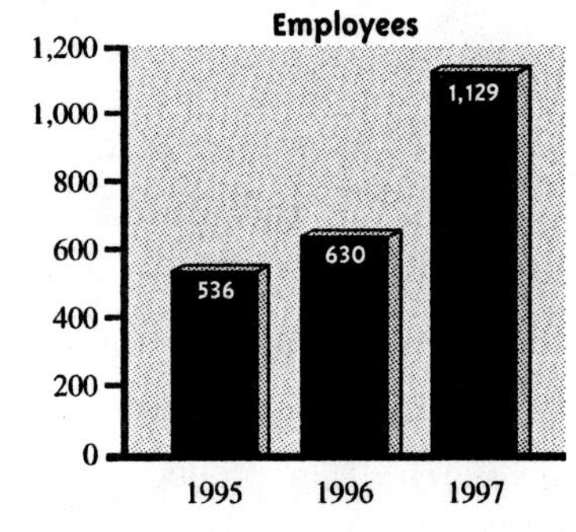

BEST 10 ↕ WORST 1	HOURS	PAY	PRESTIGE	DRESS	SATISFACTION
	BEST = SHORTEST HOURS	BEST = HIGHEST PAY	BEST = MOST PRESTIGIOUS	BEST = MOST CASUAL	BEST = MOST SATISFIED
	5	5	7	6	7

THE SCOOP

Brøderbund (Swedish for "brotherhood") was started by two brothers, Doug and Gary Carlston, in 1980 to market the computer games that Doug, a lawyer, had written in his spare time. Gary Carlston is no longer with the company, but his brother has remained to watch the company rack up industry awards from educational groups – and to search for new hits. Brøderbund regularly introduces over 70 new software titles each year. The California-based firm has made its mark in the industry by developing educational software, marketing different products for specific age groups. Kids play *Where in the World is Carmen Sandiego?* to learn about geography, and explore history with *Where in Time is Carmen Sandiego?*. Those obsessed with their heritage may trace their family history with Brøderbund 's best-selling genealogy software, *Family Tree Maker*. A game software pioneer, Brøderbund has shifted its focus to educational and personal productivity software. To maintain its wholesome family image, Brøderbund now releases its games software under a new divisional title, Red Orb. As of June 1997, Brøderbund held sway over a 4.3 percent unit share of the U.S. software market.

Brøderbund's flagship product is *Myst*, the alluring and cryptic puzzle-based graphic game. Introduced in 1993, *Myst* was one of the first products to benefit from strong word-of-mouth on the Internet and World Wide Web. For now, the company is measuring its success by the popularity of the graphic game line; in 1997, Brøderbund's financial performance faltered considerably. The company suffered a net loss of $13.4 million in 1997, a blow after a healthy fiscal 1996, which netted the company $36.7 million. Although Brøderbund continued to develop software products, in 1997 the company's best-selling product was still *Myst*, with $7 million in sales. Despite the game's continued strong popularity, revenues are declining, because the game is now sold at a discount. Brøderbund then banked on *Riven*, the sequel to *Myst* introduced in October 1997, to pull it out of its financial rut. Brøderbund spent $10 million bringing the game to market and shelled out another $10 million for marketing and advertising. The game sold more than 474,000 units at retail price in the first month on the market, but petered out to 1 million total units the following year (as compared to the 5 million copies of Myst that have sold as of December 1998).

The company's weakening stock price (in the past two years, it has dipped as low as $18.75 from a high of $76 in October 1995) gave credence to rumors of takeovers. Despite Brøderbund's troubles, the company's brand equity and creative talent made it a tempting acquisition. The Learning Company, which Brøderbund itself had attempted to buy three years ago, found the tables turned and got itself a bargain. TLC, (formerly known as SoftKey International) snatched up Brøderbund for a paltry $420 million in July 1998 when the struggling company could no longer withstand the merciless price-cutting employed by TLC and Cedant. Through the merger, however, Brøderbund can return its original focus on producing educational products – and leave the marketing to TLC. In 1998, The Learning Company announced it would be bought by toy company Mattel for $3.8 billion.

GETTING HIRED

Brøderbund accepts resumes by e-mail and regular mail. Brøderbund's job hotline and web site, at www.broderbund.com, post current openings at Brøderbund and give brief but helpful descriptions of each. Qualified candidates will be interviewed over the phone and then contacted for a company visit, insiders say. The company looks for individuals who express a passion for the software industry and a high level of enthusiasm for the particular position for which they are applying. Most new hires are placed into Brøderbund's Marin County locations.

OUR SURVEY SAYS

Brøderbund's extensive list of perks includes San Francisco symphony tickets, a happy hour every Friday, free software, and even birthday dinners. Brøderbund's employees give the company high marks for on-the-job satisfaction; they cite the "spectacular California location,"

the company's "impressive name" within the software industry, and the "laid-back, casual atmosphere" of its offices. Brøderbund management favors "fast learners" who are "driven" to push the envelope in marketing and technology innovation. Unfortunately, some employees feel that the "excessive turnover" at Brøderbund indicates that the company attributes "little value to long-term employees." One insider indicates that "Brøderbund tends to value products over employees. This makes it less than ideal for a job early in one's career."

Employees say that "there is a very young workforce that is very social and casual, as well as hard working, and slightly competitive." "It's a non-smoking building," says another employee, who adds that "the company used to allow pets, but not any longer." One insider says: "The dress code is very loose. Most days I wear jeans and a T-shirt. In marketing and finance, however, employees have to wear appropriate attire when they meet 'auslanders.'" "You couldn't tell who is the president or vice president just by looking at them," says an insider. "As far as pay goes," says one contact, "I've heard Brøderbund pays somewhere between average and low average. Benefits include good medical, 401(k), and happy hours and picnics." A longtime employee concurs that "Brøderbund is not known for its pay. They try to be competitive." Insiders say that the environment at Brøderbund is "100 percent PC (meaning politically correct)." "The workforce is extremely diversified," says one employee, who adds "there isn't a nationality that I can think of that isn't represented here. As far as the treatment of women, everything is cool. We have annual sexual harassment courses." sums up one satisfied insider at Brøderbund: "My co-workers are like extended family."

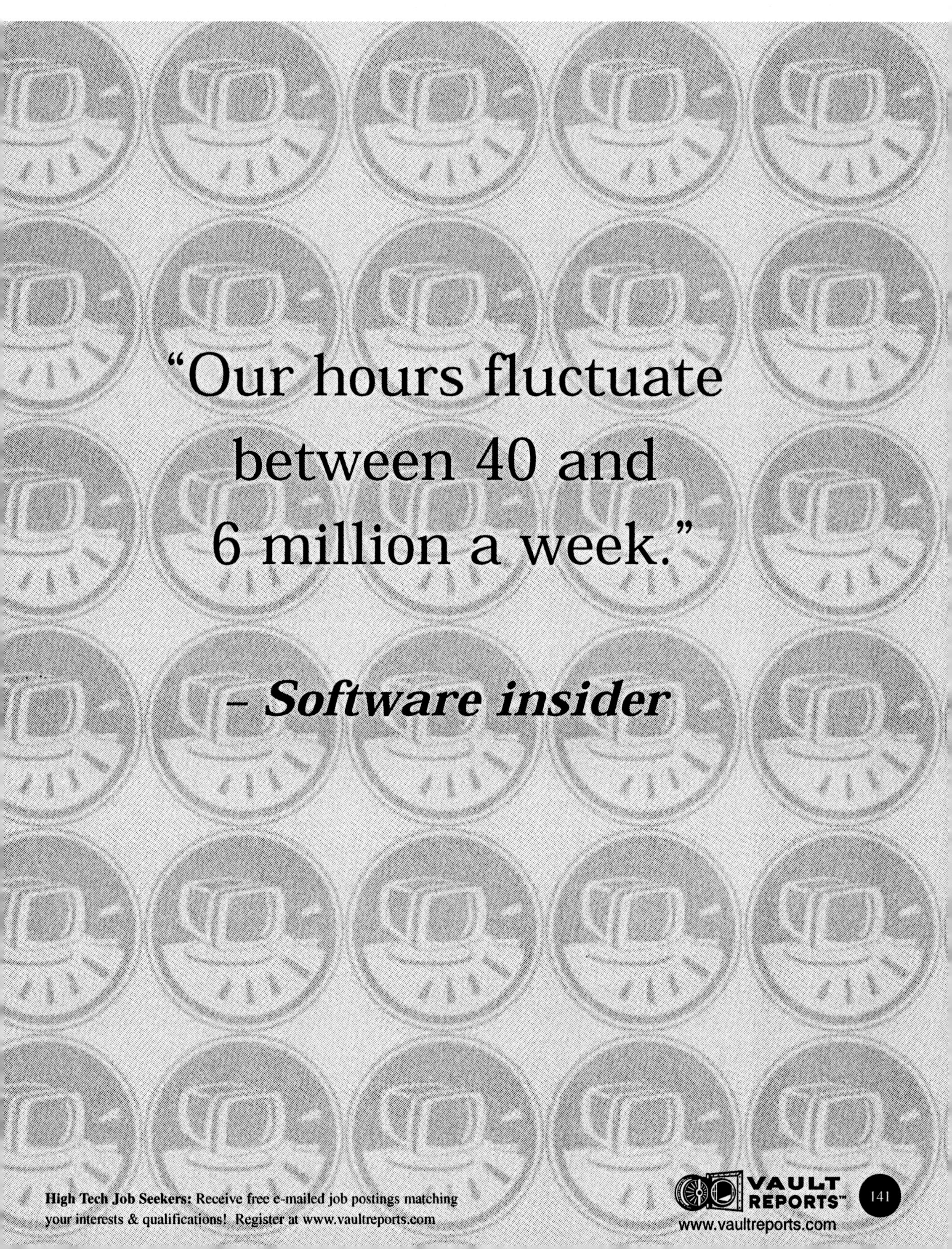
"Our hours fluctuate
between 40 and
6 million a week."
– Software insider
High Tech Job Seekers: Receive free e-mailed job postings matching your interests & qualifications! Register at www.vaultreports.com
VAULT REPORTS™
www.vaultreports.com

Cabletron Systems

35 Industrial Way
Rochester, NH 03867
(603) 332-9400
Fax: (603) 332-8007
www.cabletron.com

CABLETRON SYSTEMS

LOCATIONS

Rochester, NH (HQ)
Numerous locations throughout the U.S and the world

DEPARTMENTS

Administrative
Finance
Firmware Engineering
General Manufacturing
Hardware Engineering
Human Resources
Information Services
Manufacturing
Marketing
Product Development
Quality Assurance & Test Engineering
Sales
Software Engineering
Tech Support
Technical Documentation
Training/Course Development

THE STATS

Annual Revenues: $1.4 billion (1998)
No. of Employees: 6,900
No. of Offices: 100 (worldwide)
Stock Symbol: CS (NYSE)
CEO: Craig R. Benson

UPPERS

- Travel opportunities
- Young employees

DOWNERS

- High turnover in sales department
- Lack of diversity in some offices

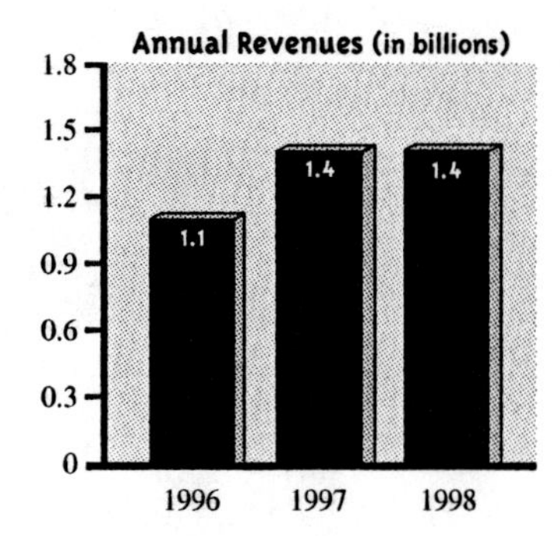

KEY COMPETITORS

- 3Com
- Ascend Communications
- Cisco Systems
- Datapoint
- Digital Equipment
- D-Link
- Hewlett-Packard
- Hitachi
- IBM
- Lucent
- Nokia
- Novell
- Xylan

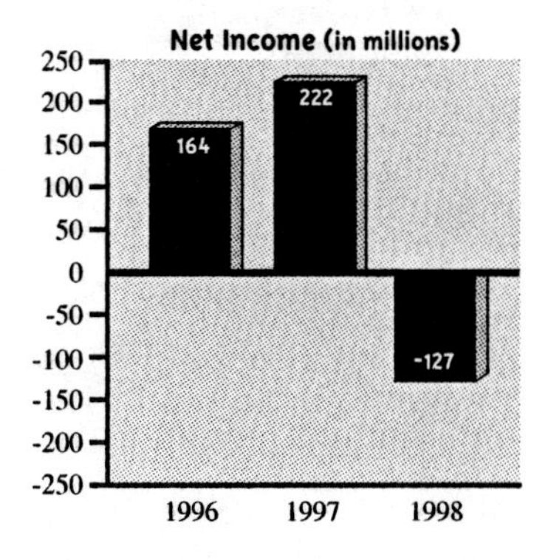

EMPLOYMENT CONTACT

Linda Pepin
Human Resources
Cabletron
P.O. Box 5005
Rochester, NH 03866-5005

Fax: (603)337-0273.
jobs@ctron.com

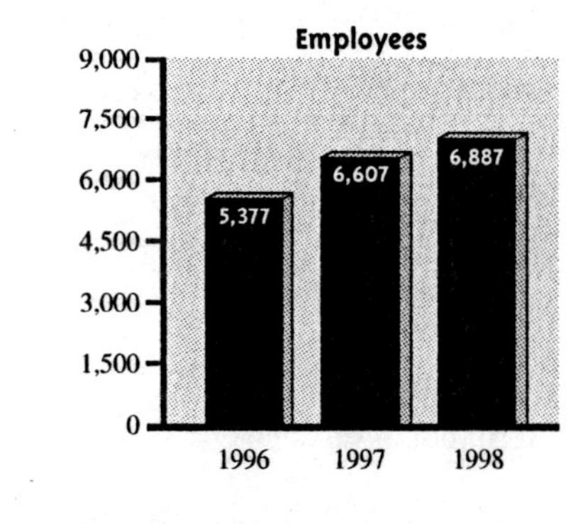

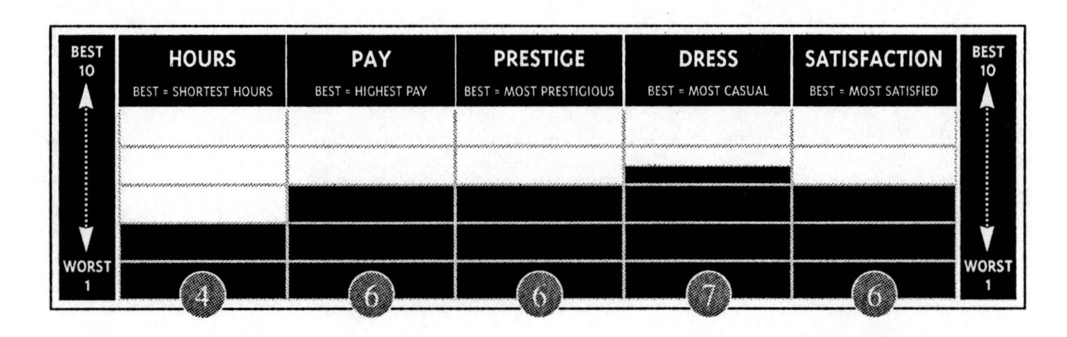

THE SCOOP

In 1983, Robert Levine and Craig Benson, a pair of wire and cable salesmen working out of a New Hampshire garage, began selling computer network cables cut to the specific lengths their customers needed. Soon after its founding, Cabletron Systems began installing networks and designing networking equipment, unveiling the Multi Media Access Center (MMAC), an intelligent wiring hub used to simplify network installation in 1988. The company went public a year later. The two networking innovators now own 13 percent each of the company they established, a combined stake worth more than $1 billion. Cabletron introduced the first SCSI-to-Ethernet adapter in 1992, enabling Mac users to connect to Ethernet LANs. Major purchases in 1995 and 1996 included Standard Microsystems, Network Express, and ZietNet.

For much of the 1990s, the company's aggressive sales force and no-frills corporate culture garnered top-notch earnings, and an aggressive acquisition policy kept it neck-and-neck with competitors like 3Com and Cisco Systems. But in 1996, the company's steady growth was interrupted by product delays, and customers began to complain about Cabletron's "overzealous," "often roguish" sales and marketing tactics. In 1997, Robert Levine stepped down as CEO and was replaced by Don Reed, who quickly began to shake things up. His first tactic was to switch the company's focus from simple product sales to becoming a "solutions provider." Cabletron expanded its product mix to include software and professional services. The next step was to streamline the direct sales force and farm out business to sales channels. Reed then positioned the company growth through expanded e-commerce operations and international business.

Reed also set Cabletron on track for new partnerships and acquisitions. He initiated the purchase of Yago Systems, Inc. and Digital Equipment Corp's networking business, which gave the company a stronghold in the router and switch markets.

But Reed was dealing with damaged goods. The company's stock fell to a five-year low, and it laid off 800 employees. After streamlining operations and instituting a strong growth strategy, Reed quit the CEO post after only seven months on the job. He handed the reins to chairman and co-founder Craig Benson. Though some doubted the claim, Cabletron execs claimed Reed's sudden departure came because he had basically accomplished what he was hired to do – create a strategy that would return the company to growth.

In June of 1998, Cabletron started something of a shopping spree, picking up FlowPoint Corp. and Ariel Corp., both producers of digital subscriber line (DSL) equipment, as well as NetVantage, a manufacturer of Ethernet switches. The company also introduced new products for wireless networks and ATM switch lines. Despite this bilking up, industry analysts speculate that Cabletron itself is now a prime target for acquisition.

GETTING HIRED

Cabletron recruits through headhunters and on college campuses. Visit the employment section of the company web site for detailed information on job openings. You can apply using an online application form, or via e-mail, fax, or snail mail.

Overall, interviewing at Cabletron "is not a very stressful process," one source said. Prospective employees endure "at least two interviews," sometimes more, depending on the situation. Each meeting lasts about 15 or 20 minutes, and "you'll get the occasional person who asks you really technical questions." The whole procedure is "somewhat relaxed, but very thorough." "They want to see if you are someone who can learn easily." Luckily, interviewers are easier on recent college graduates "because it is understood that your experience levels are lower." "For our inside sales organization, they'll make you sell something," one source revealed, "but we train all our people in-house, so we don't expect you to be an expert at interview time."

OUR SURVEY SAYS

Cabletron a leader in the "fast paced" and "highly competitive" networking industry. As one employee notes, "if you want a routine 9 to 5 job, Cabletron is probably not for you." The "loyal, hard-working" employees at Cabletron are all pretty young – "most of us are no more than 10 years out of college," and "as a result, this is a fun place to work." Engineers have a "pretty liberal" dress code, while managers and those with client contact wear suits or business casual attire. "Each office has its own personality and mix of people," reports one insider. The New Hampshire office, for example "is not the most diverse place." "There aren't many women," and one source remarks that "I haven't seen a single black person in my 14 months here." But other insiders say the Rochester facility is more diverse, and you'll "find more women in highly respectable positions." But in Rochester, the work environment "is a little more stressful," because "customers visit more frequently."

Official office hours are 8 to 5, but for engineers, "what's important is getting the work done." So during "crunch time," people work the extra hours." Sources say salaries used to be "just OK," but "compensation has gotten much better in the last two years." In addition, the company offers stock options and recently instituted a matching 401(k) plan. Though most are enthusiastic about the "excellent training programs" and abundant opportunities for advancement and travel, some are "suspicious of the rapid turnover rate." "A large number of experienced, respected people are leaving weekly," reports one source, noting that the talent drain is most conspicuous in the sales department. Engineers, on the other hand, enjoy extensive "learning and growth opportunities," and think "it's a great time to jump on board."

"Deadlines are extremely tough and pressures can be relatively high. However, when the crunch is done and we make our deadlines, things lighten up."

– ***Hardware insider***

C-Cube Microsystems

1778 McCarthy Blvd.
Milpitas, CA 95035
(408) 944-6300
Fax: (408) 944-8132
www.c-cube.com

C-CUBE MICROSYSTEMS

LOCATIONS

Milpitas, CA (HQ)

DEPARTMENTS

Engineering
Finance
Marketing

THE STATS

Annual Revenues: $337 million (1997)
No. of Employees: 750
Stock Symbol: CUBE (NASDAQ)
CEO: Donald T. Valentine

UPPERS

- Profit sharing
- Exciting DVD Products
- Tuition reimbursement

DOWNERS

- Underpayer
- Poor management

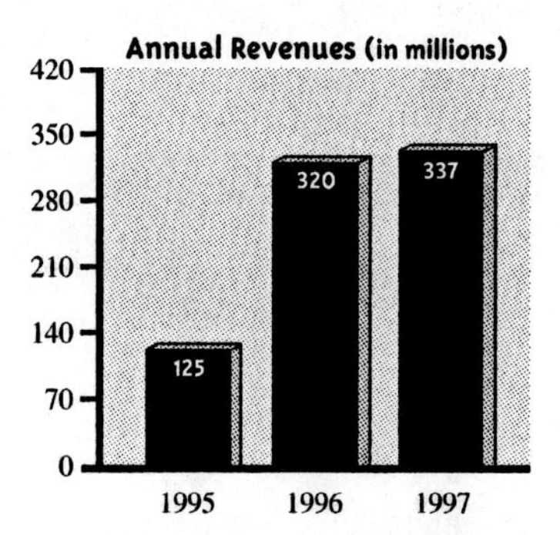

KEY COMPETITORS

- Intel
- LSI Logic
- Motorola
- STMicroelectronics
- Texas Instruments

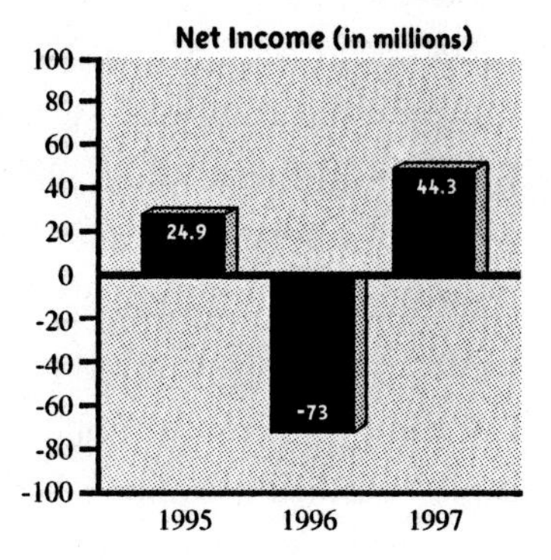

EMPLOYMENT CONTACT

Human Resources
C-Cube Microsystems
1778 McCarthy Blvd.
Milpitas, CA 95035

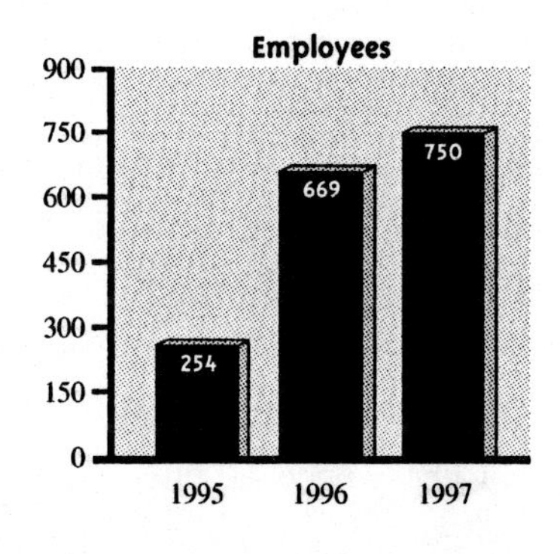

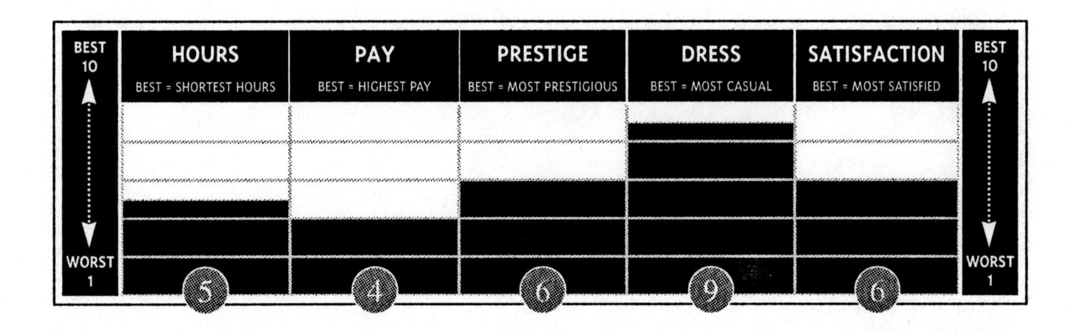

THE SCOOP

C-Cube's award-winning engineering team has Hollywood steamed. Top-notch digital video products have won the company some of the biggest clients in the digital video industry, including Apple, DEC, Samsung, and Sun Microsytems. Much to the chagrin of the folks in LA-LA land, the company's products are also favorites of video pirates in China, who use C-Cube's video compression chips to make cheap counterfeit copies of Hollywood movies. Based in Milpitas, California, the company designs chips that allow digital video to be transmitted efficiently and to be shared by a variety of products, including CD-ROMs, computers, and televisions. The growing popularity of formats like video CDs and videoconferencing has kept the company's sales on the upswing. With a broad-based clientele, C-Cube's chips can be found in products ranging from computers and camcorders to set-top cable boxes and telephone distributions systems.

Founded in 1988, the company's name represents the convergence of video technology in the consumer electronics, communications, and computer markets. (Get it? C^3?) The company's first major product came in 1990, with the introduction of a JPEG chip for the compression of still images. Two years later, C-Cube debuted an MPEG chip for audio and video compression. William O'Meara, a co-founder of microchip maker LSI Logic, was named C-Cube's CEO in 1991. O'Meara aimed the company toward the consumer electronics and communications markets. C-Cube began providing video encoding technology for Hughes Electronics' DirecTV satellite television in 1993. The company went public the following year. In 1995, C-Cube gained as clients broadcast satellite systems Alphastar and Echostar digital and electronics giant Sony, which chose C-Cube's chips for its new line of video CD players. The company bought Divicom, a leader in digital video networking products in 1996. The same year, rising decoder sales in the Asia-Pacific region caused C-Cube's sales to zoom. The company unveiled ZiVA in 1997, a new digital video disk chip that will lower the cost of digital video players.

GETTING HIRED

Employees report, "C-Cube is hiring." One contact notes, "There is a lot of growth in this company. I've been here a little more than two years, and there were 160 employees at the time I joined. Now it's close to 800." Another reports, "There are so many projects right now, and we're looking for a lot of people." Visit the "Employment Opportunities" section of C-Cube's web site for job descriptions and on-line application forms.

OUR SURVEY SAYS

C-Cube employees turn in mixed reviews for the growing company. "C-Cube is the best place in the Silicon Valley to work. I plan to grow old with this company, if time allows," one says. Another praises: "I love this place. C-Cube is a pioneer in the business of DVD, VCD, and Set-top boxes, with state of the art technology." Another reports: "C-Cube is a great company. I've been working here for two years, and believe me, this is a good experience." However, one contact reports that, "I left C-Cube a few months ago because I wasn't happy there." Another notes, "In my humble opinion, C-Cube possesses great engineering talent, but in the hands of lousy management. Though C-Cube has faced keen competition in the VCD and DVD markets, I think bad management is at the hands of the recent fall in stock price."

Though company culture varies by department, C-Cube tends to fall in line with the casual dress and flexible hours typical to the high-tech industry. "We don't have a dress code and the work hours are flexible," one reports. "On the average, we put in 45 hours a week." Employees report that hours get hectic around project deadlines. One warns: "The working schedule can get tight. You might need to work extra hours without extra pay, and may need to come in on weekends." "Working hours are flexible – as long as you meet project schedules," another reports. Though one employee says, "C-Cube pays competitively," another reports, "the pay is significantly below the market, however, the company gives incentive stock options as compensation, which is good if you join at the right time and the right price." C-Cube's perks are standard, including medical, dental and profit sharing in addition to stock options. The company offers tuition reimbursement to employees pursuing jobs related to their jobs. Employees say they are judged by performance, not gender or ethnicity. Though there are few females in engineering, employees note that, "minorities are the majority in the Engineering department." The Vice President of Engineering is Asian-American, and Gregorio Reyes, a Hispanic, sits on the company's eight-member board of directors.

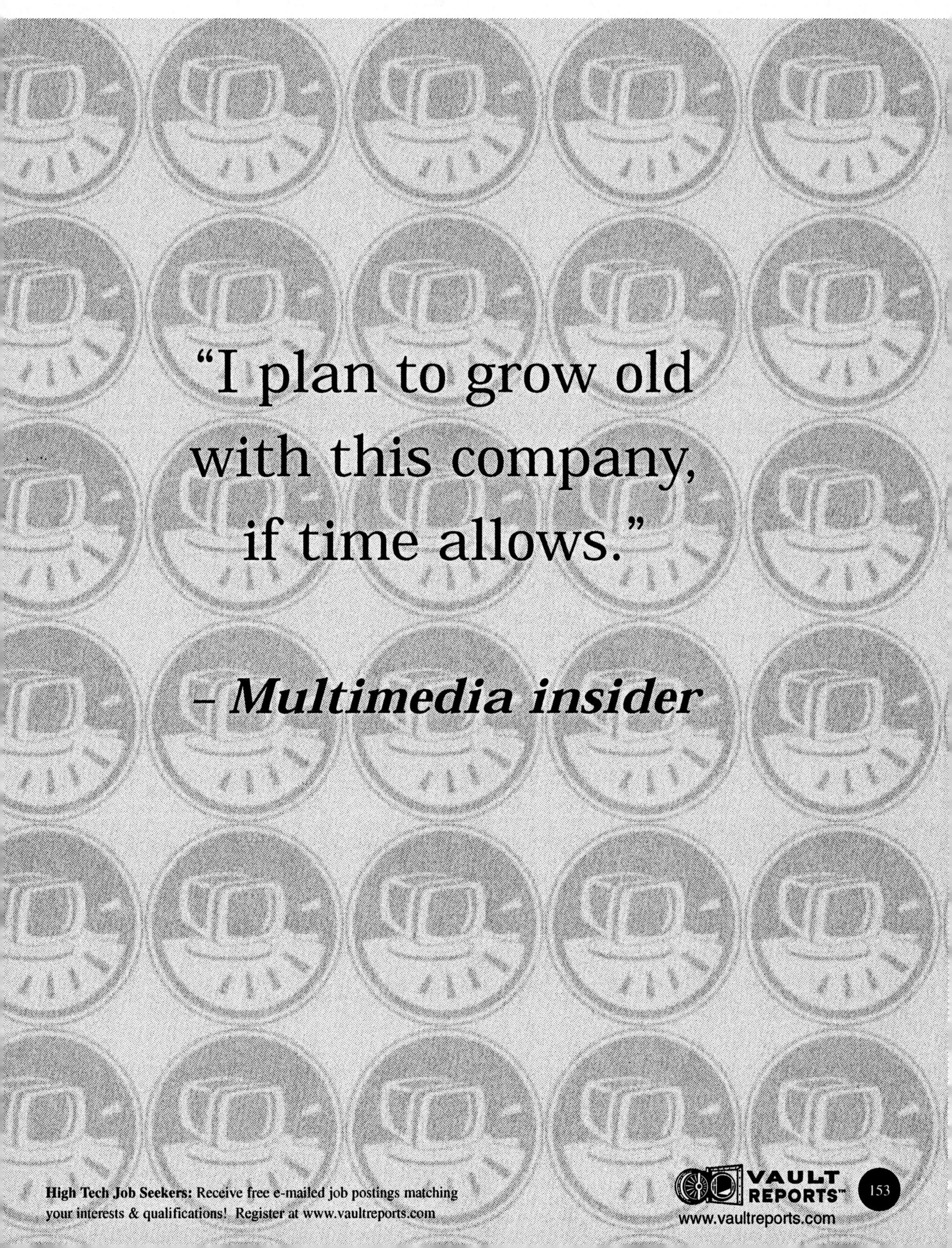
"I plan to grow old
with this company,
if time allows."
– Multimedia insider
High Tech Job Seekers: Receive free e-mailed job postings matching
your interests & qualifications! Register at www.vaultreports.com
VAULT REPORTS™
www.vaultreports.com

Cirrus Logic

3100 W. Warren Ave.
Fremont, CA 94538
(510) 623-8300
Fax: (510) 226-2240
www.cirrus.com

LOCATIONS

Freemont, CA (HQ)
Austin, TX
Boca Raton, FL
Broomfield, CO
Greenville, SC
Korea
Japan
Singapore
Taiwan
United Kingdom

DEPARTMENTS

Engineering
Finance
Human Resources
Information Technology
Manufacturing
Marketing
Quality
Research & Development
Sales

THE STATS

Annual Revenues: $954.3 million (1998)
No. of Employees: 1250 (worldwide)
No. of Offices: 10 (worldwide), 5 (U.S.)
Stock Symbol: CRUS (NASDAQ)
CEO: Michael L. Hackworth

UPPERS

- Employee stock plan
- Performance bonus

DOWNERS

- Recent company downsizing
- Limited workweek flexibility

KEY COMPETITORS

- Analog Devices
- Linear Technology
- Lucent Technologies
- National Semiconductor
- STMicroelectronics
- Texas Instruments

EMPLOYMENT CONTACT

Mr. Patrick V. Boudreau
Vice President Human Resources
3100 W. Warren Ave.
Fremont, CA 94538

(510) 226-2232
Fax: (510) 226-2270

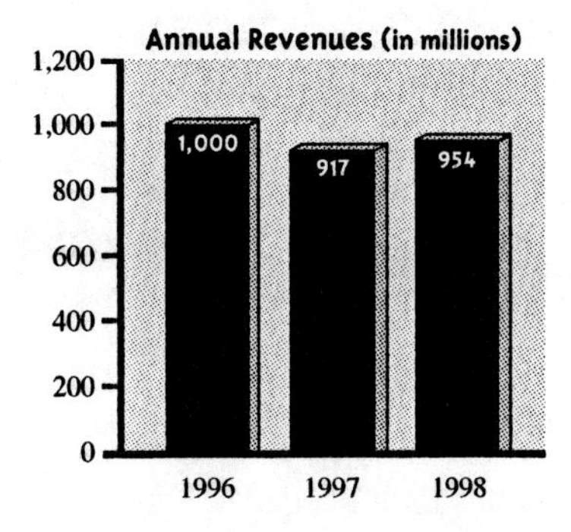

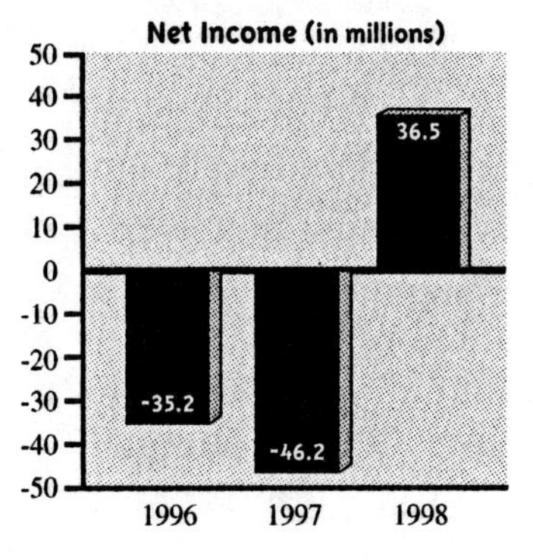

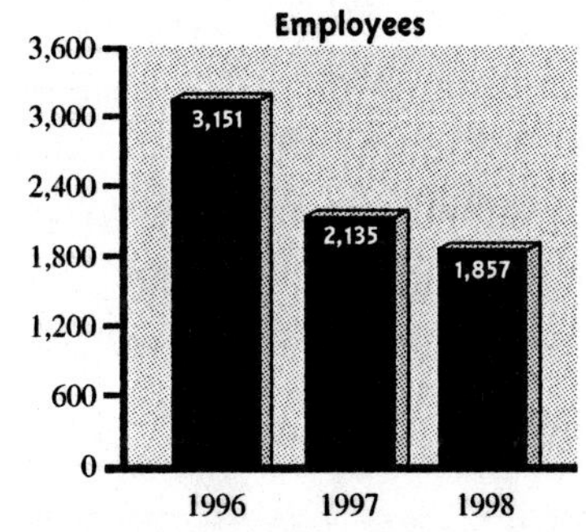
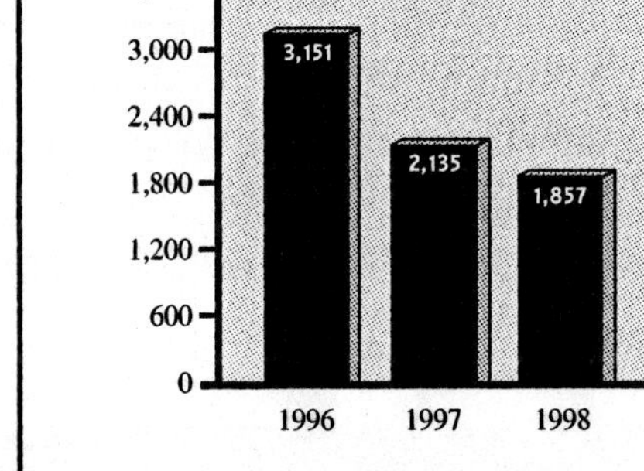

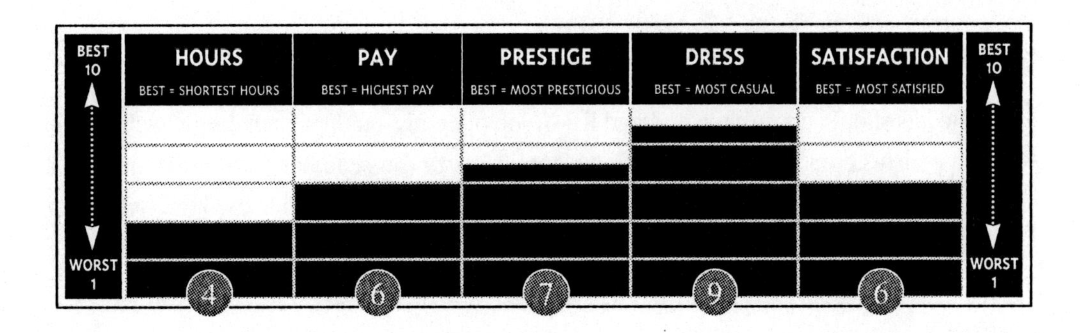

THE SCOOP

After developing advanced software for designing system-level chips, former MIT professor Suhas Patil needed an experienced business partner to help start a company to reap the benefits of his ingenuity. He found just the man in Mike Hackworth, a Northern California native and then senior vice president at Signetics. In 1984 the two men founded Cirrus Logic, a name suggested by Hackworth's daughter Lauren, after the highest clouds in the sky. In 1996, the company proved that it could live up to the name. Cirrus Logic soared to sales of more than $1.1 billion, up from $9 million just eight years before. This made Cirrus Logic the fastest Silicon Valley semiconductor firm ever to break into the billion-dollar-a-year club.

While the strong market for PC graphics chips provided a boost for the company during its early-1990s success period, demand for its 2-D graphics took a hard dive. Just one year after reaching the $1 billion mark, sales dropped over $200 million, down to $917 million. That year Cirrus Logic also gave 15 percent of its workforce the pink slip. Though the firm still has a broad portfolio of products and technologies with applications in multimedia, communications, and mass storage, Cirrus Logic is clearly undergoing a massive restructuring, including the sale of its wireless infrastructure equipment unit in 1996 for $23 million. In 1998, Cirrus Logic accused ATI Technologies of incorporating Cirrus' patented video graphics technology into its products. Though the lawsuit is pending, Cirrus Logic intends to withdraw from the graphics chip business and focus on data and communications.

GETTING HIRED

Applicants should consult the updated list of job openings on the Cirrus Logic web site, located at www.cirrus.com. This service allows applicants to run searches for specific jobs at any or all of the company's locations in the United States and Asia, and explains where to send resumes to apply for the appropriate openings. Cirrus Logic is looking for motivated individuals with strong backgrounds in technical fields, especially electrical and computer engineering; computer science; and research & development. Cirrus recruits graduating college students on campuses around the country as well. Seniors interested in a career with the company can send resumes to the head office, mailing them care of College Relations, MS 513; fax them to (510) 624-7140, attention College Relations; or e-mail them to staffing@corp.cirrus.com. Insiders tell Vault Reports that Cirrus is so eager for new employees that it runs a promotional program with free trips and other prizes given to those employees who make referrals.

OUR SURVEY SAYS

Employees find that "overall Cirrus is a nice place to work" with "competitive" pay and "flexible" hours." "I see a lot of respect and professionalism," says one insider. "Cirrus does not have as many levels of management hierarchy as some bigger companies," employees report. Insiders say that, like many high-tech Silicon Valley companies, Cirrus has "had some great times and some not so great times." Explains one employee: "The last two years have been a recuperation period for the company. We've had two downsizings and a fairly major reorganization, with the company focus turning away from the highly competitive graphics market, and turning to the sub-$1000 PC and system-on-a-chip markets. This refocusing has been very good for Cirrus Logic and this year [1998] employees will receive two bonuses." Benefits and employee stock purchasing plans are "very competitive in the market," and considered by insiders to be "the biggest perks" of the job. Another perk – "company subsidization of a very nice health club." Compensation is based on your experience "and how well you can negotiate with HR!"

As is the case almost across the board in Silicon Valley, the dress code is "fairly casual, usually comfortable jeans." Employees say that "most people have cubicles; only managers have offices." "I'm not sure how they evaluate who deserves windows," shrugs one insider. However, not all is "a sea of cubes," for "the cafeteria is very good. Kitchenettes and eating areas are provided in every building as well as outside. There are on-site basketball courts." As far as employment policies go, Cirrus has "a large number of women" and a "big percentage of minorities." However, one female employee comments that while "well over half of senior management is of Asian descent, women, on the other hand, are less well represented. It makes me wonder about the glass ceiling." Still, that insider notes that "many women engineers, have been rightfully promoted." The workweek is "not too progressive. Flex-time and telecommuting are allowed with a manager-employee agreement, but there is no company policy."

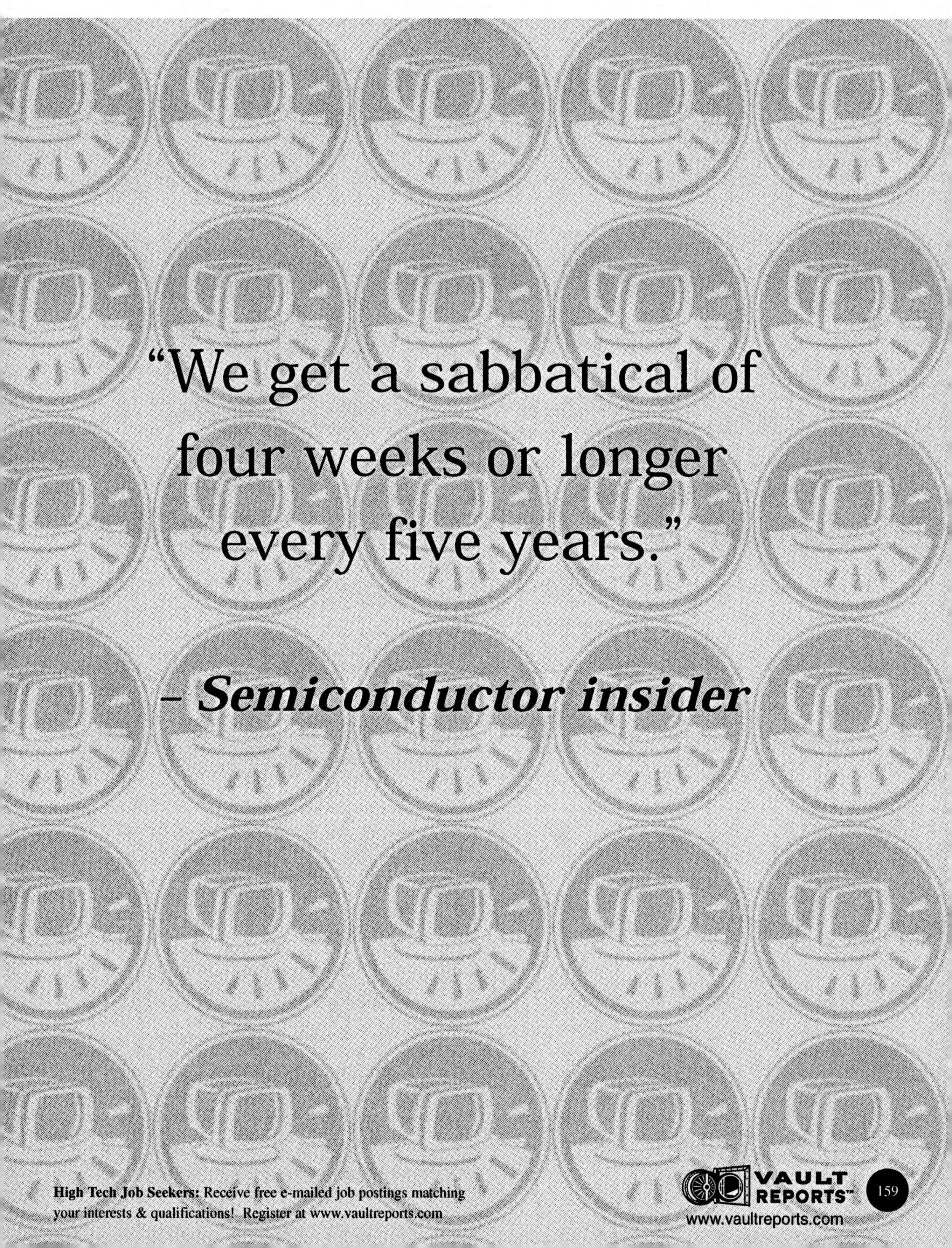
"We get a sabbatical of
four weeks or longer
every five years."
– Semiconductor insider

Cisco Systems

170 W. Tasman Drive
San Jose, CA 95134
(408) 526-4000
Fax: (408) 526-4100
www.cisco.com

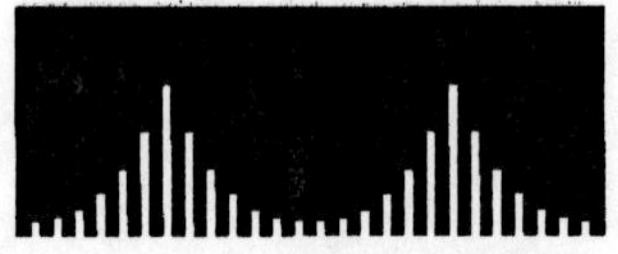

LOCATIONS

San Jose, CA (HQ)
Chelmsford, MA
Irvine, CA
Research Triangle Park, NC
Santa Cruz, CA
Additional sales locations nationwide and abroad

DEPARTMENTS

Customer Engineering
Finance and Administration
Human Resources
Manufacturing
Marketing
Sales
Software & Hardware Engineering

THE STATS

Annual Revenues: $8.4 billion (1998)
No. of Employees: 11,000 (worldwide)
No. of Offices: 5 (U.S.)
Stock Symbol: CSCO (NASDAQ)
CEO: John T. Chambers
Year Founded: 1984

UPPERS

- Widely renowned job training
- Beautiful corporate offices
- Free T-shirts
- All-u-can-eat popcorn
- Premier company in industry segment

DOWNERS

- Low pay relative to industry
- Competing for cubicles

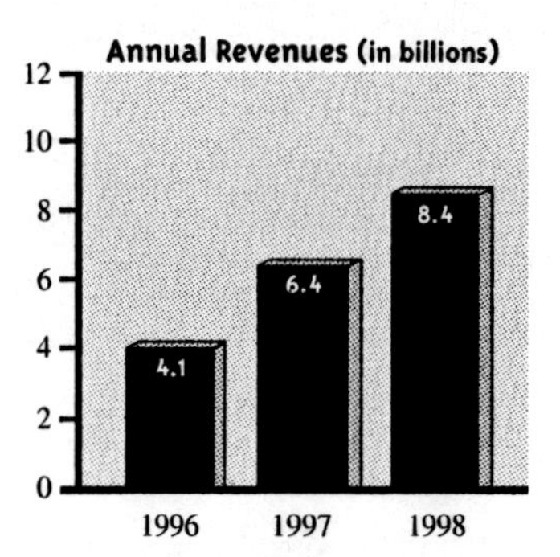

KEY COMPETITORS

- 3Com
- Cableton
- Lucent Technologies
- NEC
- Nortel

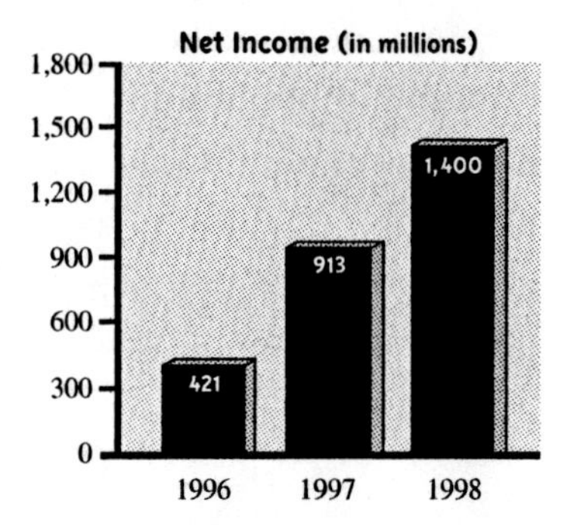

EMPLOYMENT CONTACT

Human Resources
Cisco Systems, Inc.
P.O. Box 640730
San Jose, CA 95164

jobs@cisco.com

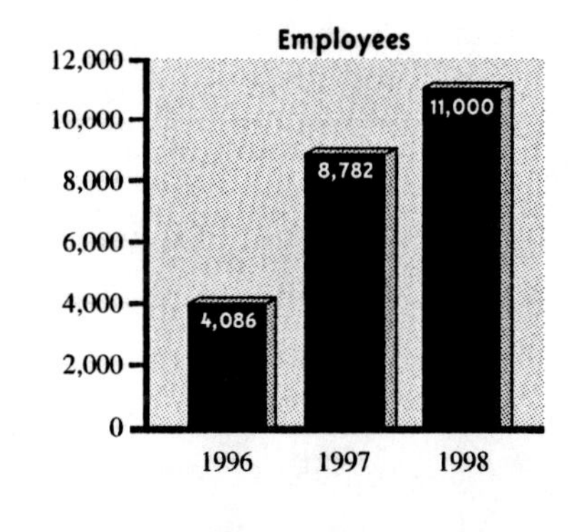

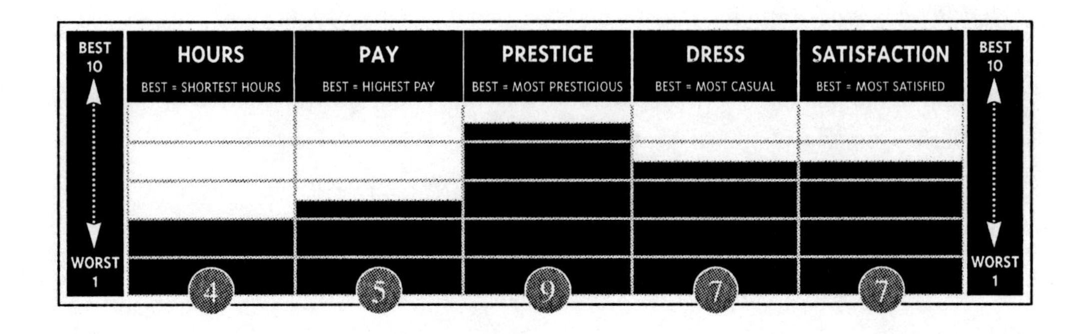

THE SCOOP

Some call Cisco Systems the third leg in the "triopoly" that also includes Microsoft and Intel. Optimistic employees already have coined a term for the synergy – "Wintelco." While Microsoft writes the software and Intel makes the powerful chips that run PCs, Cisco manufactures the routing systems and switches that help computers talk to each other via networks and the Internet. Founded in 1984 by Stanford University scientists Leonard Bosack and Sandra Lerner, Cisco has built its fortunes upon the surging demand for reliable network technology. The company sold its first router in 1986, and saw its market explode. Cisco is currently the world's leading supplier of networking products, including multiprotocol routers, bridges, workgroup systems, ethernet switches, and network management software.

Cisco's Internetworking Operating System (IOS) allows computers running from different operating platforms to work together seamlessly. The Internet explosion has translated into a booming business for Cisco, which makes more than 80 percent of the routers that serve as the Internet's backbone, as well as much of the technology that connects individual networks to the World Wide Web – routers, LAN and WAN switches, dial and other access modes, SNA-LAN integration, web site management tools, and network management software.

Cisco is prone to acquiring companies; technology expertise acquired through such acquisitions includes network management, digital subscriber line (DSL), and voice/data/video integration. Cisco Systems expects such acquisitions to play "an ongoing role in… leadership strategy." Also playing an important role: partnerships with other companies. Cisco is leery of the "go-it-alone" strategy, and has formed partnerships with Microsoft, Intel, Hewlett-Packard, GTE, Alcatel, and Dell.

Now, Cisco wants to promote its own strong brand identity. As the company says in its own literature: "A compelling Corporate Identity that expresses our personality and leadership position in the marketplace is important to Cisco. As our growth continues, the necessity to establish a consistent global presence in this expanding industry has become even more apparent. The Cisco brand is equal in value to our products and services. It cannot be compromised. And it must be enhanced."

While a mega-merger is unlikely, partnerships appeal to Cisco. In 1998, the data networking company initiated talks with telecommunications giants Lucent Technologies and Northern Telecom (now called Nortel), but was firmly rebuffed. Both companies have plans to enter Cisco's turf and develop products that compete directly against Cisco's. Lucent echoed its refusal a few months later, when it leveled a patent infringement lawsuit against Cisco, alleging a violation of eight data networking patents; Nortel bought Cisco competitor Bay Networks for $7.6 billion. Cisco remains the market leader, however, and investors have taken note: the firm hit the firm hit $100 billion in market capitalization in July 1998.

GETTING HIRED

Cisco recruits on some college campuses but also accepts applications via mail, fax, and e-mail (ASCII text). The web page also has a full on-line application apparatus. Cisco has full-time opportunities in manufacturing, engineering, customer engineering and information systems. The company's recruiting web page, located at www.cisco.com/jobs, has information on current employment opportunities as well as more general job descriptions. Typically, Cisco has recruiting seasons in the fall and the spring. The majority of Cisco employees work at the San Jose headquarters, but those interested in sales and marketing positions should indicate their geographical preferences.

OUR SURVEY SAYS

Cisco employees work in a "high energy" environment that stresses "productivity above all else." Recent hires praise the "incredible" training that they receive and relish the chance to work for a company where "intelligence, learning aptitude, and resourcefulness are more highly prized than the ability to kiss butt." *Wired* said of Cisco employees: "Nobody has this much fun going to work. All [Cisco employees] do is smile, smile, smile." An insider adds: "If you are remotely entrepreneurial, you will work crazy hours because you want to, not because you have to." Happily, though, "there is 'extra-duty' pay for salaried employees working weekends and holidays."

Any complaints? The pay is "not up to snuff," complain some employees. One insider says: "Considering the cost of living in San Jose, it isn't great – maybe better than many, but worse than many. The pay scales definitely need adjustment to reflect the extremely high cost of living, as well as the fact that most people in my department are recent college grads with big loans." However, as a bonus, "other companies will pay lots more – up to 60 percent more – to hire us away." Some employees are none too thrilled about Cisco's "non-territorial offices," its policy of not having assigned desks for most employees (except support staff, which need access to their files, and managers, with their sneaky managerial secrets) – all cubicles are first-come, first-serve basis. "There can be some competition for the best spots," says one insider. "It's like always being on line for a movie, and then the doors open." Summing up the bright side of working at Cisco, one employee says "I've learned more in eight months here than in the years I spent in college. Eventually, I'll be able to take this knowledge somewhere where I don't have to worry about making the rent every month."

"If you are remotely entrepreneurial, you will work crazy hours because you want to, not because you have to."

– Hardware insider

CNET

15 Chestnut St.
San Francisco, CA 94111
(415) 395-7800
Fax: (415) 395-9205
www.cnet.com

LOCATIONS

San Francisco, CA (HQ)

DEPARTMENTS

Administration
Advertising Services
CNET Labs
CNET Television
Content
Creative Services
Editorial
Management Information Services
Marketing
Production
Sales
Snap Online
Strategic Relations
Technology

CNET

THE STATS

Annual Revenues: $33.6 million (1997)
No. of Employees: 581
No. of Offices: 1
Stock Symbol: CNWK (NASDAQ)
CEO: Halsey M. Minor

UPPERS

- Young, bright co-workers
- Diverse environment

DOWNERS

- Can be overly political
- Hemorraging money
- CEO known to yell

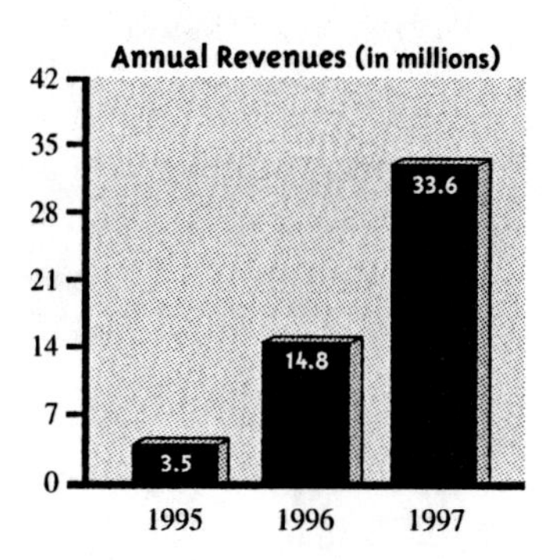

KEY COMPETITORS

- CMP Media
- IDT
- International Data Group
- Mecklermedia
- Software.net
- Time Warner
- Wired Ventures
- Ziff-Davis

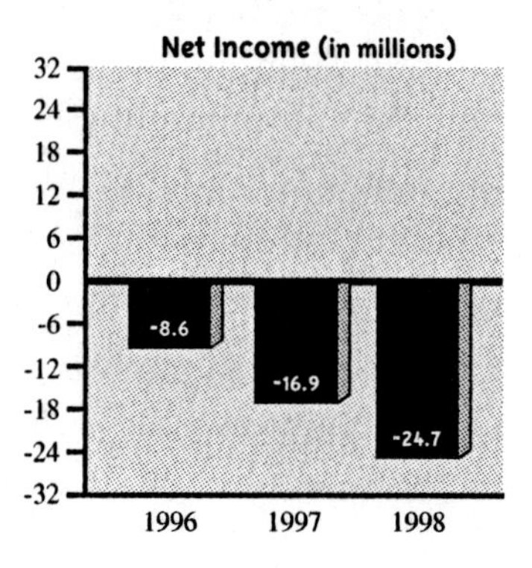

EMPLOYMENT CONTACT

Heather McGauhy
VP Human Resources
CNET
15 Chestnut St.
San Francisco, CA 94111

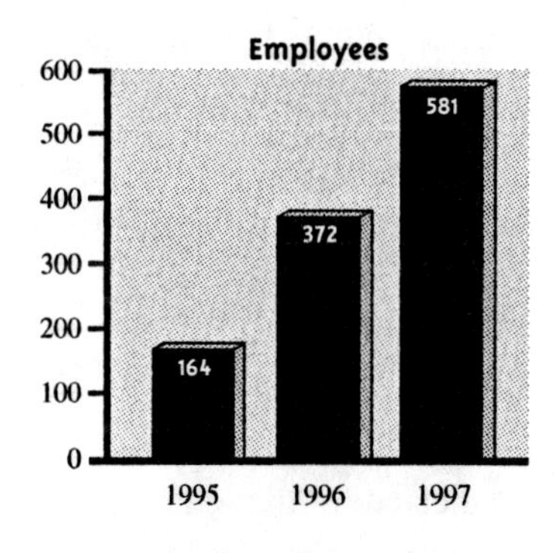

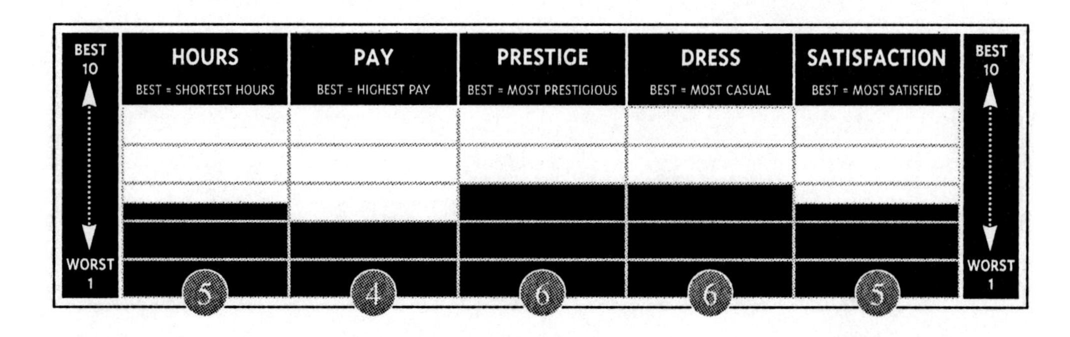

THE SCOOP

CNET keeps the world informed about the most important developments in the world of computers and the Internet. It is one of the few companies that publishes original content on the Internet, a decidedly expensive strategy when successful competitors simply regurgitate information already in print. Halsey Minor, a former executive recruiter, founded CNET in 1992, with $5 million in venture capital from Microsoft co-founder Paul Allen. Minor's idea was to develop a cable network devoted to computers and new technology, with a companion project on the Internet. In April 1995, the company launched a weekly half-hour program on the USA Network, and the cnet.com site, which featured news, product reviews, and helpful tips. Today, CNET has four tech-focused TV series, including *CNET Central.* It also operates a variety of Internet sites, which are devoted to a variety of topics, ranging from web design to online gaming to downloadable software.

CNET's brightly colored, user-friendly sites get the best ratings in their categories, and have evolved into high-tech staples. Those in the business trust CNET to provide up-to-the-minute information; those in the know look to CNET's product reviews before they buy. It's no surprise then, that advertisers love CNET's sites. Still, despite rapidly growing revenues, the company has yet to turn an annual profit.

In 1995, Minor spent $25 million to create Snap!, a search engine/directory site. Minor attracted much criticism for this expenditure, mostly because CNET was forced to play catch-up with the likes of established names like Yahoo! and Excite. In 1998, however, Minor was vindicated for his search engine move, when NBC laid out $26 million for a 4.9 percent stake in CNET. As part of the deal, NBC purchased a 19 percent stake in the Snap portal, and has an option to buy a 60 percent controlling stake for an additional $32 million. Industry analysts expect Snap to fare better now because of NBC's marketing and advertising muscle.

GETTING HIRED

CNET wants employees with a real love for technology and the possibilities of the Internet. Visit the company web site at www.cnet.com for information on the company and a list of current job opportunities. Snail mail, e-mail or fax your resume to the Human Resources contact indicated. The company does not accept phone inquiries.

OUR SURVEY SAYS

The work environment is "very young and relaxed" and employees are "some of the smartest people you could ever meet." The company is "very diverse, and we have a large percentage of women and minorities all the way to upper management." Be careful what you say around the office, however – one source describes several employees having problems with superiors. In fact CEO Halsey Minor has been known to bawl out an employee or two on a stressful day. Another insider reports that "a few careers have been killed by politics through e-mail."

Dress is business casual, though "upper management seems to dress a bit more formally." In most departments, typical hours are 9 a.m. to 6:30 p.m., although people in tech positions usually work 50- to 70-hour weeks. Employees enjoy typical industry salaries and "all the usual benefits – 401(k), stock options, and so on."

Compaq

20555 State Highway 249
Houston, TX 77070
(281) 370-0670
Fax: (281) 374-1740
www.compaq.com

COMPAQ

LOCATIONS

Houston, TX (HQ)
Numerous other U.S. and international locations

DEPARTMENTS

Administration
Business Planning
Customer Service
Hardware Engineering
Information Management
Internet Services
Manufacturing/Operations
Sales & Marketing
Software Engineering

THE STATS

Annual Revenues: $24.6 billion (1997)
No. of Employees: 80,000
No. of Offices: 600 (worldwide)
Stock Symbol: CPQ (NYSE)
CEO: Eckhard Pfeiffer

UPPERS

- Gym memberships
- Management bonus program
- "Family festivals"
- Merchandise discounts

DOWNERS

- Long workdays
- Rigid bureaucracy in some locations
- Looming layoffs in wake of Digital aquisition

KEY COMPETITORS

- Dell Computer
- Hewlett-Packard
- Apple Computer
- Gateway
- IBM
- Micron Electronics

EMPLOYMENT CONTACT

Compaq Comp Corporation
Attn: U.S. Employment
MS 110415
P.O. Box 692000
Houston, TX 77269

Fax: (408) 285-6938
careerpaq@compaq.com

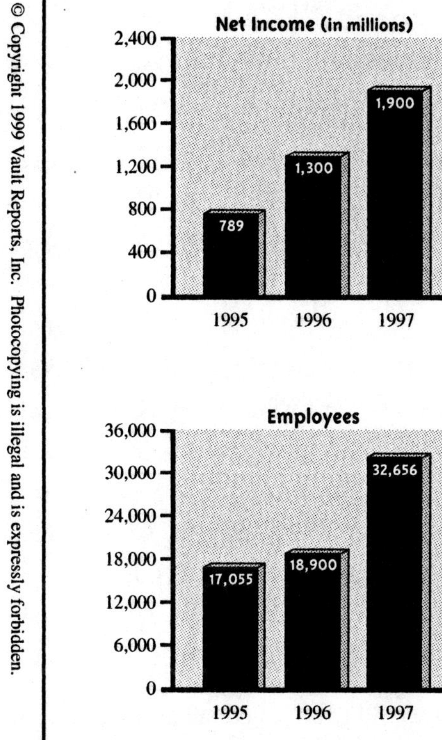

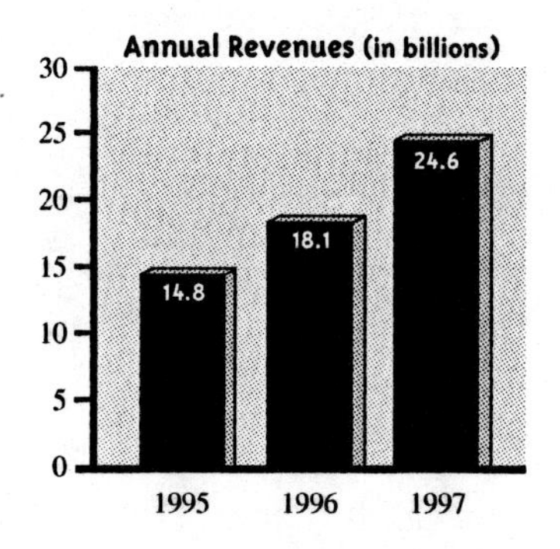

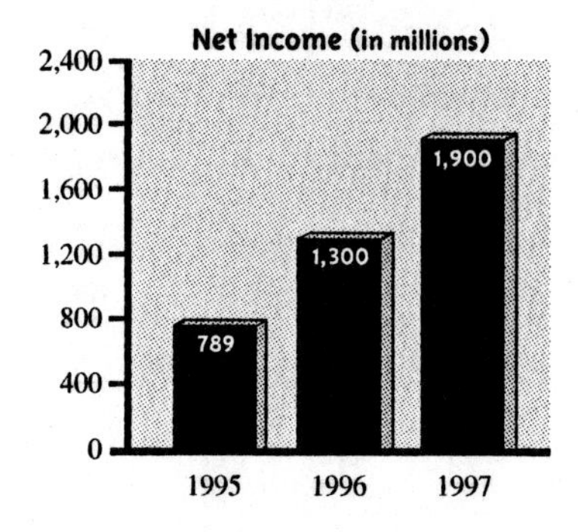

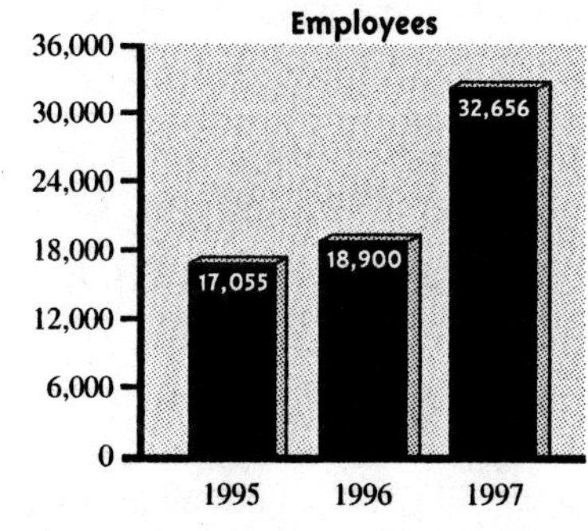

HOURS	PAY	PRESTIGE	DRESS	SATISFACTION
BEST = SHORTEST HOURS	BEST = HIGHEST PAY	BEST = MOST PRESTIGIOUS	BEST = MOST CASUAL	BEST = MOST SATISFIED
3	8	9	7	10

BEST 10 — WORST 1

THE SCOOP

The world's leading PC designer and manufacturer has one thing to say to those who are astonished by their explosive sales: "We're not finished yet." With over 12 percent of the world's PC market under its control in 1996, Compaq is now poised to dominate the network server market and is pushing ahead into the networking and workstation markets. Completely compatible with Windows NT, the spread of Microsoft's dominion is good news for Compaq CEO Eckhard Pfeiffer and his firm.

Founded by Texas Instruments executive Joseph "Rod" Canion, Compaq boasted $111 million in sales in its second year. The reason for the company's fast success was simple: the newest technology in a crisp, IBM-compatible package. The company kept its overhead costs low and boosted its sales by granting exclusive dealership rights to stores across the country. Recently, Compaq has had to fend off new, smaller competitors in the PC market. To do so, the company has developed mid-priced portable computers and cut its prices on some of its best-selling models. Meanwhile, Compaq has expanded its sales force and purchased Tandem Computers as part of its push to reach the $40 billion sales mark by the year 2000. So far, Compaq has had notable success. The preeminent Wintel computer vendor, Compaq had $25 billion in sales in its 1997 fiscal year. Compaq has also moved strongly into the server market. By 1998, Compaq expects to gain $10 billion in revenues from the sales of servers and data processing systems. Compaq servers will run Walt Disney's new high-tech entertainment center, and a Compaq data processor runs NASDAQ, the red-hot New York stock exchange.

How will Compaq compete when it needs to service all its servers and fancy data processors? Until recently, Compaq only had 8000 sales and service people, a twentieth the number of IBM. But in June 1998, Compaq acquired Digital Equipment for the impressive sum of $8.5 billion. The resultant Compaq has a combined annual revenue of $38 billion and edges out Hewlett-Packard as the second-largest computer company. After paying $5.4 billion to finance the merger, Compaq announced that it would eliminate overlap by gradually cutting 17,000 jobs worldwide – mostly on Digital's end – through plant closings and consolidations. Compaq's new strategy is to keep inventory low and product variety high in hopes that it will rival IBM in size and reach.

GETTING HIRED

Compaq accepts resumes through both e-mail and regular mail. Applicants should consult the company's career web page, located at www.compaq.com/jobs. The web page enables applicants to construct a resume online, provides a link to a list of career opportunities, and offers a calendar of Compaq's upcoming recruiting visits around the nation. Insiders say Compaq looks for "self-driven, friendly, customer-oriented, friendly workaholics." Once in the door, say our contacts "expect to be in a position for 18 months before becoming eligible to move to your next job at Compaq." You can e-mail your resume to Compaq at careerpaq@compaq.com. Plain ASCII text, please.

OUR SURVEY SAYS

A "universally-respected" industry giant, Compaq has created a "high energy," "fluid" work place for its employees. So-called "fast-trackers" in the company sometimes work "intense, long weeks," but they are rewarded with a pay scale that is "substantially above average" for the computer industry. Added to the compensation is a profit-sharing plan that insiders say paid 6 percent of a full-time employee's salary each of the last four years. "Middle management and above" receive stock options. "Compaq has a culture that employees have grown to love," says one insider. "It pays its people well and rewards successful ones. It is definitely a team effort, with everyone pulling together." This "team" atmosphere has some very concrete manifestations: "Almost everyone lives near the main campus." Says another: "Most of Compaq is on a first-name basis."

As for diversity issues, employees report that there is a "high percentage of women and also a significant percentage of employees of Asian, Hispanic, and African American descent." However, one female insider reports "mentoring relationships which are easier between older white males in power and younger white males in whom they see themselves reflected a generation ago." "At Compaq we've made a lot of progress but there is still more to do," that insider says.

Employees at Compaq's Houston headquarters call the campus a "technological mecca" that provides perks such as an "an outstanding internal web site" and intramural sports. Internal job postings, meanwhile, create "frequent" advancement opportunities within the company. "We have very comfortable and attractive facilities" at the Houston headquarters, says one insider. "We are now building a gym and central cafeteria facility on site. It will be called Compaq Commons. There will be a conference and convention center there too." Crowded employees at this rapidly expanding company are also pleased to note that "they are building two new 10-story office buildings, as we are at overcapacity now."

Employees dress casually in the Houston heat. "Dockers is a well-known name around here, and Fridays are generally known as blue jeans and T-shirt days. Some shorts tend to slip in during the summer too." Despite a bit of crowding, employees still enjoy the main campus (or, as it's known to techie employees, CPQMS). "I have a lovely office that faces a woodland, and I can see deer at night," says one insider. Another contact praises the also praises the locality: "There are several nice and reasonably priced subdivisions nearby." Other bonuses include athletic facilities and a plethora of eating choices. "For lunch, we have the Olive Garden, Shipley's Donuts, and a few fast food restaurants." Also, "we have a running track that winds through the campus. We have showers too." One employee adores Compaq's "check-out policy" which allows you to "take home the latest in Compaq tech for your personal use. When it's obsolete, just return it and get a new one."

"The campus is pretty posh, with a huge lawn in the middle, a very nice cafeteria, and a big indoor atrium with a café."

*– **Hardware insider***

Computer Associates International

One Computer Associates Plaza
Islandia, NY 11788
(516) 342-5224
Fax: (516) 342-5329
www.cai.com

COMPUTER ASSOCIATES

LOCATIONS

Islandia, NY (HQ)
New York, NY • Culver City, CA • Aurora, CO • Fort Lauderdale, FL • Atlanta, GA • Lisle, IL • Bloomington, MN • Princeton, NJ • Cincinnati, OH • Irving, TX • Reston, VA • Bellvue, WA • Toronto, Montreal • Ottawa, Calgary • Vancouver, Canada

DEPARTMENTS

Client Services
Development
Programming
Project Management
Sales
Technical Services

THE STATS

Annual Revenues: $4.7 billion (1998)
No. of Employees: 11,400 (worldwide)
No. of Offices: 160 offices in 43 countries.
Stock Symbol: CA (NYSE)
CEO: Charles B. Wang

UPPERS

- Free breakfast
- On-site gym and child care
- Paid medical benefits

DOWNERS

- Blemished corporate image
- HQ in Long Island, NY
- Long, rigorous interview process

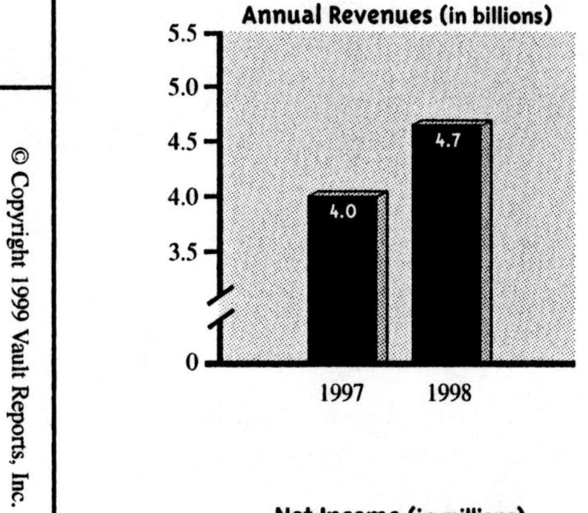

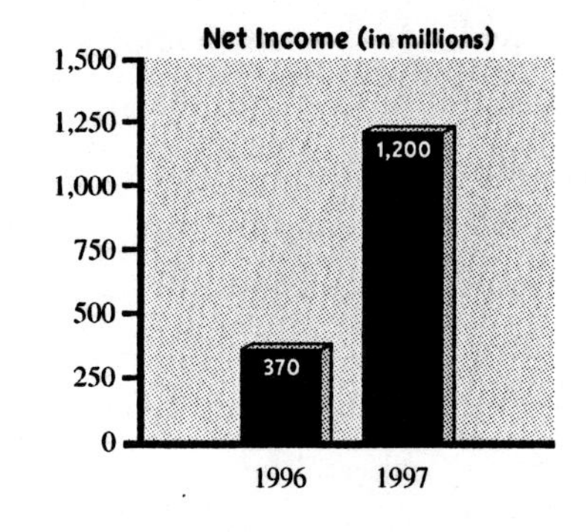

KEY COMPETITORS

- Amdahl
- Hewlett-Packard
- IBM
- Oracle
- Microsoft
- SAP
- Sybase
- Symantec

EMPLOYMENT CONTACT

Deborah Coughlin
Human Resources Dept./NET
Computer Associates International, Inc.
One Computer Associates Plaza
Islandia, NY 11788

resumes-usa-r1@cai.com

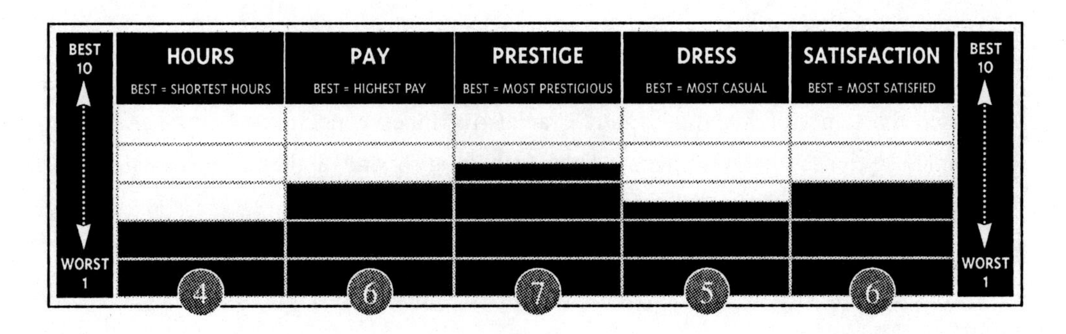

THE SCOOP

Computer Associates International is the third-largest independent software company behind Microsoft and Oracle. Charles Wang founded a U.S. subsidiary of CA, which was originally a Swiss company in 1976, bought out his partners in 1980 and took Computer Associates public in 1981. The multibillion-dollar technology company began by making mainframes and managing data center systems and entered the software business in 1994. Today it is a major force in distributed enterprise management (client/server) networks. Its clients include more than 90 percent of Fortune 500 companies. Commonly referred to as the vulture of the software industry, the company has a 20-year history of acquiring ailing competitors and ruthlessly cutting costs – and jobs – in the assimilation process. CA has also succeeded by creating partnerships, developing strong products, and integrating key technologies. Now offering 500 different products, CA has the ability to deal with technology ranging from COBOL to the latest NT, on any platform. Industry insiders (some grudgingly) refer to CA as "the plumbers," because they offer so many options, enabling businesses to integrate their existing applications with new technologies, and "manage the network mess."

Today the company is concentrating on further diversifying its product mix and improving its notoriously poor client services. In 1997, CA launched Unicenter TNG ("The Next Generation"), which allows access to each component of a company's internal and external links from a central console. One of the most desirable aspects of the program is its 3-D monitoring capability, which allows users to see how problems affect the entire network. Sales of the product have skyrocketed since its release, and revenues are expected to continue growing, as more and more companies establish their own Intranets.

The company tried to strengthen its client services business in 1998 by acquiring Computer Sciences Corp. in a hostile takeover. CA publicly pursued the consulting/computer service company for three weeks before it finally withdrew, citing "ugly mudslinging tactics" and accused the target of making negative racial overtones concerning Wang's Asian heritage. Computer Sciences likewise accused the technology giant of illegal business practices and "economic extortion."

Though CA may never attempt another hostile takeover, Wang and his company seem to have recovered just fine. CA has recently been gloating over the fact that it has effectively beaten out rival IBM's Tivoli Systems for a new partnership with Microsoft. As one industry analyst put it: "CA's the preferred partner… it's the one Microsoft's getting behind." (IBM's Tivoli division is CA's main competitor for the small-business market.) Microsoft will bundle The Real World Interface, a portion of CA's Unicenter TNG, with a new version of Windows NT set for release in early 1999. The web-based enterprise management application offers a 3-D view of different sources, giving users a cohesive view of their IT environment. This deal further cements the relationship CA has been trying to build with Microsoft over the past few three years. In February of 1998, CA's Korean subsidiary announced a deal to bundle CA's data backup, storage and virus protection applications with Microsoft Korea's network operating system, and in April, Bill Gates gave the keynote address at the CA 1998 conference in New Orleans. CA has also forged pacts with Intel, Tandem Computers (owned by Compaq), and Hewlett-Packard to bundle software and develop new products.

Though CA has traditionally marketed its network management software to big corporations, it is now eyeing the growing small-business market. Bundling the Real World Interface with Windows NT was a good first step, as that Microsoft product is widely used by smaller companies. The company plans to use independent resellers, systems integrators and consultants to serve small business customers – a cheaper alternative to using its own direct sales force. In addition, CA has launched its own Professional Services organization, which is expected to employ 3000 people by the end of the year. The company's latest software offerings include Harmony, a strategy for information management, and Jasmine, a multimedia object-oriented database and development environment. CA has not done as well in desktop management – though the company launched an Open Desktop Management Initiative, and is looking into partnerships with Intel and Microsoft.

CA has also been expanding internationally. In 1996, CA acquired South African Dimension Data Group's share of their joint software distribution and servicing center in sub-Saharan Africa. Betting that the political climate had settled down, CA decided to take control of the business, called Computer Associates Africa. The next year, CA stepped into India – the company is expanding $100 million dollars to build a technology center, complete with dormitories and several offices. CA is expected to continue with its acquisitive strategy to gain a foothold in India; and is looking throughout the rest of Asia for possible targets.

GETTING HIRED

CA recruits on college campuses, especially at SUNY Stony Brook, a university near its Long Island, New York headquarters. The company also posts job listings on its web site, in major metropolitan newspapers, and on billboards along the Long Island Expressway. In addition to participating in job fairs around the country, CA holds several of its own on site each year. Insiders say the company is concentrating most of its current hiring at company headquarters, and intends to double the size of the Islandia office.

The interview process generally starts with a phone interview, followed by a meeting with the hiring manager. Successful candidates generally come in for another round with people on the tech side (if that's where they are applying). Insiders say the process sometimes take a long time – one tech employee reports waiting four months for an offer. For marketing or sales, the process may not be as rigorous, but do include "several rounds of interviews starting with middle managers and ending with a one of the VPs." Says one insider: "They tend to do some things the old fashioned way, with a formal approach unlike a lot of the Silicon Valley-type companies."

OUR SURVEY SAYS

Though it tends to fight dirty on the corporate playing field, on the inside, CA is "a true meritocracy," where "everyone has a chance to go as far as they want to." One employee on the technical side reports, "I have found my raises accurately track what I did in the previous year," and "my responsibilities have increased as I have asked for them to be." Further, they advise that "it's important to be very proactive."

Employees say Computer Associates is very open to women and minorities, which is common for the industry, and point out that "the CEO and COO are both minorities." One employee adds that "the overall attitude is positive towards advancement of anyone with talent that can generate revenue." If they can't, however, they're out. Some insiders – salespeople, at least – say "there is a medium to high turnover rate" for those who do not meet their quotas.

Computer 'Associates' also enjoy the "team mentality" within the company, and its "second family" feel. One insider, however, makes you wonder whether there isn't a little human-type programming going on at CA: "No matter the country where you live or the language you speak, if you are a man or a woman, there is only one important thing: CA. Remember these words, because I'm sure you'll hear them. Remember..."

CA offers an extensive array of perks, including free breakfast, tuition reimbursement, fully-paid medical benefits, and a 'discretionary distribution' – through which the company matches up to 8 percent of your base salary in a 401(k) plan. CA maintains "first-rate" corporate fitness facilities, and on-site child care centers in its offices all over the world. Perks like this gain a great deal of publicity for a company commonly criticized for its surly business demeanor. Computer Associates is consistently ranked high on "top places to work" surveys in publications around the country. The dress code in the corporate offices is pretty straightforward: "If you meet clients, you wear full business attire; if you don't, you wear business casual." The lucky employees in development and tech support outside the NY headquarters say they wear jeans and T-shirts on a regular basis.

Cypress Semiconductor Corporation

3901 N. First St.
San Jose, CA 95134-1599
(408) 943-2600
Fax: (408) 943-2796
www.cypress.com

CYPRESS SEMICONDUCTOR

LOCATIONS

San Jose, CA (HQ)
Austin, TX • Beaverton, OR • Bloomington, MN • Colorado Springs, CO • Nashua, NH • Seattle, WA • Woodinville, WA • Round Rock, TX • Starkville, MI • Cork, Ireland

DEPARTMENTS

Accounts Payable
Administrative
Customer Service
Design
Engineering
Finance
Marketing & Sales

THE STATS

Annual Revenues: $544.4 million (1997))
No. of Employees: 2,770
No. of Offices: 14
Stock Symbol: CY (NYSE)
CEO: T. J. Rodgers
Year Founded: 1982

UPPERS

- Tuition reimbursement
- In-house fitness facilities
- Automotive center with discount gasoline, car wash and detailing

DOWNERS

- Flamboyant, domineering workaholic CEO
- Falling stock prices
- Recent downsizing

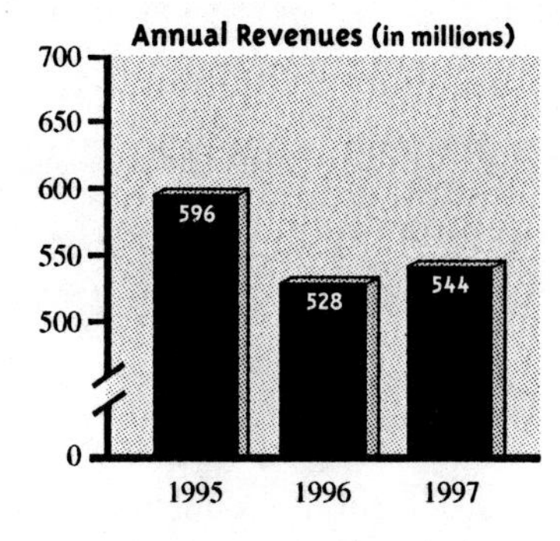

KEY COMPETITORS

- Advanced Micro Devices
- Alliance Semiconductor
- Atmel
- Integrated Circuit Systems
- Integrated Device Technology
- Intel
- Lattice Semiconductor
- LSI Logic
- Micron Technology
- National Semiconductor
- Phillips Electronics
- Samsung

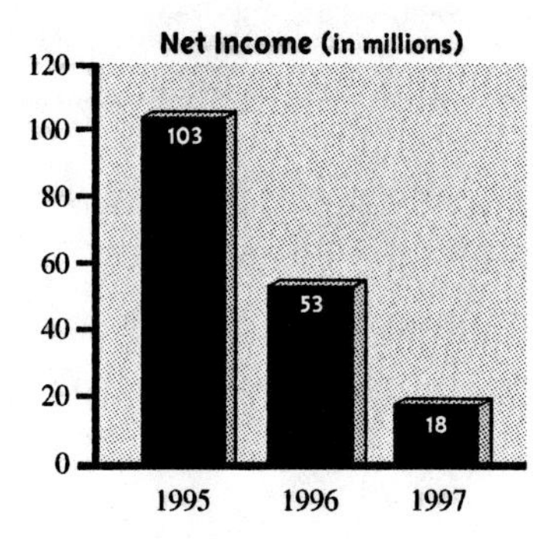

EMPLOYMENT CONTACT

Joyce Sziebert
Vice President of Human Resources
Cypress Semiconductor Corporation
3901 N. First St.
San Jose, CA 95134-1599

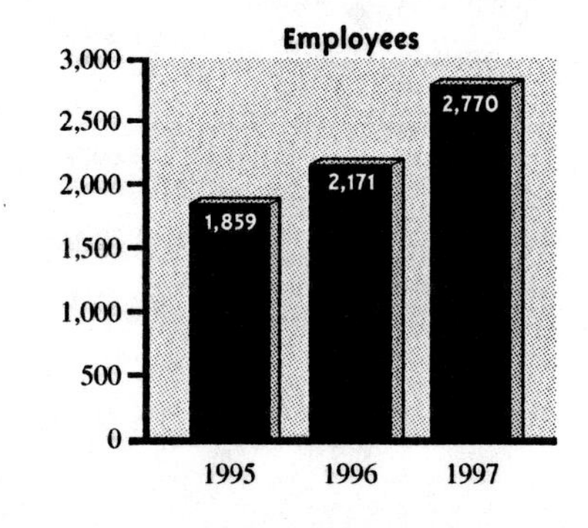

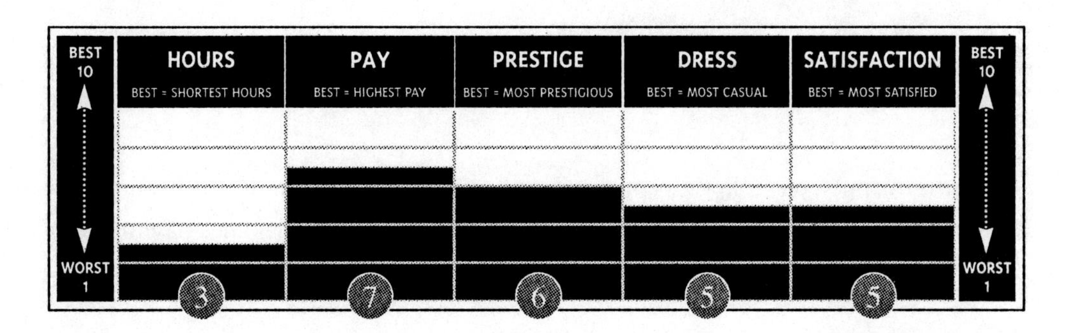

THE SCOOP

Cypress has long been a leader in the microchip-making business. It manufactures a wide range of products including SRAM (static RAM used in cellular phones, networks and computers), memory and data communications chips, timing devices, and PC chips. Its customers are major computer, networking and telecom companies, including AT&T, IBM and Motorola. About 10 percent of the company's revenues comes from military contracts.

T. J. Rodgers, Cypress Semiconductor's bombastic CEO, thrives on media attention. His memo paper is reportedly distinguished by the header: "From the Desk of God." He has publicly spoken out against corporate welfare, and politically correct hiring policies – though he was one of the biggest proponents of the recent move to increase the annual visa quota for highly-trained foreign engineers to work for American high-tech companies.

But industry observers say Cypress' recent performance has been "mediocre," and some suggest Rodgers should make fewer speeches and return to the hands-on management style that initially made his company a success. Like everyone else in the semiconductor business, the company has suffered as a result of falling chip prices and market saturation. Rodgers refuses to quit the SRAM business, despite the fact that Cypress' stock prices have fallen steadily over the past two years. In fact, the company is expected to introduce new "value added" SRAMs in the near future. In 1998, like most computer chip makers, Cypress Semiconductor went through a reorganization, laying off 100 workers in Texas consolidating operations in its Minnesota plants, and closing its test facility in Thailand.

GETTING HIRED

Cypress recruits on college campuses, and posts openings on its web site, located at www.cypress.com. Our sources say the company is going through some "painful" changes at the moment, "so there are not too many job openings."

Insiders admit that "the interview process is rather rigorous." Expect to face "a wolf pack" – four to six people who "try to test your ability to cope with a little bit of stress to see if you can remain focused." "Naturally, extensive technical questions are involved," and candidates are advised to "learn about T. J. as well as his company." Make sure you're familiar with the company's history and current financials as well. "It's demanding," said one employee, "but it makes receiving an offer feel like a great accomplishment."

OUR SURVEY SAYS

"It's really a one man show here," remarks one employee, "most everything is run by T.J.," the "extremely outspoken," "driven" CEO. Sources say his attitude "permeates throughout the company," creating a "unique culture." One insider describes it as a "work-hard, play-hard, little-nonsense-tolerated environment." "The entire semiconductor industry is very aggressive," explains another, "but Cypress is more aggressive than most."

Sources say Cypress' numerous in-house amenities were set up by the CEO "to make his life easier," benefiting employees as well. Cypress maintains cafeterias, fitness facilities, a hair salon, dry-cleaning service ("it's expensive but they deliver to your desk"), and an automotive center that offers discount gasoline, car washing and detailing (done while you work), and monthly oil changes. There is also a tuition reimbursement program, and workers can take classes broadcast live from Stanford University via satellite. Also, "Cypress gives employees $1200 every two years towards the purchase of a computer and peripherals for a home office." Unfortunately, there is no childcare center. It's "something a lot of parents here would love," say insiders woefully. Though there's no company matching for the 401(k) program, Cypress does offer a "good stock option plan" that will be beneficial once the company recovers from the recent downturn in the SRAM and semiconductor markets.

While their cars are being detailed, "hard working" employees drink their free coffee in "unusually small cubicles." Not fun, considering the "very long hours" most put in. There are some lucky employees who "keep things at eight hours a day," but no one is exempt from "one hard fast rule – you must be in by 8 a.m. Our CEO reminds us via e-mails when he notices people coming in late."

Cypress hires "only the best," and for that reason, the company does not "skimp on salaries," which employees describe as "middle to high." The company also employs "an incredible amount" of minorities. It's no secret that T. J. Rodgers "actively lobbies the U. S. government to increase the number of visas awarded to immigrant workers." A number of company executives are minorities, including the VP of R&D and the CFO. Though female employees "are often the minority in their respective groups," most "seem very happy with their situations." Dress varies by department: "Marketing people always wear suits and ties, engineers wear mostly jeans." Some engineers at the company's California HQ even report wearing shorts and sandals to the office.

Recent college grads say they've been given "challenging projects at a much faster rate" than they would have at companies. "Things are lean here," says one recent hire, "so if you demonstrate a willingness to take on more projects you will constantly be given challenging work." Plus, "since most of the teams are very small, you get your hands into everything – that's the best way to learn." "It's easy to get recognized for your good work," adds one source. And once you've proven yourself, "there is room for progress and it is encouraged at all levels." As one engineer sums up: "This is a very good place for electronics engineers to learn and grow."

"The culture places more emphasis on performance than appearance."

– ***Semiconductor insider***

Dallas Semiconductor Corporation

4401 S. Beltwood Pkwy.
Dallas TX 75244-3292
(972) 371-4000
Fax: (972) 371-4956
www.dalsemi.com

LOCATIONS

Dallas, TX (HQ)

DEPARTMENTS

Administrative
Marketing
Engineering
Finance
Information Technology
Manufacturing
Quality and Safety
Sales
Technical Support

THE STATS

Annual Revenues: $368.2 million (1997)
No. of Employees: 1,500
No. of Offices: 1
Stock Symbol: DS (NYSE)
CEO: C. Vin Prothro
Year Founded: 1984

DALLAS SEMICONDUCTOR

UPPERS

- Flexible hours
- Promising products

DOWNERS

- Hiring freeze
- Few women and minorities at high levels

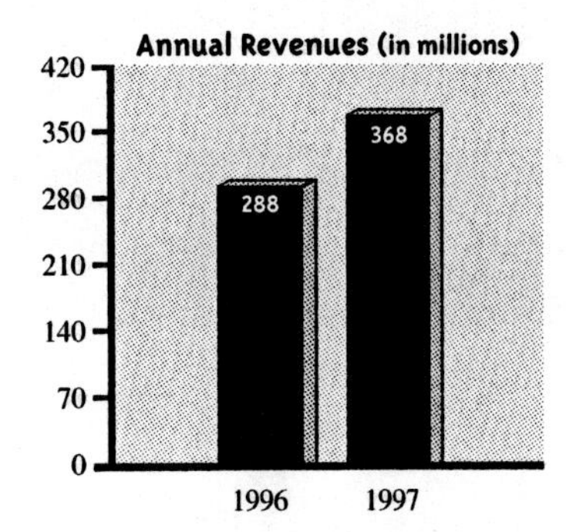

KEY COMPETITORS

- Alliance Semiconductor
- Analog Devices
- Cirrus Logic
- Cypress Semiconductor
- Linear Technology
- Lucent
- Micron Technology
- National Semiconductor
- Rockwell International
- Siliconix
- STMicroelectronics
- Unitrode
- Xicor

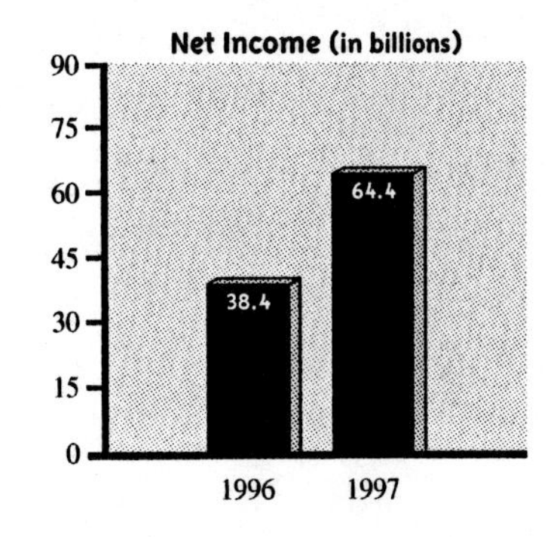

EMPLOYMENT CONTACT

Gay Vencill
Staffing Department
Dallas Semiconductor
4401 S. Beltwood Parkway
Dallas, TX 75244-3292

Fax: (972) 371-6337
recruiter@dalsemi.com

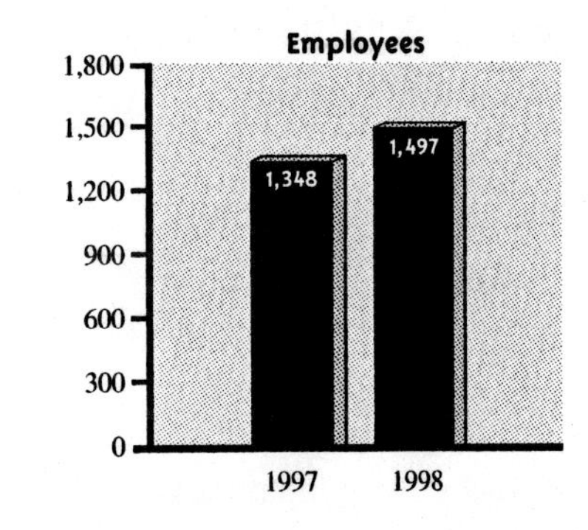

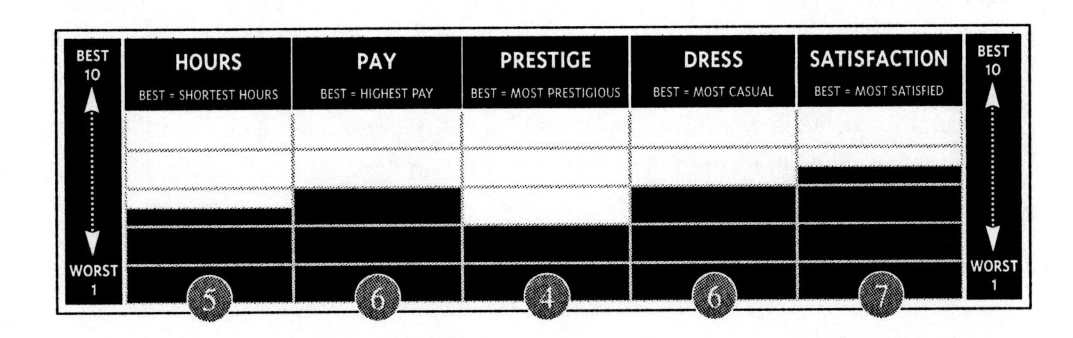

THE SCOOP

Dallas Semiconductor was created by four top execs from the now-defunct DRAM chip maker Mostek Corp. With a very small series of products and an even smaller set of customers, Mostek was forced to shut down in 1981. CEO Vincent Prothro and his colleagues learned from their mistakes, and went on to create a much more diverse business, calling it Dallas Semiconductor. The company has introduced more than 260 products in 14 years, including devices used in computers, telecommunications systems, electronic security, and industrial equipment. DS also has a much wider customer base than its predecessor – serving more than 10,000 customers, as compared to the five served by Mostek. For this reason, the company has managed to stay healthy during the recent volatile period in the chip industry. According to observers on Wall Street, DS' broad base should enable it to weather the storm.

The company's present strategy is to concentrate on the production of devices with higher profit margins and longer shelf lives. These will include new chips for high-speed Internet access and its iButton technology. Introduced in 1992, the iButton is a small device that looks like a watch battery – it is comprised of an eight-bit microprocessor, a co-processor, and memory. It encrypts and stores information, and can be used in a variety of different ways. The iButton makes Internet transactions safer between two of its users. Widespread use of the iButton in personal computers could lead to increased commerce over the Internet, because credit card numbers would be transferred without worry.

The technology is already in use in countries such as Turkey, Argentina and Russia, as a memory device that stores cash electronically for small transactions. It has been employed in mass transit systems, parking meters, gas pumps, and vending machines. Here in the U.S., the Postal Service is in beta testing of the iButton, and plans to use it to replace the postal meters that companies currently lease to stamp their mail. Eventually, business customers will be able to refill their meters over the Internet. iButtons may also be installed in wallets to be used instead of cash and in watches or bracelets as security passes and identification bands. They are also used as a high-tech "Medic-Alert bracelets," to store important medical information in case of emergencies.

DS expects this technology to surpass the performance of the Smartcard, which has had limited success in North America. The iButton is more durable, contains a real-time clock to monitor transactions, and contains security provisions to prevent unauthorized access to stored information. The company released an updated version, called the Crypto iButton, in March 1998.

GETTING HIRED

Dallas Semiconductor is in the midst of a "temporary hiring freeze," but you can check out job descriptions at the company web site, located at www.dalsemi.com. Sources say to expect a thaw sometime in 1999. Resumes and cover letters may be sent to the staffing department via e-mail or regular post. DS also recruits on college campuses.

The interview process varies from one department to another, but in general, prospective hires meet with employees from a specific group. There is usually "one escort – whose responsibility is to take you around to the different people and most likely take you to lunch." For engineering positions, "the interview process is a bit more on the stressful side," and includes "several technical sessions with several people." But sources say interviewers "are more interested in finding out how candidates go about solving problems they do not know the answers to." Another source notes that "since we basically hire right from college, we don't expect a person to have a large skills matrix from which to draw."

OUR SURVEY SAYS

DS employees say they love working in this "highly technical," "very entrepreneurial" environment where "the CEO, CFO and upper-level managers are still very much involved in the everyday activities of the company." Insiders say "there is no micromanagement," and appreciate the fact that their "ideas are considered very seriously." "There's not really a rat-race mentality," explains one contact, "because the company has been growing quickly and there are always opportunities for people to move up if they're ready." Though the whole industry has been hard hit by the Asian fiscal crisis, employees feel pretty secure that DS will ride it out. "They have a very conservative financial and marketing plan," one insider remarks, "and they stick to it." Instead of laying off workers, "DS simply scaled back production and cut expenses by making factory operations more efficient."

The company's Dallas HQ is on a 40-acre campus with more than 20 buildings – "depending on what your job is, you might make a habit of going to at least five or six of those every day." Work hours are flexible, and "Dallas tries to work around any reasonable time conflicts – i.e., school, family issues, etc." Official work hours for engineers are 8 to 5, but in general "the latest they come in is 10, and no one leaves before 2 p.m." The corporate culture is "relaxed for the high-tech talent," and the dress code is comparable. Those in research and development are "definitely not suit and tie." "Jeans are fine," and one source reports seeing co-workers in shorts. Employees in the corporate and marketing departments dress more formally.

Employees enjoy "really great" benefits, including 401(k) with 6 percent matching, profit sharing, stock options, and relocation assistance. One source who's been around the high-tech loop says "I have been treated with so much more respect and freedom here that I have stayed for three years." Salaries are "comparable to other companies in the industry."

"I found that women and minorities were not treated especially well," reports one source from the Dallas HQ. Though there are many female engineers and several female managers, "there are no women above the director level." As for ethnic minorities, sources say they "interact with people from all over the world every day." However, "there is only one minority SVP" in the Dallas office. Considering there are more than 1500 employees, "that ratio is not the best."

"The entire semiconductor industry is very aggressive, but we are more aggressive than most."

– ***Semiconductor insider***

Dell Computer

One Dell Way
Round Rock, TX 78613
(512) 338-4400
Fax: (800) 224-3355
www.dell.com

LOCATIONS

Round Rock, TX (HQ)
Austin, TX
Bracknell, UK
Hong Kong
Kawasaki, Japan

DEPARTMENTS

Administration
Human Resources
Information Systems
Manufacturing Operations
Product Development
Product Support
Project Management
Sales and Marketing

THE STATS

Annual Revenues: $12.3 billion (1998)
No. of Employees: 16,000 (worldwide)
No. of Offices: 32 (worldwide)
Stock Symbol: DELL (NASDAQ)
CEO: Michael S. Dell

UPPERS

- Employee discounts on computers
- "Dell Discounts" on shopping and dining
- Company gym
- Extensive social events

DOWNERS

- Frequent shifts in company organization
- Little access to stock options

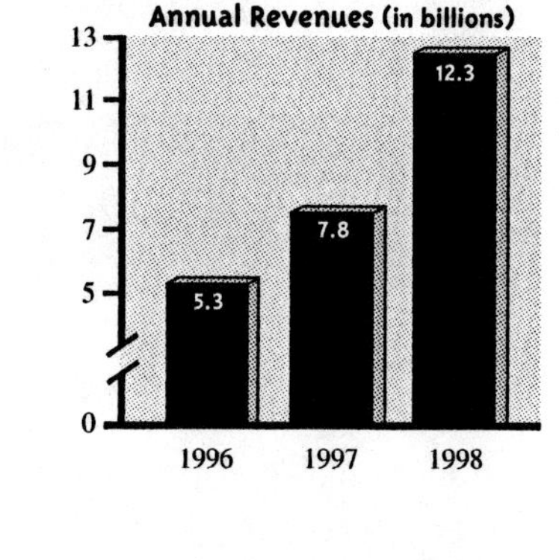

KEY COMPETITORS

- Compaq
- Gateway
- Hewlett-Packard
- IBM
- Micron Electronics

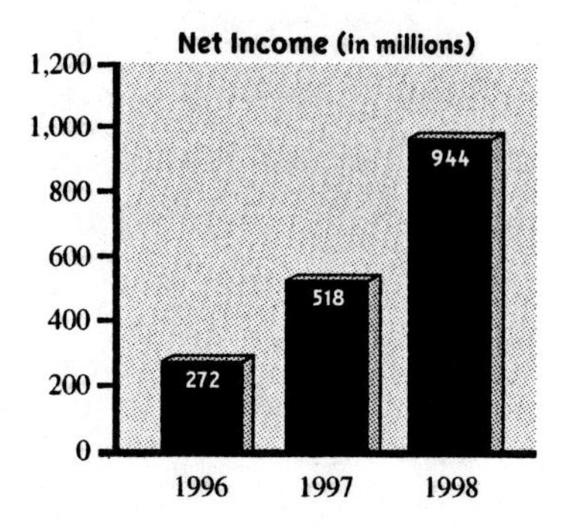

EMPLOYMENT CONTACT

Human Resources
Dell Computer
One Dell Way
Round Rock, TX 78613

(512) 728-4747
Fax: (512) 728-0571
careers@us.dell.com

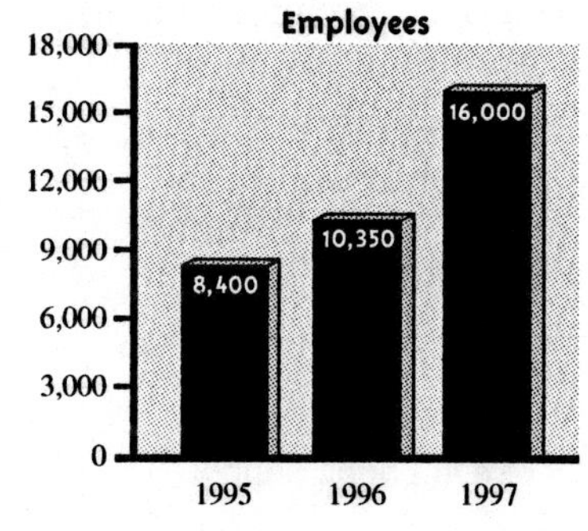

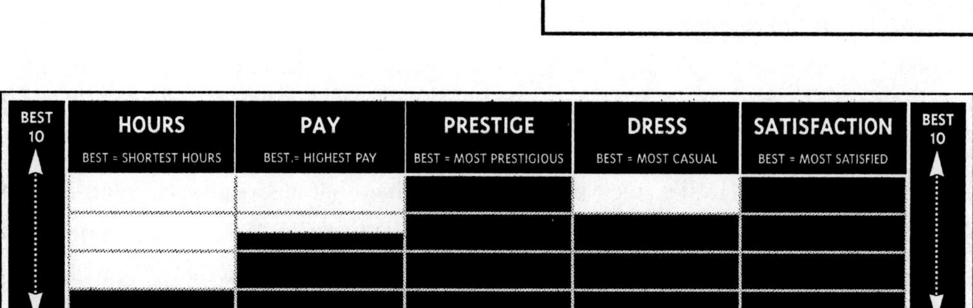

THE SCOOP

There's gold in them there hardware. Dell Computer's success is amazing by any measure – even more so when you consider the company is less than 15 years old. Based in Austin, Texas, Dell is now the leading direct-seller of computers in the world. In 1998, Dell's sales totaled over $12 billion. As recently as 1992, sales were under $1 billion annually. Dell started out making desktops and now manufactures a line of desktop computers, notebooks, workstations and servers. Desktops still account for 78 percent of this rapidly growing company's business. Dell sells primarily to small and medium-sized businesses and government agencies, though increasingly the company is targeting the consumer market. Dell still sells more than 90 percent of its products to businesses and government entities. Corporate customers include Ford Motor Company, Boeing, and international giant Deutsche Bank.

The business is relatively simple. Roughly 50,000 calls, most of them orders, come into Dell's toll-free numbers each day. Though Dell assembles a computer specifically for each order, making and shipping a PC usually only takes Dell 36 hours. Dell builds all of these computers at plants in Austin, Texas; Limerick, Ireland; and Penang, Malaysia. In the early 1990s, Dell entered the retail market by striking deals with retail chains to sell its computers at mail-order prices. Though many were skeptical when Dell began selling computers online, the company now sells $6 million worth of equipment on the Web daily. Dell prides itself on its customer service. Custom-building computers according to each order means Dell can respond rapidly to the changing needs of the marketplace. Dell was the first PC maker to offer direct, toll-free technical phone support and next-day, on-site service.

Michael Dell, the founder of his eponymous company, is the richest man in Texas. It's no wonder – the company's stock is going gangbusters. A hundred dollars of Dell stock purchased in 1990 would have been worth over $20,000 by September 1997. A $10,000 stake at the initial public offering in 1988 would by that time have been worth over $1 million. In the past three years, Dell's stock has risen faster than that at Coca-Cola, Intel, or Microsoft. While Dell's stock began to fluctuate through the autumn of 1997, it did so not because of serious negative indicators but because analysts couldn't imagine that the stock could keep rising. However, Dell continues to grow: In the first two quarters of 1998, the company posted its 15th & 16th consecutive quarters of record sales. The company is also hiring like gangbusters: From 1995 to 1997, the company nearly doubled its workforce.

Recently, Dell has been increasing its presence outside the U.S., particularly in Latin America and Asia. Dell's international expanision is all the more remarkable given the current global financial crisis. While major U.S. high-tech companies have been scaling back their investments in these two volatile regions, Dell has been aggressively pushing forward. In 1998, the company began a new mail-order service in Hong Kong, Japan, and Singapore, a new Asia/Pacific Customer Center in Malaysia, and direct-sales operations in South Korea and Taiwan. Dell also began production in 1998 at a new factory in Xiamen, China. Company officials announced plans to expand Dell's Latin American market share (a region where it lags far behind IBM, Hewlett-Packard, and Compaq) by opening a new plant in Brazil.

GETTING HIRED

Dell's employment web page, located at www.dell.com/careers/index.htm, provides information on positions available in each department, including specific experience and degree requirements. Applicants may e-mail a resume but should be sure to type "resume" as the message's subject line; Dell will then route the resume directly into its applicant tracking system. The company conducts some campus interviews, primarily in the South and Southwest, although it fills few jobs this way. Dell looks for "people with a high level of motivation, determination, and the ability to be a team player," and apparently finds many with these qualifications: according to insiders, the company hires from 100 to 200 new employees each week. For many jobs like customer support, Dell will hire through temp agencies around Austin.

According to insiders, most positions require three interview sessions. The first session serves "mainly to see if you are the right kind of person and have the relevant experience to perform the position." This interview will be with a human resources recruiter. The following interviews will be with managers or subject experts (depending on the field). Remember that they are looking for "the right people technically as well as socially." "The biggest function of the interview is to try to determine if you would fit in to the culture," says one insider. "With the fast pace and multitasking required, we need team players who are also self-starters."

While Dell reportedly does not utilize brainteaser questions like Microsoft, you must be familiar with the Dell Direct Business Model. The Dell model means low inventory, just-in-time manufacturing, built-to-order products, and direct customer relationship with manufacturer. There are many benefits to this system. By cutting out the middleman, Dell can afford to sell its products well below retail. Since Dell builds as it goes, it keeps very low inventories, allowing the company to switch to the latest technologies faster than firms with pre-built stocks in warehouses. Understand this model. Think about it; don't just regurgitate it. In an industry that is essentially a commodity business, this model explains Dell's success in the face of so many others' failures.

OUR SURVEY SAYS

Dell prides itself on a "flat" corporate structure that encourages each worker to contribute "innovative" ideas, and its employees say they appreciate this "openness" and "absence of hierarchy." This "unstructured," "decentralized" environment allows Dell's "young," "energetic" employees to "gain responsibilities quickly and get a chance to prove yourself." "Managers that respect you" establish good relations. "There is much less corporate politics than in many other environments, which is a refreshing change," says one MBA intern. Dell is not a company for those who like to take things slowly, because "everything moves quickly." "Working at Dell is like jumping out of an airplane with your hair on fire," says one employee. "Everything about Dell is fast. We call it velocity," explains another.

Combined with Dell's "explosive" growth, the "meritocratic" promotion policy enables "talented management" to "rise rapidly." "There are some tremendous opportunities if you want to work hard and think 'out of the box,'" says one insider. Employees say they are generally satisfied with their compensation packages, adding that "the pay is fair, swell perks (401(k)/employee stock purchase plan) are great." A profit-sharing plan "comes out to about an extra 8 percent of your base pay per year," according to one insider. Other perks employees mention include discounts on computers and various establishments in Austin (including rent in some apartment complexes).

Employees also like the casual, "family-like" atmosphere, which includes a "lax" dress code. While dress codes differ by department, some employees can come to work in shorts and sandals while others go with "business casual dress." "There are even a couple of people with purple hair," says one. Insiders say there are "many minority employees, especially Asians and Asian-Americans." One technical support worker in Austin says "women are given opportunities that I did not see at my former employer." All agree that "performance" and "not gender or skin color" moves employees up the career ladder. "For what it's worth, I'm gay, and I felt relatively comfortable interacting with the folks at Dell," says one employee. "The EEOC could take lessons from Dell," says another.

The overall level of employee satisfaction is astonishing. Numerous employees say exactly the same thing: "Dell is the best job I've ever worked." Most have "absolutely no complaints," but obviously no company is perfect. As one technician says, "The only bad thing is that there are some 'Dell ways' of doing things. Since a lot of areas are new, when you hit those new areas, things can slow down." Minor criticisms aside, when contacts say things like, "I have found the place I intend to retire from. The only way I'll leave is kicking and scratching," it must be a pretty good place to work.

Electronic Arts

1450 Fashion Island Blvd.
San Mateo, CA 94404
(650) 571-7171
Fax: (650) 286-5137
www.ea.com

LOCATIONS

San Mateo, CA (HQ)
Austin, TX
Baltimore, MD
Chicago, IL
Dallas, TX
Louisville, KY
Maitland, FL
New York, NY
Seattle, WA
London, UK
Paris, France

DEPARTMENTS

Administration
Game Development
Information Technology
Marketing
Product Support
Sales

THE STATS

Annual Revenues: $908.9 million (1998)
No. of Employees: 2,100 (worldwide)
Stock Symbol: ERTS (NASDAQ)
CEO: Lawrence F. Probst III

UPPERS

- Stock options
- Company gym
- Free beer on Fridays
- Free computer games

DOWNERS

- 100-hour workweeks
- Low pay for entry-level employees
- Few women, especially in technical departments

KEY COMPETITORS

- Broderbund
- Cendant
- Lucas Arts
- Nintendo

EMPLOYMENT CONTACT

Human Resouces
Electronic Arts
209 Redwood Shores Parkway
Redwood City, CA 94065

resumes@ea.com

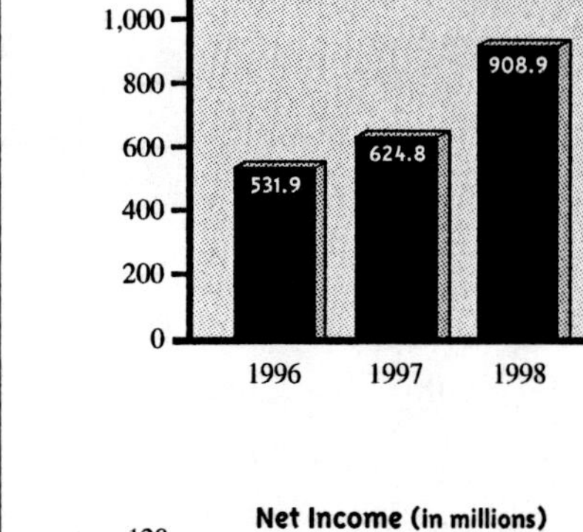

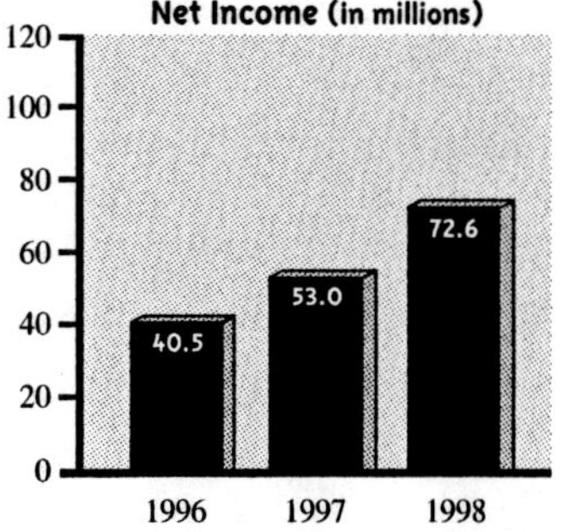

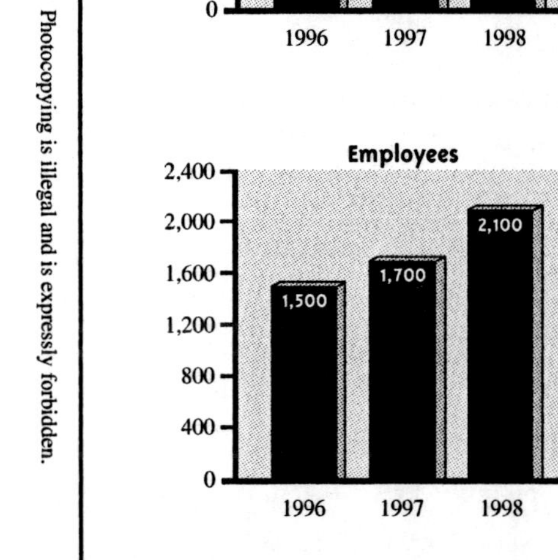

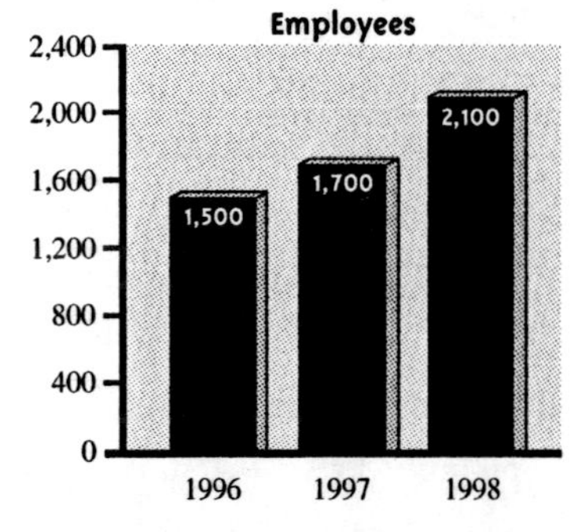

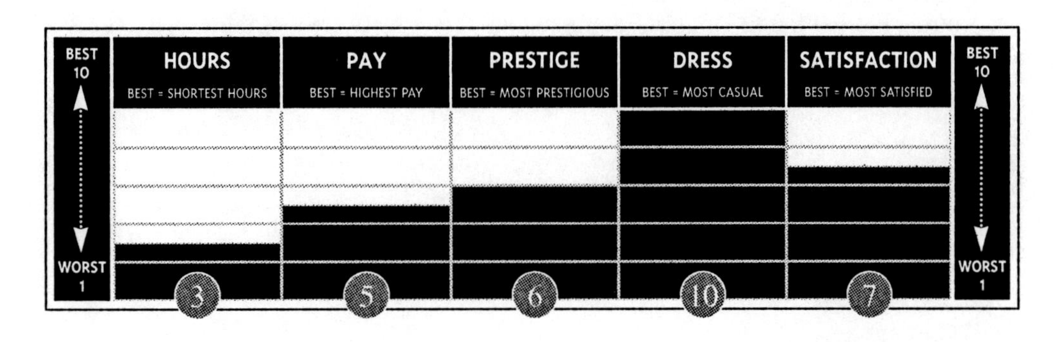

THE SCOOP

With 120 software titles, Electronic Arts' offerings range from laid-back virtual golfing to gory battles with virtual aliens. The San Mateo, California-based entertainment software giant operates like a movie studio, with producers coordinating the work of artists, writers, animators, sound engineers, musicians, set designers, and programmers. With some of the most popular products in the business, including *NHL Hockey* and *Wing Commander III*, Electronic Arts is shooting for the top position in the growing entertainment software market.

The company was founded in 1982 by Apple Computer's techno-wiz Trip Hawkins. EA began designing games for the Sega Genesis video game system in 1990, and acquired software publisher Origin Systems in 1992. The same year, EA formed a joint venture with a unit of JVC to market EA products in Japan. International acquisitions throughout the 1990s have boosted sales. Although 40 percent of the company's sales in recent years have come from software for the Sony Playstation System, EA began offering the first of several sports games it will produce for the Nintendo 64 system in 1997. Anticipating N64 usage to double in 1998, the company intends to release up to eight games for the N64 in 1998. Revenue from Nintendo 64 is expected to make up 10 percent of Electronic Art's revenues in 1998, up from only 1 percent in 1997. EA has also gotten a boost from its 1997 acquisition of Maxis, the producer of the popular *SimCity* titles. The acquisition of Maxis represented a coup for EA, whose strength has traditionally been in cartridge games, since Maxis is a player in the personal computer market. EA has also moved into the pay-to-play online gaming market. In September 1997, EA introduced an online version of *Ultima* – the classic role-playing game by subsidiary Origin. While competitors have avoided Internet gaming thus far, EA is already planning to offer another title, *Wing Commander*. EA continued steamrolling in 1998, acquiring Florida-based Tiburon Entertainment, Swiss software distributer ABC Software, and two video game units of Virgin Interactive Entertainment.

GETTING HIRED

Visit the "Job Opportunities" section of EA's web site for the lowdown on job openings, locations, and requirements. Send or fax resumes to human resources, or e-mail resumes via the web site.

OUR SURVEY SAYS

Working at Electronic Arts is a lot like playing the games it produces – lots of fun and very intense. "Because they let us have so much fun, we work like maniacs," one employee says. Another employee reports that "about three times a year, I'll work 80- to 90-hour weeks for a month. Two years ago, I frequently worked 100-plus hours a week to get a product to ship, though I've lost some steam since then." Employees offer mixed reviews on salaries. Though one contact reports that, "the pay is fine, I'm happy with my salary," another EA insider says "the work hours suck and EA salaries are the lowest in the industry, unless you're an exec, in which case you make 70 times the salary of the average employee here." As expected, the worst pay and hours go to employees hired straight out of college. "If you don't have industry experience, the pay and hours are horrible until you prove yourself to [Electronic Arts]," one such recent hire says. "Granted, no matter where you work in the computer industry, you'll be treated like a slave for the first year or two, but I would recommend working somewhere else for one or two years, then transferring to EA once you have some experience under your belt."

Pay and work hours notwithstanding, employees report excellent opportunities for advancement in this growing field. Like most tech-driven media companies, the culture is "laid-back" despite the "long hours." A non-existent dress code and "free beer and junk food on Fridays" are a few reasons why employees say working at EA is "really fun." Co-workers are described as "the best and the brightest in the industry," and "a good mix of creative and technical talent." The structure of Electronic Arts encourages congeniality; the company is divided into "studios" of approximately 100 to 200 employees. Employees say women and minorities are treated well, though they are not well represented in many departments: "There are many women in the administrative part of the organization, there are some, but less, in the artistic ranks, and almost none in the technical ranks. It's not unusual for there to be fewer women than men in technical environments, but EA is really lopsided," say insiders. Perks include an excellent benefits plan, free gym access, generous vacation time, and, of course, "free games."

"The company tends to value products over employees. This makes it less than ideal for a job early in one's career."

– ***Software insider***

Excite

555 Broadway
Redwood City, CA 94063
(650) 568-6000
Fax: (650) 568-6030
www.excite.com

LOCATIONS

Redwood City, CA (HQ)
Austin, TX
Chicago, IL
Westminster, CO
Australia
Japan
London

DEPARTMENTS

Administration
Content Production
Engineering
Finance and Accounting
Human Resources
Management Information Systems
Marketing and Business Development
Network and Systems Operations
Product Management
Sales

THE STATS

Annual Revenues: $50.2 million (1997)
No. of Employees: 700
No. of Offices: 10 (worldwide)
Stock Symbol: XCIT (NASDAQ)
CEO: George Bell

UPPERS

- Stock options
- Free soda
- Flexible hours
- Telecommuting options
- Masseuse

DOWNERS

- Still in the red

KEY COMPETITORS

- America Online
- Geocities
- Infoseek
- Lycos
- Microsoft
- NBC/Start.com
- Yahoo!

EMPLOYMENT CONTACT

Yvonne Agyei
Human Resources Manager
Excite
555 Broadway
Redwood City, CA 94063

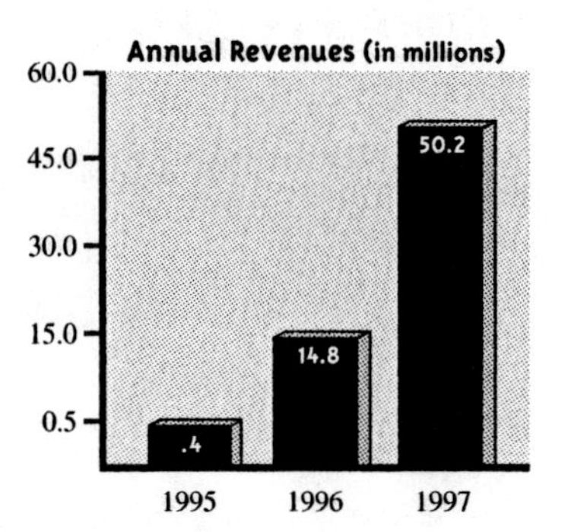

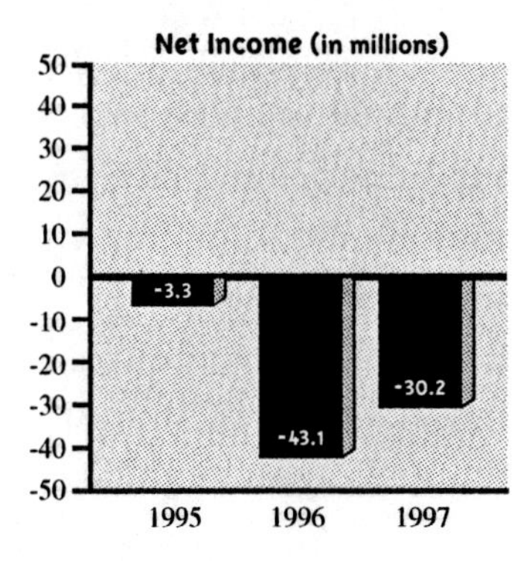

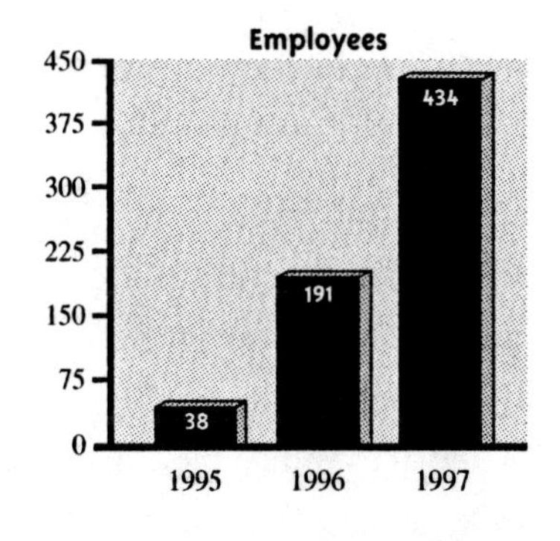

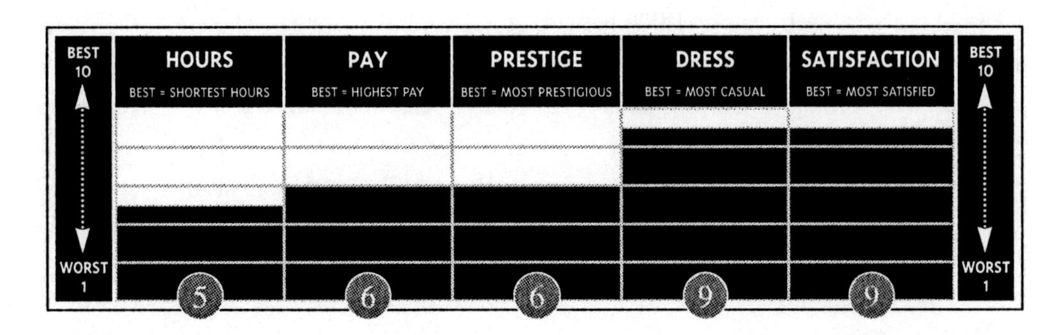

THE SCOOP

Serving an estimated 3 million web crawlers a day, Excite's popular portal service lets users browse the Web according to concepts, not just single keywords. The company's humble beginnings date back to the winter of 1993, when six Stanford graduates met for burritos at Rosita's Taco Shop. The group formed Architext Software in mid-1994, to fulfill a $100,000 contract for an online project. In 1995, the company introduced its NetDirectory site reviews and NetSearch indexing service, and released EWS, the companion software to its search service. The same year, the firm purchased the CityNet, an online resource for local and regional information. The company took the name Excite in 1996, and debuted Personal Excite, a customizable search tool the same year. Recent acquisitions, including America Online's WebCrawler, online Internet guide Magellan, and online advertising services MatchLogic and Classifieds2000, have raised Excite's profile on the Web. In 1997, financial services software powerhouse Intuit purchased 13 percent of Excite; together the two firms launched a financial information site called the Excite Business and Investing Channel. For the recreationally inclined, Excite contracted a gaming company to add Java-based games like hearts, chess, and checkers.

CEO George Bell has spent $75 million building market share, and Excite has certainly carved out a place for itself in the portal wars: the company's revenue in 1997 was $50.2 million, making it second only to Yahoo!. Though Excite's net income for the year was in the red, losing $17 million, Excite is banking on a profit late in 1998, according to analyst predictions. In the meantime, the company continues to seek out marketing partnerships and useful acquisitions. For example, Excite and Cybermeals have signed a $15.5 million agreement whereby Cybermeals will be the exclusive provider of takeout and meal delivery services to Excite users. Cybermeals's 11,000 networked restaurants will be promoted on the Excite site. Excite has also signed an agreement with Prodigy to provide co-branded content to Prodigy subscribers. Prodigy users will be able to create a custom default web page, and have access to Excite-topical channels. The arrangement will effectively take Prodigy out of the content business. Rival Yahoo signed a similar agreement with MCI Worldcom in April 1998.

Excite believes in cooperative competition, as evidenced by the two-year mutual technology and revenue deal it made with rival navigation firm Netscape in May 1998 (a deal that may be off following the AOL-Netscape merger). Included in the $70 million package is prime Excite visibility on the Netscape site, a move that has propelled Excite's sales. Excite inked a three-year deal with AT&T WorldNet Service in June to provide Internet access (with www.excite.com as the start page, of course) and various voice communications services.

Excite's activity has drawn some unwanted attention. Earlier, in May, the portal spurned a $1.7 billion bid by oil and fish-meal company Zapata Corp. The Walt Disney Company dispelled rumors that it would acquire a 40 percent share of Excite when it contracted the company to include content from its sites, later purchasing a 43 percent share of Infoseek instead.

GETTING HIRED

Visit the "jobs" section of Excite's web site, www.excite.com, for details on job openings and an online application form (which asks applicants to explain why the company should hire them). The site allows job seekers to search for open positions by department. Resumes can also be faxed or sent via snail mail to the company's human resources. Most openings are in Excite's Redwood City, California headquarters.

OUR SURVEY SAYS

A young and growing company, Excite boasts employees who are, well, excited. "There's never a dull moment here at the office," one employee says. "I love working here." Another reports, "Excite is an excellent place to work. I recommend it highly. You can work on exciting projects, and take on as much responsibility as you can handle." Work hours can be long, but are flexible. One employee notes: "Sometimes I come in late and leave late, sometimes I come in early and leave early. In general, I do work more than 40 hours each week, but I truly enjoy what I do, and it is worth it." Salaries are described as "meeting or exceeding industry standards." There is no need to don dress clothes for work at Excite, where employees typically sport jeans and shorts. Employees report excellent treatment of minorities and women, noting "There are people of all types, genders, races here at Excite." Stress busters? You bet – they include "a masseuse that comes in once a week, yoga classes, a corkscrew slide, BBQs and pizza lunches, free sodas and cheap Odwallas (a fruit drink popular in California)."

Gateway

619 Gateway Drive
North Sioux City, SD 57049
(605) 232-2000
www.gateway.com

LOCATIONS

North Sioux City, SD (HQ)
Hampton, VA • Kansas City, MO

Ireland (European HQ)
Overseas offices in Australia • Austria • Belgium • France • Germany • Japan • Luxembourg • Malaysia (manufacturing facility) • The Netherlands • Switzerland • The United Kingdom

DEPARTMENTS

Engineering
Finance
Information Systems
Manufacturing
Marketing

THE STATS

Annual Revenues: $6.3 billion (1997)
No. of Employees: 13,300
No. of Offices: 50+ (worldwide)
Stock Symbol: GTW (NYSE)
CEO: Ted Waitt

UPPERS

- Profit sharing
- Ample advancement opportunities
- Company credit union
- Cute boxes

DOWNERS

- Some rural locations
- Confusion from rapid corporate growth

KEY COMPETITORS

- Apple Computer
- Compaq
- CompUSA
- Dell Computer
- Hewlett-Packard
- IBM
- Micron Electrons

EMPLOYMENT CONTACT

Human Resources
Gateway
619 Gateway Drive
North Sioux City, SD 57049

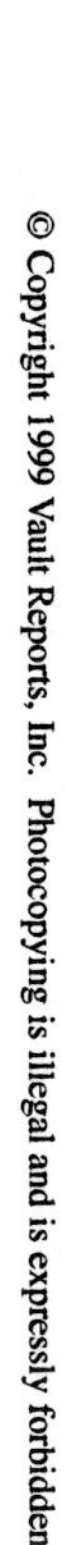

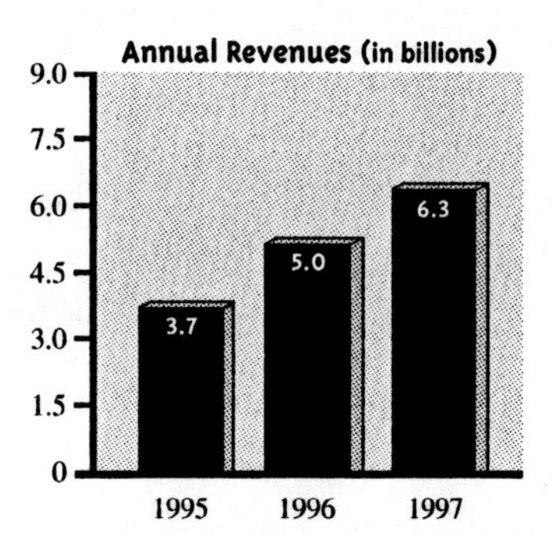

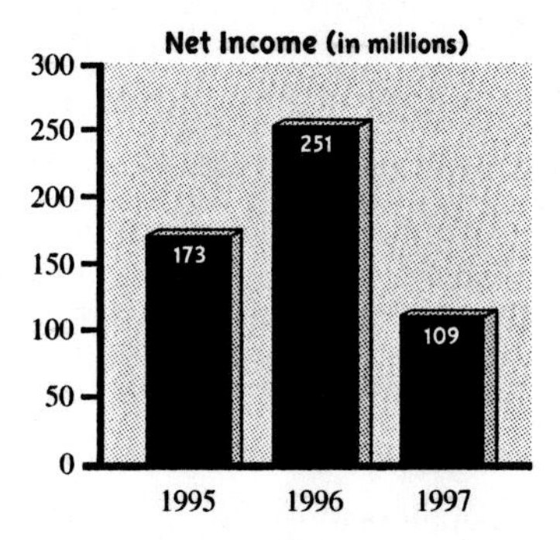

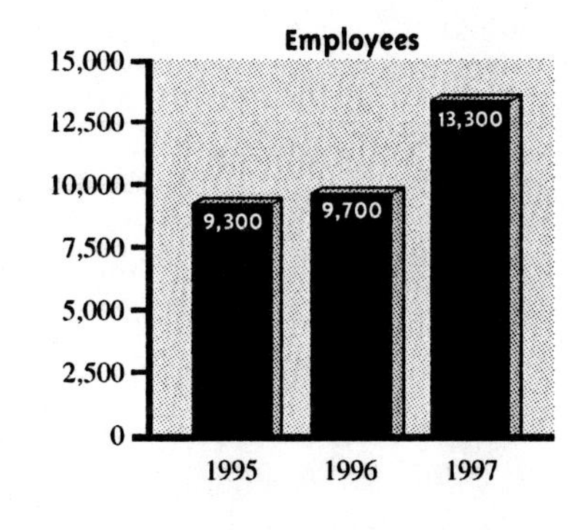

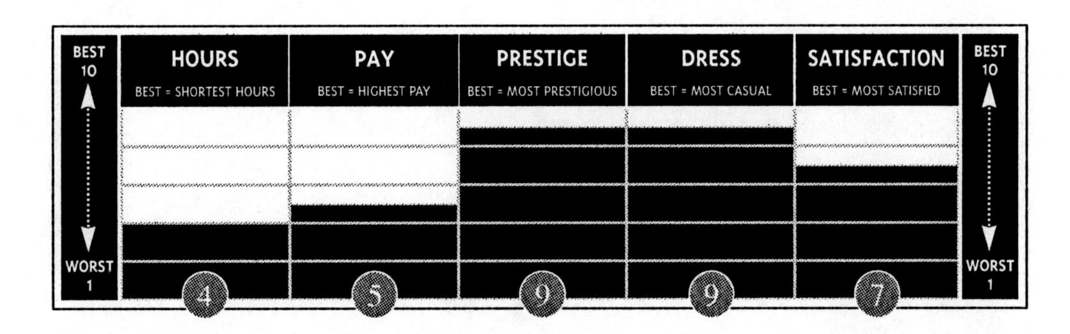

THE SCOOP

Gateway combines Midwestern thrift and common sense with the latest in computer technology. Started by college dropout Ted Waitt in 1985, Gateway (formerly Gateway 2000) has grown into one of the nation's largest direct marketers of personal computers. Based in North Sioux City, SD, Gateway emphasizes its heartland location by printing all of its packaging and promotional material with black and white cow spots. The previous four generations of Waitts were all cattlemen; Ted essentially does the same thing as his forefathers – he simply switched commodities, from beef to computers.

Gateway bases its corporate philosophy on the belief that retail customers will buy personal computers much more readily if they can bypass large markups, avoid a confusing array of dealers, and receive superior service directly from the manufacturer. By marketing Gateway computers as a simpler and less costly alternative, Waitt has steadily steered the company to the forefront of the market. At the same time, Gateway remains the No. 2 direct marketer of computers overall, behind Michael Dell and his eponymous company. Dell's philosophy – not to mention his incomplete college education – is very similar to Waitt's. The biggest difference lies in their customer bases. Dell is tops in corporate and government markets, whereas Gateway sticks mainly to home computer buyers and small businesses. In an effort to continue its growth, Gateway has begun aiming at larger businesses.

Gateway went public in 1993 with a listing on NASDAQ. In that same year, Gateway opened a manufacturing and service facility in Ireland. In 1994 it added retail showrooms in France and Germany. Although roughly 85 percent of Gateway's business is restricted to the U.S. and Canada, major efforts are underway to expand the company's potential customer base. Gateway expanded into Australia by purchasing 80 percent of that country's largest computer maker. The company also opened a manufacturing facility in Malaysia and a sales and support operation in Japan. While sales to the Asia/Pacific region total only $236 million right now, those figures should rise in the coming years.

Perhaps Gateway's greatest asset is its ability to market itself as a folksy, eccentric alternative to other "faceless" producers. In addition to the cow-like boxes, there is Ted Waitt's own approachable public persona, with his ponytail, blue jeans and sandals. It is not unfair to liken

Gateway to the car company Saturn, both upstarts in established industries relying on quirky advertising and quirkiness (though Saturn, of course, is a division of car monolith GM). Waitt proved his commitment to his own way of doing business when Compaq offered $7 billion to take over Gateway. Waitt turned Compaq down flat.

GETTING HIRED

Gateway places most of its new employees in its North Sioux City, SD headquarters. At its web site, applicants can submit their resumes via fax or regular mail, or use an online resume form available on the site. While many positions require degrees in technical fields such as computer engineering and web development, Gateway also offers entry-level opportunities in marketing and sales.

If human resources personnel like the look of your resume, you can expect a phone interview. The duration of these interviews varies from "about five minutes" to "over half an hour." Employees say "they asked simple DOS and Windows questions, also simple questions about memory." If the phone interview goes well, the company brings you to its North Sioux City headquarters to talk in person. The interview process takes "a single day" and consists of a session "with an HR representative, then a technical session – either panel format or individual interviews with about three different developers – and finally an interview with a supervisor or manager."

OUR SURVEY SAYS

Employees rave about Gateway's "fast-paced" corporate culture and say the company's recent "explosive" growth makes them proud. "Gateway creates a climate in which we all feel like we are contributing to its success," says one employee. This growth, however, has engendered some temporary "confusion" about corporate lines of communication and authority. "One area many people find frustrating is the lack of organization in some areas," according to one insider. "Gateway has grown so rapidly in the past 12 years that policies and standards haven't quite kept up."

Some employees, however, find this "hectic" atmosphere quite "exciting." Gateway has "growing pains," they say, only because the company has "accomplished so much in an extremely short time." They are still pleased that the company is "down home and informal." Employees display the corporate values like a badge of honor. Well, actually, it is a badge: "We all wear a 'Gateway Values' laminated badge attached to our ID badges that lists our corporate values." Through these values, the company retains the "attitude of the Midwest. One strange thing is that there are a lot of family members working with each other."

While pay varies widely, employees "love the generous stock options" that they earn after one year, and find that the "monthly profit-sharing checks are great," and "come in especially handy around Christmas time, when the checks tend to be larger due to increased sales volume." On a deeper level, the profit sharing makes them "feel like owners and not just employees." Also, "after one year of employment, you begin earning stock options, which are free to you and can be exercised anytime within the next 10 years." Other perks that come with working at Gateway include "special deals on computers and software" and "discounts at local businesses." Gateway's vacation plan isn't the hottest, though: "five days the first year, 10 the next, 15 after three years," according to one insider. The catch? "Sick days are included in those vacation days."

Surprise! Few employees mention living in South Dakota as a perk. Instead they find it "bearable" and "relatively inexpensive," but "at times isolated" and "undeniably in the middle of nowhere." However, one transplanted Californian gives the 120,000-strong community that spans Iowa, Nebraska and South Dakota rave reviews. "The locale is surprisingly scenic, not anything like I initially pictured Iowa. We have a brand new art center, theater, modest symphony, numerous parks, and assorted minor league sports. If you're into the outdoors, the area can be particularly appealing, with hills, trees, river bluffs, hiking and biking trails, camping, fishing, hunting, horseback riding, all just a few minutes from the main parts of the city."

Some insiders say that Gateway, "isn't for everyone." But most contacts assert that they love the firm and some that they "couldn't go anywhere else." Insiders say there are "no minority problems." One insider remarks, "I work with Hispanics, people who came from India, blacks, Native Americans, vegetarians, you name it. We respect each other and value our diversity."

All those groups can wear what they want as well, since "about the only attire not allowed here is torn, tattered clothing and T-shirts with objectionable messages." "If you see someone here in a suit and a tie, he's probably a visitor," according to one Gateway vet. But informal does not mean lax. Employees say Gateway will "push you to your true potential. If you're not afraid of thinking and working outside of 'normal' schools of thought, you'll fit right in."

Hewlett-Packard

3000 Hanover St, MS 20APP
Palo Alto, CA 94304-1181
(415) 852-8473
Fax: (415) 852-8138
www.jobs.hp.com

LOCATIONS

Palo Alto, CA (HQ)
Andover, MA • Atlanta, GA • Boise, ID • Colorado Springs, CO • Corvallis, OR • Lake Stevens, WA • Rohnert Park, CA • Roseville, CA • San Diego, CA • Santa Rosa, CA • Spokane, WA • Vancouver, WA • Wilmington, DE • As well as sales offices around the country and offices worldwide

DEPARTMENTS

Finance
Factory Marketing
Learning Products
Marketing
Information Technology
Procurement
Research & Development

THE STATS

Annual Revenues: $42.9 billion (1997)
No. of Employees: 127,200
Stock Symbol: HWP (NYSE)
CEO: Lewis E. Platt

UPPERS

- Generous pay
- Flexible scheduling options
- Profit sharing
- Merchandise discounts
- Free sporting events tickets
- True fitness

DOWNERS

- Big-company bureaucracy
- Infrequent employee interaction
- Cubes
- No individual glory

KEY COMPETITORS

- Apple Computer
- Compaq
- Dell Computer
- Gateway
- IBM
- Micron Technology
- Packard Bell
- Sun Microsystems
- Xerox

EMPLOYMENT CONTACT

Employment Response Center
Hewlett-Packard
3000 Hanover Street, MS 20APP
Palo Alto, CA 94304-1181

resume@hp.com

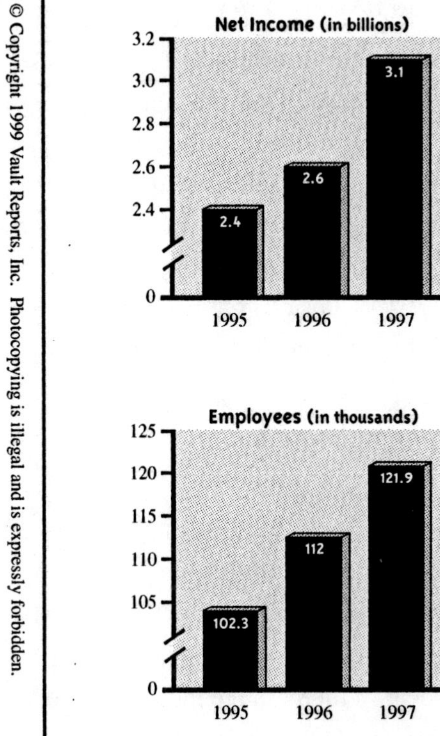

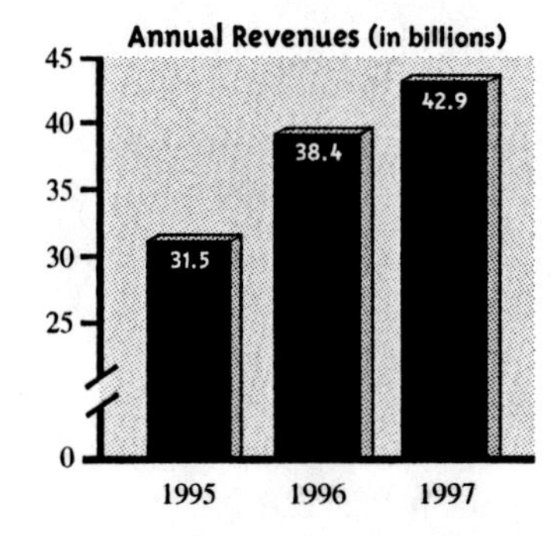

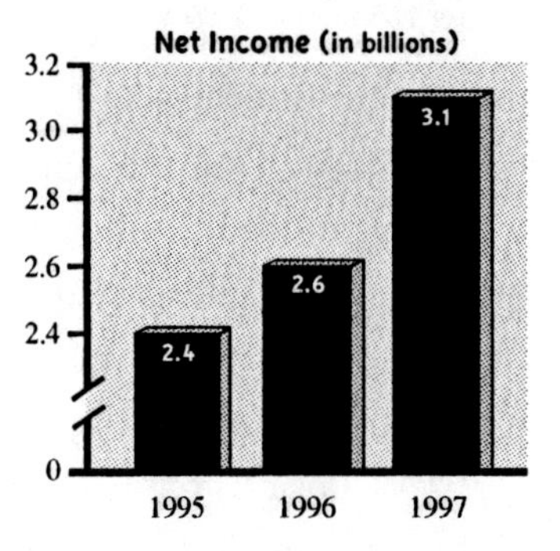

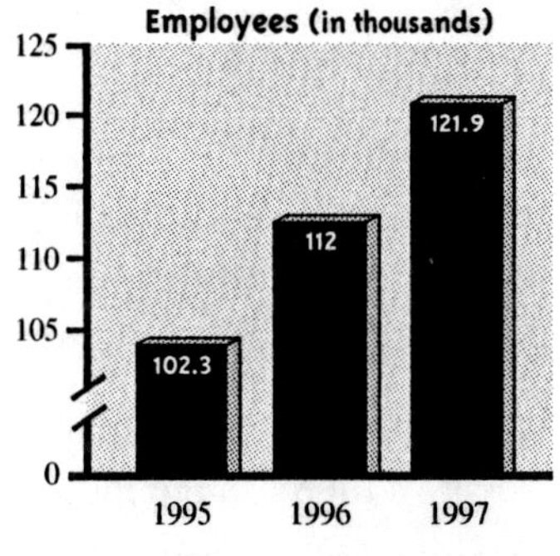

HOURS	PAY	PRESTIGE	DRESS	SATISFACTION
BEST = SHORTEST HOURS	BEST = HIGHEST PAY	BEST = MOST PRESTIGIOUS	BEST = MOST CASUAL	BEST = MOST SATISFIED
4	7	9	6	8

BEST 10 — WORST 1

THE SCOOP

Founded by a pair of engineers in a garage in 1938 with just over $500 in capital, today Hewlett-Packard is a large and diverse company with broad interests stretching across the technological spectrum. While mainly known now as a computer company, HP still makes products in its older fields such as measurement and testing equipment and medical electronics. Products from these older divisions include everything from stethoscopes to atomic clocks. But the computer industry is clearly the company's bread and butter now, accounting for $35.5 billion of its $42.9 billion in annual revenues. Hewlett-Packard ranks as one of the top five providers of personal computers, servers, printers, and both systems integration and contract programming services. The company's newfound success as a PC producer is probably its most impressive recent achievement. HP has gone from the 11th-largest vendor of PCs in 1993 to the third-largest worldwide, with 5.5 percent of the world market share in 1997.

Competitive pressures spurred HP's jump into the PC market. Though PCs had never been a big part of Hewlett-Packard's sales, company executives feared that HP would quickly be marginalized without a solid market position. (The sad fact is that companies that do not sell personal computers are unlikely to score contracts for integration and support services selling hardware helps firms get a foot in the door, allowing them to vend other services.) IT consulting happens to be a highly profitable area, and one in which HP enjoys an exceptional reputation. Although the recent jump into the top three is certainly a good sign for Hewlett-Packard's future in the industry, the picture is not entirely rosy. The company's laptop sales, for example, have been extremely sluggish; HP sells more high-priced servers than laptops annually, even though the market for laptops is much larger overall.

Despite its recent successes, HP executives aren't stopping to smell the roses. Instead they are pushing forward, planning joint ventures with fellow technology titans Microsoft and Intel. Hewlett-Packard and Microsoft have formed an alliance to promote Microsoft's Windows NT operating system. The two companies are making joint sales calls, and HP engineers are working to improve the reliability of Microsoft's system so it meets the rigorous standard for installation in Hewlett-Packard machines. HP and Intel's alliance, though it centers on a tiny chip, is also causing a huge buzz. The companies are collaborating on a new microprocessor called the Merced that will reportedly run 10 times faster than the Pentium. When the Merced is ready – it's projected to be on the market in 2000 – Hewlett Packard's computers will be the first to contain the souped-up microprocessor. The firm hopes HP computers equipped with glitch-free Microsoft operating systems and ultra-speedy Merceds will sell like hotcakes.

The company is also looking to develop Internet software that will enhance electronic commerce. In May 1997, HP introduced "Domain Commerce" for businesses to identify and differentiate Internet users who visit their web sites. The firm's "OpenPix Image Igniter" will allow Internet users to see the products that retailers and catalog companies place in their electronic showrooms in greater quality and detail. HP is also working with a Canadian bank to create an "electronic bank of the future" where financial transactions and services are performed online.

While Hewlett-Packard is known for reliable hardware, HP's Software & Services Group is perhaps most responsible for the company's sterling reputation and rabid customer loyalty. In professional and support services, HP invariably scores high marks for customer satisfaction in both consulting and technical support. As Hewlett-Packard moves into mass-market computer product lines and faces cutthroat competition from the likes of IBM, Compaq, and Dell, the firm must rely on its good name to stand out in this crowded field and attract inexperienced new buyers.

GETTING HIRED

With more than 100,000 employees, Hewlett-Packard is always in need of young, talented people to replenish its ranks. HP recruits at major universities twice a year, once in the fall and once in the spring. Applicants submit a resume and interview once. HP also interviews at seven diversity conferences: NBMBAA, NSHMBA, AISES, NSBE, SHPE, CGSM and SWE. The company then enters all resumes and interviewer remarks into a database ("our automated applicant information system") to which managers across the world refer whenever they have job openings. HP sends all interviewees a letter confirming that their information is available online. Then the wait begins. Prospective hires are usually called for consecutive on-site interviews with several managers. Those found to fit both the job and the company culture will then receive an offer.

For job hunters who are out of school or missed the recruitment rounds, the HP web page also features a detailed list of job openings, with page-long descriptions and a list of qualifications for each opening. These listings include jobs in all departments, from R&D to personnel to finance. When applicants decide they are qualified for and interested in a position, they may fill out an application form for each job, which include space to paste resumes. Applicants can apply for as many positions as they like over the Web. Job seekers who apply to jobs through HP's employment web site are encouraged to apply for specific job openings. When a job seeker applies to a specific opening, hiring managers worldwide can have the resume on their desktops one working day later.

In a corporation as large as Hewlett-Packard, job qualifications vary widely among departments and levels. As at most high-tech companies, HP seeks applicants with undergraduate degrees in engineering and computer science for research and development and other hands-on fields, but jobs in finance and marketing are available for business and accounting majors. Hewlett-Packard expects candidates in technical fields to have a strong working knowledge of C/C++ programming and UNIX development environments. Beyond the entry-level positions, advanced degrees are almost essential for new hires, whether that means an MBA for the business side or a PhD in electrical engineering for product development. As one of the top companies in the country and a leader in its field, Hewlett-Packard can pick and choose the best applicants, but its focus on teamwork and communication strongly affects its hiring practices. Antisocial engineers who get along better with their harddrives than their neighbors should probably seek employment elsewhere. HP is just as eager to avoid misanthropes as cutthroat competitors. Hewlett-Packard has a "good-guy" reputation for a reason. Corporate conquistadors should seek jobs elsewhere since they just wouldn't fit in.

OUR SURVEY SAYS

All of HP's company literature emphasizes the importance of teamwork at Hewlett-Packard. When you recall that two friends working side-by-side in a garage founded the corporation, it isn't hard to understand why. Potential employees should find out if this is their ideal work environment. Says one engineer, "the key is to find out if you fit into HP's low-key, nice-guy and team-oriented work environment." One employee insists "the key word is informal." With all of the emphasis on teamwork and the "HP Way," one could get the idea that conformity is the rule, but as a procurement manager with 16 years experience at the company puts it, "it's a good environment for people with all types of skills and backgrounds."

Expect to make some cash if you score a job at Hewlett-Packard. Silicon Valley is known for compensating its employees out of its huge profits. While the range varies, employees say it "pays among the leaders in the industry." Top-level engineers pull down over $100,000 a year plus "two profit-sharing checks and other benefits." The only places likely to pay higher salaries are "hotshot startups," but you "trade a little bit of money for a lot of security." Compensation is "based on relative performance"; expect the serious producers to see much more serious money, especially in the profit-sharing checks.

Just remember, you're not going to get the big corner office. In fact, no one has an office at Hewlett-Packard. As one employee describes the situation, "even the CEO has a cubicle. While gopher-holing isn't everyone's idea of a good time (one employee repeatedly mentions how much "cubes suck"), there is a rationale: "It encourages the open door policy." Obviously, where there is no door, it is difficult to close one. Other employees say the cubicle-heavy offices have a "minimalist, modern feel."

Hewlett-Packard employees have nothing but praise for their co-workers, saying that they "are the best, but not in an arrogant way." Another employee insists, "I don't see office politics at the lower levels that I hear of at other places." Indeed, "behaving in an overtly ambitious manner is not supported." One insider remarks, "I have never seen individual engineers trying to grab the glory." As pleased as HP employees are with the quality and integrity of their co-workers, they say "the office atmosphere deters interaction between them." In fact, at a company where job satisfaction is extremely high, the lack of social contact between workers is one of the biggest complaints.

“Take home the latest in tech for your personal use. When it’s obsolete, just return it and get a new one.”

– ***Hardware insider***

IBM

New Orchard Road
Armonk, NY 10504
www.ibm.com
(914) 765-1900

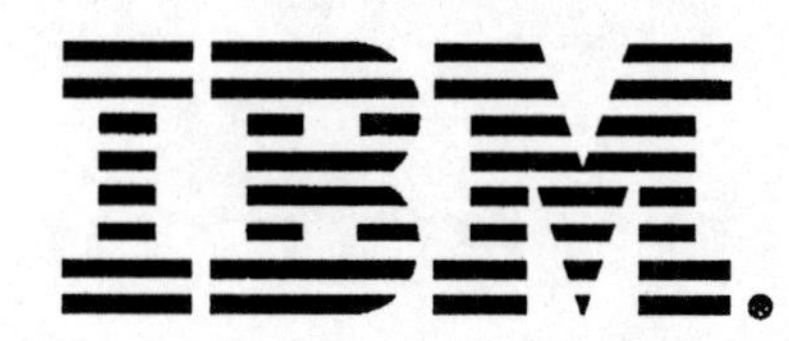

LOCATIONS

Armonk, NY (HQ)
Offices around the world

DEPARTMENTS

Accounting
Consulting
Finance
Hardware Engineering
Information Technology
Internet
Market Research
Network Analysis
Programming
Research & Development
Software Engineering

THE STATS

Annual Revenues: $78.5 billion (1997)
No. of Employees: 269,465 (worldwide)
No. of Offices: 18+ (worldwide)
Stock Symbol: IBM (NYSE)
CEO: Louis V. Gerstner

UPPERS

- Tuition reimbursement
- Stock purchase plan
- Loosened dress code
- Health benefits to partners of gay and lesbian employees

DOWNERS

- Remnants of conservative culture
- Lack of guidance from upper management

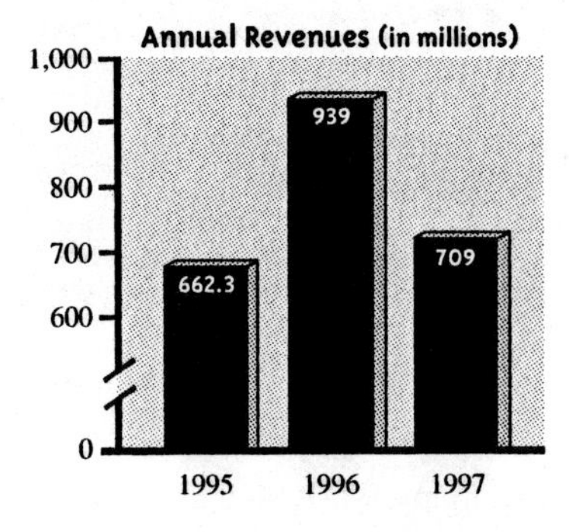

KEY COMPETITORS

- Apple
- Compaq
- Dell Computer
- Gateway
- Hewlett-Packard
- Hitachi
- Micron Technology
- Microsoft
- Sun Microsystems

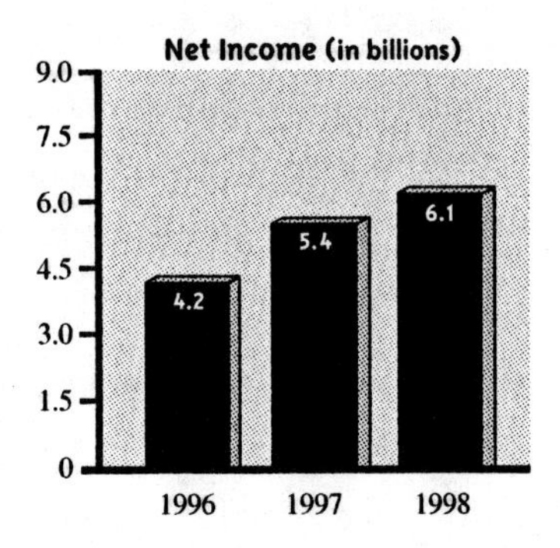

EMPLOYMENT CONTACT

Human Resources
IBM
New Orchard Road
Armonk, NY 10504

Fax: (800) 426-6550

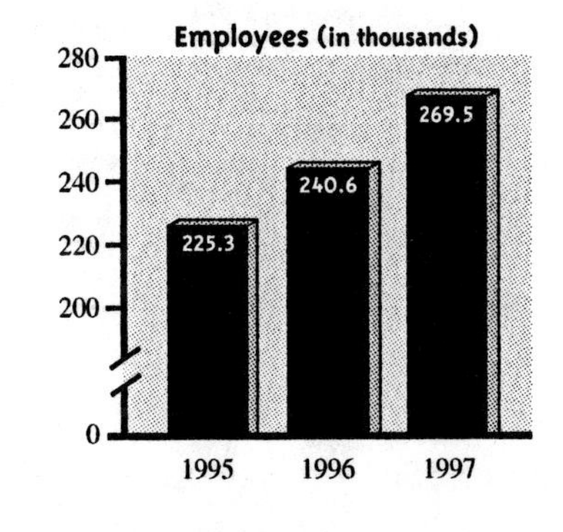

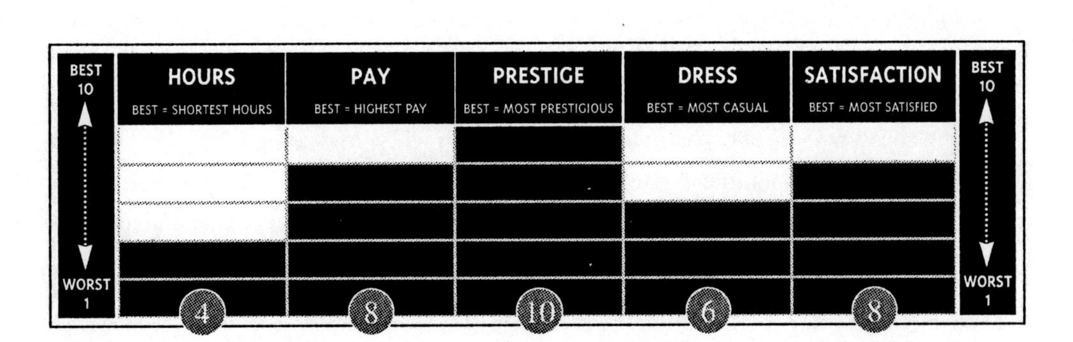

THE SCOOP

International Business Machines (IBM) began as the Computing-Tabulating-Recording Company, a floundering office machine firm rescued by National Cash Register salesman Thomas Watson in 1914. Watson turned the company around by securing government contracts during World War I; by 1920, annual revenue had tripled. Four years later, the company took its present moniker and quickly established its dominance in the office machine market, selling its tabulators, time clocks, and electric typewriters domestically and abroad. IBM introduced its first computer in 1952 and maintained control of about 80 percent of the market throughout the 1960s and 1970s.

The subsequent "PC revolution," however, found IBM unable to compete in the face of the shift to smaller, more open systems. To adapt to the new conditions, the company changed its focus to big computers, semi-conductors, software, and professional services in order to secure the top spot in the computer world. With its recent acquisitions of Lotus and Tivoli Systems, IBM now plans to extend its dominance to the business software and Internet markets.

Just a few years ago, IBM was floundering. When crafty CEO Louis Gerstner took the helm of the firm in 1993, IBM was lumbering toward a $8 billion loss. Many observers were clamoring for the firm to split up. Instead, Gerstner turned IBM's focus to meeting customer needs. Big Blue's acquisition of Lotus in 1995 and Tivoli Systems in 1996 both increased sales and allowed IBM to supply a wide variety of software, instead of just software for mainframes and other proprietary software. Since then, IBM has jumped into the software market full force. Software now accounts for about 18 percent of all IBM sales, making Big Blue, not Microsoft, the biggest software company in the world. IBM has more developers working in Java than Sun Microsystems, which wrote the computer language in the first place. The increased focus on software, including voice recognition technology, has made IBM more flexible; the firm no longer writes software solely for its own hardware. The company is also moving ahead with its encryption technology, which will allow companies to verify customers purchasing products online. IBM's new mainframe, S/390 G5, is expected to run 1040 millions of instructions per second, 15 percent more than what analysts were expecting.

Once-costly IBM is even delving into the sub-$1000 PC market, introducing a personal computer that sells for $999. IBM sells computers directly to customers, but, unlike its nimbler rivals Gateway and Dell, also sells computers through distributors or "middlemen," who take a slice of profits and demand compensation when surplus inventory sits in warehouses. IBM is also looking to sell off its printer division, which analysts say could fetch up to $2 billion.

GETTING HIRED

During a typical year, IBM hires more than 15,000 new employees in everything from accounting and data warehousing to Internet applications and software development. Applicants should consult IBM's web site, located at www.empl.ibm.com, to find out more about opportunities within each of its various business lines. At the web site, applicants can also construct a resume online and send it directly to the company; IBM scans all resumes, whether electronic or paper, into a central database where they circulate for six months.

OUR SURVEY SAYS

IBM is "trying to reengineer itself to compete in the new millennium" by shifting its emphasis to "individual empowerment and entrepreneurial thinking." Observing that even the IBM's famous "unwritten dress code" is gone, employees describe the "new IBM" as "more relaxed and informal." Many are excited about the organizational changes, but some say they "would appreciate more guidance and input" from upper management. Others worry that IBM's changes have come "at the expense of some of its longtimers." For new employees, however, the shift in IBM's corporate culture will translate into "spectacular new opportunities" at an "exciting moment in the history of one of the most storied companies in America." One insider is positive about the changes at the company: "The culture has changed from command-and-control to highly entrepreneurial. Things move fast and change often. IBM employees now have a great deal of latitude to deploy resources to opportunities while drawing upon the talents and technology of a $76 billion company. Most of the old IBM bureaucracy is gone, replaced with an environment where performance and personal contribution are what get rewarded."

Employees think their company is cutting edge indeed. "IBM is the leader in more areas and technologies than anyone else in the most important and exciting industry on the planet," concurs one insider. Changes have shaken up the traditional Big Blue, however. Reports one insider: "The dress code is business casual. Dress is selected based on what your customers expect as normal. Inside IBM, nobody much cares anymore about white shirts, dresses and all that" – this at a firm once famous for its unspoken white-shirt-dark-suit dress code. As befits a worldwide power, IBM has generous and far-sighted perks and policies." As you'd expect from a forward-looking company, IBM has an egalitarian structure. "There's no differentiation between men's and women's jobs. There are several women in the executive ranks in IBM. IBM is a worldwide company, with employees in every country, so there's a lot of cultural and ethnic diversity among the employees." Perks? Yup. "Special perks include free Internet access, ThinkPad laptops, support for mobile employees, and a benefit package you would expect from a company like IBM."

Still, IBM retains a bit of the old gentility – which may not be to everyone's liking. "I have to say that, depending on who your manager is at IBM, you may or may not have a good experience," says one insider. "Generally, the newer people who don't come from a large corporate background (IBM has been hiring a lot of younger, more aggressive types) may not reflect the old IBM culture. They get frustrated with IBM's 'politeness' and eventually leave, but not before annoying a lot of people."

Infoseek

1399 Moffett Park Dr.
Sunnyvale, CA 94089
(408) 543-5000
Fax: (408) 734-9350
www.infoseek.com

LOCATIONS

Santa Clara, CA (HQ)

DEPARTMENTS

Client Application
Human Resources
Product Marketing
Public Relations
Sales
Software development

THE STATS

Annual Revenues: $34.6 million (1997)
No. of Employees: 171
No. of Offices: 1 (U.S)
Stock Symbol: SEEK (NASDAQ)
CEO: Harry Motro

INFOSEEK

UPPERS

- No dress code
- Flexible hours
- Full medical benefits
- Free sodas, toys, T-shirts

DOWNERS

- No job training

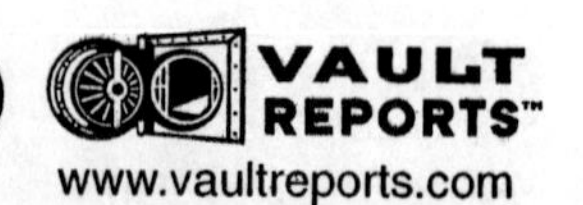

KEY COMPETTORS

- America Online
- Excite
- Lycos
- Snap!
- Yahoo!

EMPLOYMENT CONTACT

Ginny Tolan
Director of Human Rsources
Infoseek
1399 Moffett Park Dr.
Sunnyvale, CA 94089

Fax: (408) 734-9403
jobs@infoseek.com

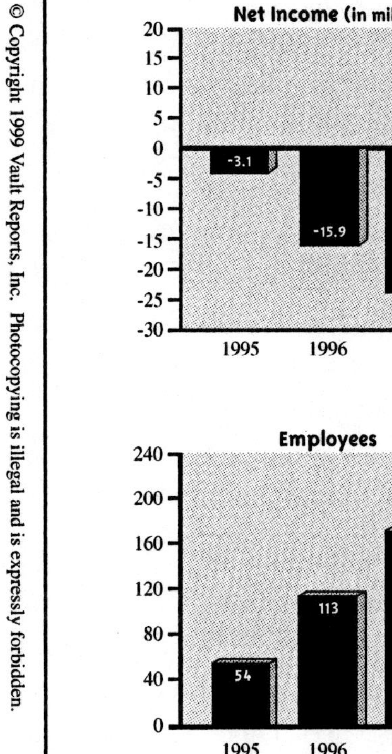

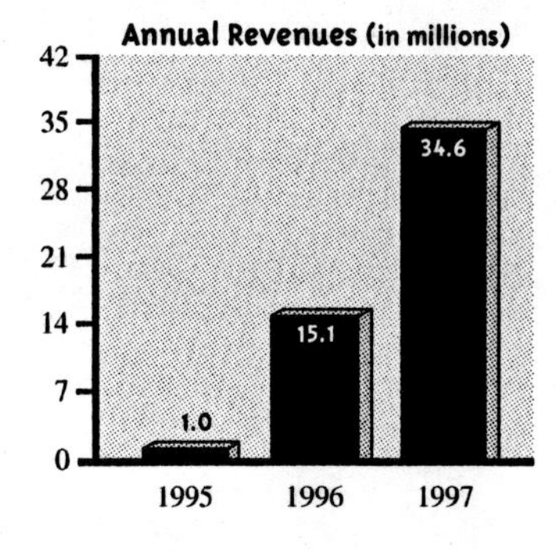

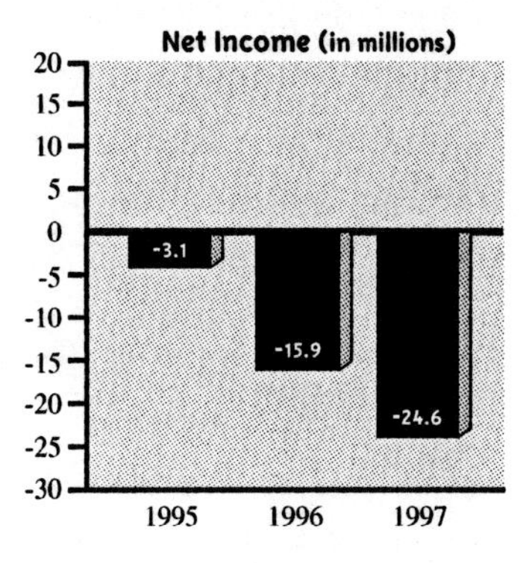

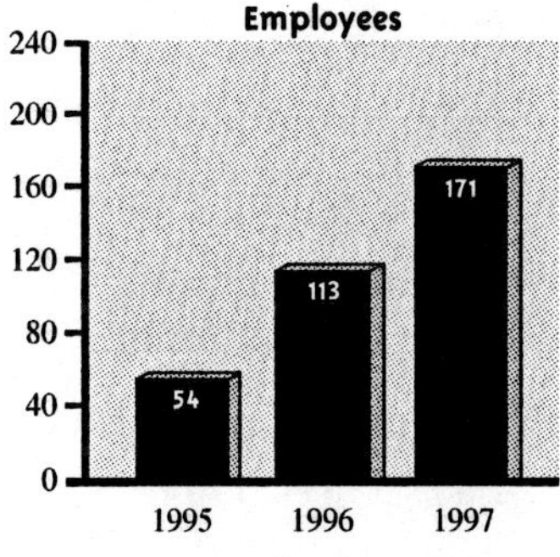

BEST 10 / WORST 1	HOURS	PAY	PRESTIGE	DRESS	SATISFACTION	BEST 10 / WORST 1
	BEST = SHORTEST HOURS	BEST = HIGHEST PAY	BEST = MOST PRESTIGIOUS	BEST = MOST CASUAL	BEST = MOST SATISFIED	
	6	6	7	10	8	

THE SCOOP

Build a better search engine, and the Mouse will beat a path to your web site.

In 1994, Steven T. Kirsch, a high-tech entrepreneur who had founded a computer mouse company in 1982 and a software publishing company in 1986, developed an idea for an Internet search engine that would provide users with personalized news. The search engine, Infoseek quickly became one of the most popular free search engines on the Web. And in 1998, the company joined the media big boys when the original mouse king, Walt Disney, bought 43 percent of the company.

The Disney deal gives Infoseek security in the new media industry, which has in its brief existence shown itself to be a place where market leaders can disappear and fail in the blink of an eye. In buying 43 percent of Infoseek for $70 million, Disney handed over control of its Starwave web site development company to Infoseek. As a result, Infoseek will share its subscribers with Disney and feature content from the Disney.com and Disney-owned ABC and ESPN (both sites designed by Starwave). With Disney's help (and cash), Infoseek announced its plans for a new portal site called Go Network, scheduled for launch in 1999.

But Infoseek isn't putting its energy just into its partnership with Disney. In September 1998, the company unveiled a redesign of its site, which includes a sidebar for news headlines and software downloads, as, like chief competitors Excite, Yahoo and Lycos, Infoseek seeks to transform itself from merely a search engine to a portal – a site where web surfers go for myriad services, from stock quotes to e-mail to shopping.

GETTING HIRED

The interviewing process at Infoseek varies depending on the department to which you are applying. As a general rule, however, "it's safe to say that you will meet wide range of potential colleagues," insiders tell us. There is a preliminary screening of resumes and references prior to the interview. Infoseek job listings are posted on its web site at www.infoseek.com. Despite its casual environment, Infoseek does look for candidates with strong academic records and experience in web site creation and maintenance. Also, Infoseek often tenders offers to its summer interns. Most of the openings listed on its web site call for at least a year of experience in high-tech.

OUR SURVEY SAYS

Infoseek is a "hyper-casual place with fun people," and no dress code. One engineer admits to "walking around barefoot most of the time." Although the atmosphere is "very laid back," there are "occasional periods of very demanding work," since the company is "understaffed and growing quickly." The work hours, though, are "flexible in the extreme," with some employees working "noon to midnight and/or from home. The pay is "pretty much in line with similar companies in the Valley," with the "standard array of extras, like 401(k), ESPP, medical and dental," and the not-so conventional perks like "free sodas, occasional parties, toys, and more T-shirts than you can wear in a month." One engineer, however reports that "pay is OK but not quite as good as one might expect in this market."

One respondent attests that the greatest aspect of Infoseek is the people: "They're by far the smartest group I've ever worked with, including the CEO." An engineer agrees: "They are by far the smartest group of engineers I've ever worked with." Adds a source: "No place is perfect, but the team of folks they have is top notch." "All in all, the teams there are very tightly knit," says one insider. However, there is a "tendency toward office politicking," which a new hire "should watch out for." The key to succeed at Infoseek, a source says, is "to find the power players in the organization and ally them to accomplish stuff" and to "speak loudly." Finally, an insider notes: "The industry dynamics are wild. Everything can change here."

Intel

P.O. Box 1141
Folsom, CA 95763
(408) 765-8080
www.intel.com

LOCATIONS

Santa Clara, CA (HQ)
Albuquerque, NM • Folsom, CA • Fort Worth, TX • Phoenix, AZ • Portland, OR • Sacramento, CA • Salt Lake City, UT • Seattle-Tacoma, WA • Munich, Germany • Shanghai, China • Swindon, England

DEPARTMENTS

Finance • Human Resources • Information Technology • Integrated Circuit Engineering • Integrated Circuit Manufacturing • Marketing • Materials and Planning • Operations • Sales • Software Engineering • Systems Hardware Engineering and Manufacturing

THE STATS

Annual Revenues: $25.1 billion (1997)
No. of Employees: 63,700
Stock Symbol: INTC (NASDAQ)
CEO: Craig R. Barrett
% Minority: 28
%Male/%Female: 68/32

UPPERS

- Stock options a go-go
- Tuition reimbursement
- "Intel University" training program
- Paid sabbaticals
- Fitness facility

DOWNERS

- Long workdays
- Massive bureaucracy

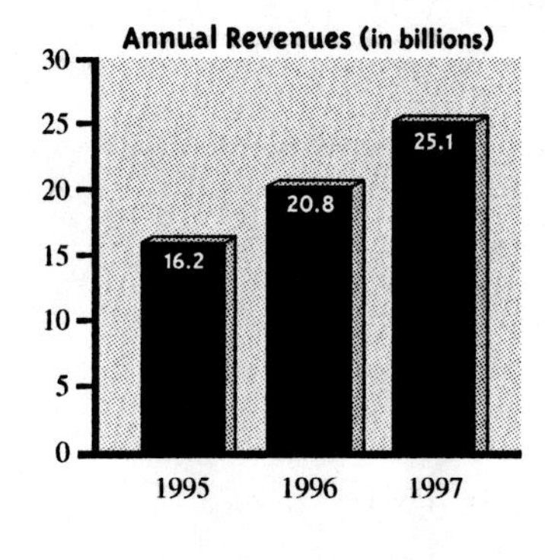

KEY COMPETITORS

- Advanced Micro Devices
- Cyrix
- LSI Logic
- NEC
- Rise

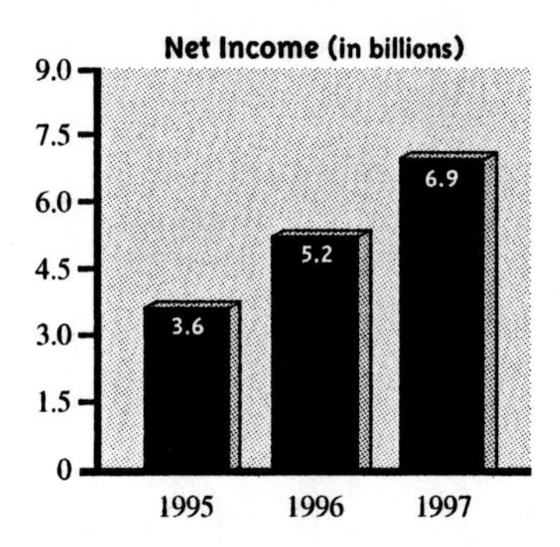

EMPLOYMENT CONTACT

Human Resources
Intel
P.O. Box 1141
Folsom, CA 95763

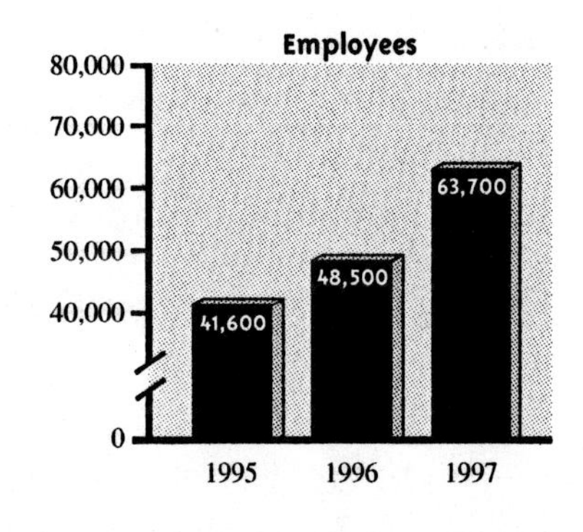

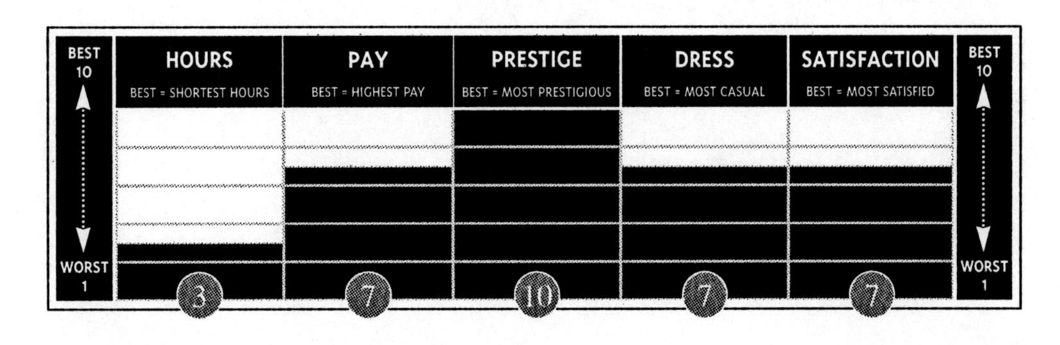

THE SCOOP

Love to root for winners? Look no further than the Silicon Valley juggernaut Intel. Its microprocessors – also known as chips – power an estimated 89 percent of personal computers worldwide. The firm is just as dependent on its microprocessors as computers around the world are on its products; Intel garners 80 percent of annual revenues from chip sales. As consumers clamor for more computing might and sharp programmers (like those at Microsoft) concoct increasingly intricate software, Intel supplies the chips that will make computers run those new-fangled programs (and the old ones, too) faster and better.

Intel's products have become increasingly complex and widespread. The firm now earns 60 percent of its revenues from outside America, with new chips like the Pentium II and the Pentium II Xeon chip. Coming up – something called "multilevel memory," which permits Intel engineers to fit even more transistors onto microprocessors (making them more powerful) and the Hewlett-Packard-Intel collaborative effort called the Merced, a chip that should be three times more powerful than the speediest Pentium II to date. Can anything shake this microprocessor powerhouse? Nothing's impossible – the boom in sub-$1000 PCs means that lower-priced competitors like AMD and Cyrix are profiting – but in the meantime, Intel's one of the most valuable companies in America.

Fierce competition has cut into Intel's profits, forcing the company to slash its prices in order to compete with the cheaper microchips that are powering the under $1000 computers that are flooding the market. In June 1998, Intel announced that commercial distribution of its highly touted Merced chip will be delayed until the middle of year 2000. The Federal Trade Commission has also filed an anti-trust suit against Intel, charging the company with using its near-monopoly to stifle competition by withholding technical information from its rivals.

GETTING HIRED

Intel is an active recruiter of new college graduates; the company typically hires more than 1800 recent graduates each year. Although a high-tech company, Intel also offers a range of business positions in sales, marketing, finance, and operations. Select new hires may qualify for either the company's "Graduate Rotation Program for Engineers" or the "Technical Sales Engineering Program." The company accepts resumes by both regular mail and e-mail and then scans them into its database. For a list of current openings, as well as a campus recruiting schedule, consult Intel's web site.

Intel does the bulk of its recruiting for its technical areas – Internet technology, hard engineering, and software engineering. Most positions in marketing and finance are offered to MBA grads, but undergraduates with technical majors are also considered. Intel recruits extensively at campuses and at job fairs; its internship program is also a sort of meta-recruitment endeavor. "Don't get stressed out," says one Intel insider. "All the interviewees are placed in little cubicles. There are about 40 other people being interviewed at the same time and the managers just keep going from person to person." After your interview, you'll be ranked on criteria like technical ability, analytical ability, communicative skills, and personal qualities. The last question on every interviewer ranking form: "Would you want to work with this person?" So remember: smile! If you're offered a job at Intel, you'll get a call anywhere from 24 to 72 hours after your interview.

Most Intel technical hires are electrical engineering, computer engineering, or computer science majors. Qualifications vary; prospective software engineers would be well-served by a thorough knowledge of C++ or multimedia software; those who aim at integrated circuit engineering might want to brush up on their VLSI circuit design knowledge, or behavioral modeling and logic synthesis. However, Intel also considers students who study chemical engineering, industrial engineering, materials science, mechanical engineering, physics, and computer information systems. While MBAs applying for financial analyst positions do not necessarily need technical knowledge, it will assuredly come in handy.

OUR SURVEY SAYS

Intel has a very particular culture – on the one hand emphasizing egalitarianism and meritocracy, on the other hand offering its workers featureless cubicles and a healthy dose of paranoia. Insiders say "Intel is a flexible meritocracy. They increase responsibilities of people who show they can do the work and want to do more. Raises and promotions are also based on meritocracy." Don't expect to relax at this firm. "Intel is a little more uptight compared to other big Silicon Valley companies. There is more discipline and more of a business-oriented focus. That is one reason why the company is so successful." And don't expect a lot of hand holding, either. "There is no executive training programs where you are eased into the culture. I loved it, because you have the freedom to take risks and make things happen, but many people found it a difficult environment." The upshot of all this independence and discipline is that "Intel expects you to speak your mind and express your viewpoint. But once a decision is made, you must back it 100 percent and do what you can to make your project a success. Office politics and project sabotage is not tolerated at Intel." Fortunately, says a former Intel insider, you should be able to get your point across. "Intel culture is something I miss very much. Your success depends on how well you can convince people of your point and your ability to communicate. There is not as much of a problem with a stagnant 'old boy networks' as at other firms."

Intel's meritocracy and discipline has its downsides – paranoia and bureaucracy. Intel insiders want to give you a tip: "Word of advice – meetings suck. They are just an excuse for the managers to justify their stock options. Avoid." An engineer assesses the firm through its motto output. "Intel has a couple of company slogans. One of them is 'A great place to work,' which is also one of our values, and another is from [Chairman] Andy Grove, 'Only the paranoid survive.' Yes, we are No. 1, but there is always someone who wants to take us down. So put those two together and maybe you can get an idea of what Intel is like."

Though Intel is a king in its market, don't expect posh digs at Intel offices. "Everyone sits in a cubicle, except in the labs and the conference rooms. In fact, there are very few doors. That makes it easier to have our 'Open Door' policy." One manager concurs "The offices at Intel are cubicles. No one has an office, not even Andy Grove. This promotes a very open environment. I like it, but I know that not having an office drives some people crazy." Don't even count on keeping a stable cubicle. One source reports "My office in Fulsom is growing so fast that our cubicle size was reduced, from 9 x 9 to 6 x 9. This is called compression. It's a little cramped." Another Intel insider mentions "By walking through the offices, you would have a hard time figuring out where one departments ends and another one starts." One employee tells us "We all sit in rows and columns. I call it the Cube Farm." As for other office attributes, say insiders, "there is a fitness facility on-site, which is handy because I don't have time to go anywhere else."

Intel doesn't shun sociability. "Each quarter the department has a quarterly event off-site. Each summer the division has a company picnic. Each winter the division (and sometimes the site) has a formal ball. When on a cross-functional team, the division sponsors team-building events, like cross country skiing." There is such thing as a free lunch at Intel – we hear that "some meals are provided when there are meetings during lunchtime, or late. Every Friday, during the 8 a.m. meeting, continental breakfast is served. Hard-working Intel employees can earn some time off – "every seven years, employees earn a sabbatical of eight paid weeks off in addition to regular vacation time." As for health and retirement benefits, "there's the full medical coverage (free), dental plan (free), vision care (free), life insurance (free), the 401(k). One employee thinks "the greatest, absolute best perk Intel gives me is stock options and the ESOP [employee stock option plan]. That means cash, and sending my kids to college."

Employees are thrilled by their total compensation, but are well aware that salaries alone are "competitive, but on the low end of competitive." One Intel insider lists various bonuses available at Intel. "Individuals can be recognized by other individuals for outstanding work with bonuses up to $100. Each division honors people with bonuses each quarter for outstanding contributions. All employees receive semi-annual profit-sharing bonuses." One employee says "because of its powerhouse status, Intel doesn't use base salary to attract top talent. It only pays at industry average. That's been a source of gripes for me and my colleagues. But over time, the complaints taper off as we watch the stock climb." A happy Intel employee reports "Intel is not the highest paying company out there, as far as salary goes. But the whole package, or "total compensation," as they call it, is quite substantial. We got four bonuses this year. One was a $1000 'thank you' bonus, one was for 15 days of pay, one for 17 days of pay, and the last was for about five percent of yearly salary. Can't beat that."

“Because the company has been growing quickly, there are always opportunities for people to move up if they’re ready.”

– ***Semiconductor insider***

Intuit

Box 7850, MS 247
Mountain View, CA 94039
(415) 944-5000
Fax: (415) 944-2700
www.intuit.com

Intuit

LOCATIONS

Mountain View, CA (HQ)
Alexandria, VA • Aurora, IL • Fredericksburg, VA • Pittsburgh, PA • Rio Rancho, NM • San Diego, CA • Tucson, AZ • Alberta, Canada • Boulogne, France • Cherksey, United Kingdom • Ismaning, Germany • Ontario, Canada • Tokyo, Japan

DEPARTMENTS

Administration
Customer Service/Technical Support
Documentation
Engineering
Finance
Human Resources
Information Systems & Technology
Internet/World Wide Web
Marketing/Sales
Operations

THE STATS

Annual Revenues: $592.7 million (1998)
No. of Employees: 2,900
No. of Offices: 15 (worldwide), 11 (U.S)
Stock Symbol: INTU (NASDAQ)
CEO: William Harris, Jr.

UPPERS

- 401(k) plan
- Flexible scheduling options
- Profit sharing
- Attractive Internet deals

DOWNERS

- Crunch times getting software out

KEY COMPETITORS

- IBM
- Quarterdeck
- Oracle
- Microsoft

EMPLOYMENT CONTACT

Human Resources
Intuit
Box 7850, MS 247
Mountain View, CA 94039

www.intuit.com/careers
intuitcareers@intuit.com

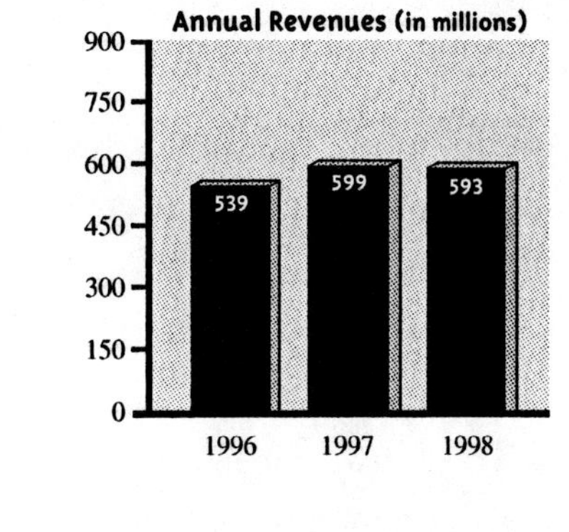

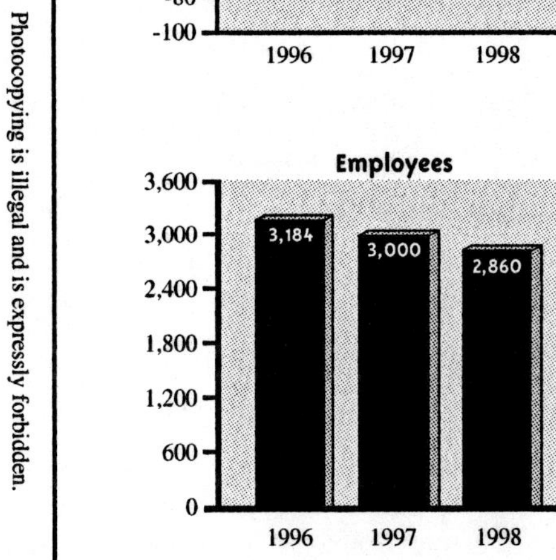

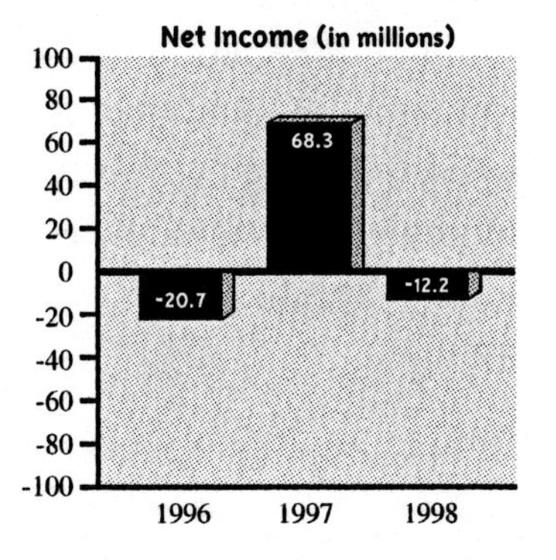

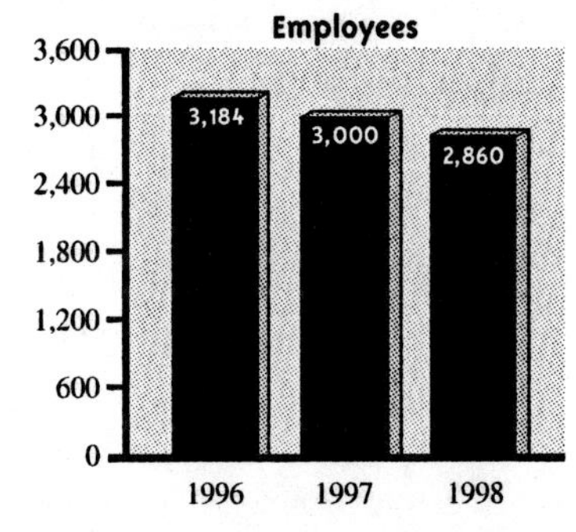

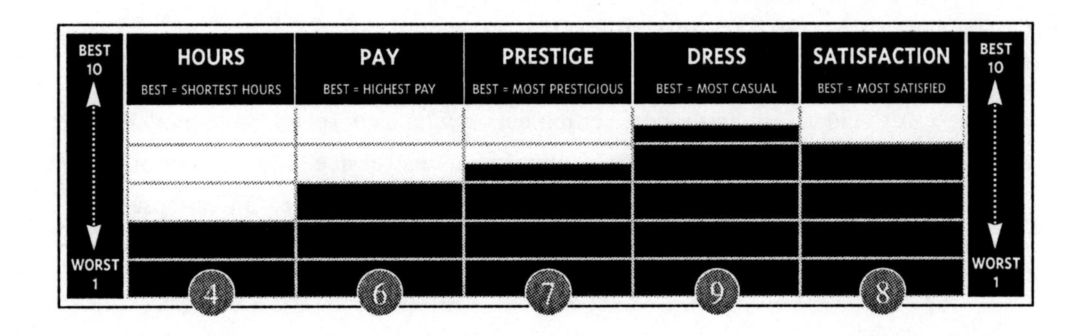

THE SCOOP

Founded in 1983, Intuit has become the largest manufacturer of accounting, personal finance, and tax software in the U.S. Founders Scott Cook and Tom Proulx recognized that there was a demand for a user-friendly personal finance software package and met that demand with Quicken – currently America's top-selling personal finance software product. The company also makes an accounting software package, QuickBooks, which is the leading choice of small businesses. Intuit's 1994 acquisition of the National Payment Clearinghouse has made the company a presence in the electronic banking industry. With the company's 1995 purchase of Milky Way KK, a Japanese software company, Intuit entered the second-largest PC market in the world.

Intuit is one of the few companies that has thus far consistently won battles against Microsoft in the software wars. Quicken, its star product, distinguishes itself by its blissful simplicity – its design replicates a simple checkbook, and is among the easiest of home finance products to use. To counter Quicken's popularity, Microsoft introduced its own product, Money, in 1991. Despite its attempt to undercut Quicken by price – Microsoft even gave away Money occasionally – Quicken remained by far the most popular financial software program, with roughly 75 percent of that market. In April 1995, Microsoft opened its own checkbook and offered to buy Intuit for the lavish sum of $2 billion, even promising to give away its own financial software offering, Money, in order to retain the semblance of competition. But in the face of displeasure from American antitrust watchdogs, Microsoft dropped its bid.

Intuit can leverage the loyalty of Quicken users to its advantage. The company runs a popular web site, Quicken.com, which is accessible through Excite, the web portal. (Intuit invested a $40 million stake in Excite, the second-largest web browser after Yahoo, and owns 19 percent of that company.) The Intuit-Excite web site has eight units: investments, chat, home and mortgage, banking and borrowing (which helps consumers search for the best rates on loans, credit cards and CDs), insurance, retirement, saving and spending (complete with debt-reduction planner), and tax planning. Intuit claims its service is much more inclusive than competing sites like Yahoo! Finance and Microsoft Investor, since, unlike those sites, it does not concentrate primarily on stock trading. Perhaps more appealing to the average web consumers, Inuit offers free services that cost money at other sites. Intuit plans to make its money by becoming a sort of web landlord, renting out space on the site to financial services firms and banks, and charging up to $500,000 for the privilege. Intuit tells banks that association with the Quicken web site will be a benefit to them, but banks have shown lukewarm interest; many fear that financial sites like Quicken.com could draw business away from bank web sites.

Intuit has looked to other partnerships to increase its Internet business. Through a December 1997 deal with CNNfn.com (the web site for the CNNfn financial news channel), the two companies set up a co-branded area on CNNfn.com, utilizing finance news from CNNfn and personal finance software from Intuit.

In February 1998, Intuit made another strategic alliance by agreeing to supply America Online with content for its personal finance channel; most of the content will come from Quicken.com. AOL has more than 14 million subscribers. Combined with already utilizing Quicken.com, and the viewers of other Intuit partners – Excite with 2.5 million viewers, and CNNfn.com with 7 million – Iniut will now be able to reach up to 30 million potential customers at one time. Under the February agreement, Intuit will shell out $30 million to AOL over the next three years and promote AOL as the primary online service for Intuit software users. Net deals still account for barely 5 percent of Intuit's available assets, but analysts say that Intuit is well-positioned to set a profitable Internet strategy in motion.

GETTING HIRED

Applicants should submit resumes to the particular Intuit location in which they are interested. The company's recruitment web page, located at www.intuit.com/int-human-resources/career-opportunities, lists the addresses of the locations that regularly hire new employees. Intuit's corporate headquarters at Mountain View houses the company's major business units, in addition to its finance, human resources, and operations functions, and is home to about 700 of the company's employees. The hiring process for MBAs involves three rounds of interviews, described by insiders as "casual."

OUR SURVEY SAYS

Intuit promotes a culture of "hard work and hard play," insiders say. The company stresses teamwork, and employees remark that their ability to complement each other's strengths and weaknesses is one of Intuit's leading assets. Employees describe the office atmosphere as "casual and sometimes quite playful," with "no dress code – written or unwritten – to speak of." While employees often work long schedules when a project nears completion, they say that Intuit "recognizes that its employees have lives outside of the office." "At Intuit," says an insider, "the firm considers its employees 'internal customers' and those who call for support 'external customers.'"

Many employees praise Intuit for its generosity. Says one developer: "The company has generous stock option and stock purchasing plans. I would say that as far as salary goes, it is in line with the industry. If you are involved in product development, there are also ship bonuses." "We often rely on a 'roving help desk' to help us assist our customers technically," says a tech support representative. "It is a very intense environment considering we must ship new tax software each year without fail. We work hard at Intuit but we also play hard." Intuit helps its employees play: "We have private offices, an on-site cafeteria and gym. Every Friday evening at 5 we have a social in the outside courtyard, serving beer, wine, food and the like." Intuit is kind in other ways, too. One employee reports: "I recently became a single parent and the company has bent over backwards to allow me to have as flexible a work schedule as possible. I work from home for several hours a day." Concludes another insider: "I do not personally know anyone who is dissatisfied at Intuit."

"This is a huge company, and I think sometimes you have either to sink or swim. You can easily get swept up and passed over."

– Software insider

Lotus

55 Cambridge Parkway
Cambridge, MA 02142
(617) 577-8500
Fax: (617) 693-1909
www.lotus.com

Lotus®

LOCATIONS

Cambridge, MA (HQ)
Atlanta, GA • Austin, TX • Chicago, IL • Cincinnati, OH • Cleveland, OH • Dallas, TX • Denver, CO • Detroit, MI • Hartford, CT • Houston, TX • Los Angeles, CA • Miami, FL • Minneapolis, MN • Mountain View, CA • New York, NY • Philadelphia, PA • Pittsburgh, PA • Phoenix, AZ • Richmond, VA • Rochester, NY • San Francisco, CA • Seattle, WA • St. Louis, MO • Washington, DC • Wayne, PA • Additional locations worldwide

DEPARTMENTS

Consumer Support
Corporate Marketing
Finance & Operations
Information Systems
Internet Services
Professional Consulting
Research & Development; Sales

THE STATS

Annual Revenues: N/A
No. of Employees: 6,000
No. of Offices: 65 (worldwide), 27 (U.S.)
A Subsidiary of IBM
CEO: Jeff Papows

UPPERS

- Flexible scheduling options
- Tuition reimbursement
- Birthday off
- Sabbatical program for long-term employees

DOWNERS

- Competition from Microsoft
- No longer independent

KEY COMPETITORS

- Intuit
- Microsoft
- Oracle
- Sun Microsystems

EMPLOYMENT CONTACT

Human Resources
Lotus
55 Cambridge Parkway
Cambridge, MA 02142

SELECTED PRODUCTS
1-2-3
cc: Mail
C++API
Domino Application Server
eSuite
FastSite
LotusScript
NetObjects Fusion
Notes
R5 Preview
Sametime
SmartSuite

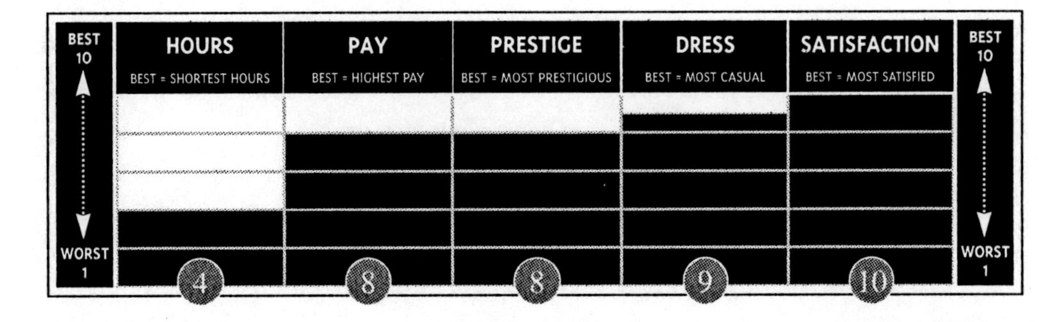

THE SCOOP

Lotus went into business in 1982, when the IBM PC ruled the world and Lotus' 1-2-3 was the only spreadsheet in town. The introduction of Microsoft's Excel late in the 1980s was a major blow to business, but Lotus has now won a new niche by marketing itself as the web-savvy software company. (Microsoft has met with limited success on the Web.) The current product list includes Java Desktop, cc:Mail, Asset Manager, a variety of IBM software, and the myriad of applications related to Lotus Notes and Domino, the company's two top sellers.

Though most recent press attention has been devoted to Lotus' web-related software, the company has also been developing intra-office communications software. Lotus' products have become so popular that tickets to Lotusphere '97, the annual showcase of new products and presentation of ongoing projects, were sold out six weeks before the actual event. Acquired by IBM in 1995, Lotus has managed to maintain an autonomous corporate identity, and has won industry awards for superior customer service.

GETTING HIRED

Lotus lists current job openings at its web site. Each listing includes an e-mail address to which applicants can send a cover letter and resume (in ASCII format only). Lotus also accepts resumes submitted via fax or regular mail at its Cambridge, MA headquarters. The web site also includes a listing of nationwide recruiting events.

For MBAs who join the company in its financial or corporate consulting departments, the company offers The Lotus Consulting Academy, a two-week residential training program based in Cambridge that provides a crash course in Lotus' products. Those who land technical positions at Lotus take note: "You will, of course, start at the bottom, taking small programming projects, but as you learn more you'll move up. If you're good, you'll move up fast." Says another insider: "On the technical side, you can advance to being a senior architect, or even a Lotus Fellow." "Lotus Fellows are sort of independent researchers. They just work on whatever they think is interesting. Most have PhDs."

OUR SURVEY SAYS

Policies such as "generous" tuition reimbursement, "frequent" company parties, and "outstanding" insurance coverage foster long-term, satisfying Lotus careers, employees tell Vault Reports. Employees appreciate the "informal," "laid-back" office culture. "Lotus is very casual, no dress codes," says one employee. Working parents also praise Lotus' "widespread efforts" to accommodate their particular scheduling needs, such as the option available to some employees to work at home. "Hours are extremely flexible," says one insider. "I usually come in at 11 and leave at 9. My boss, on the other hand, comes in at 7 and leaves at 4."

Lotus was one of the first U.S. employers to extend same-sex partner benefits to employees on its health and life insurance plans. The company also boasts "an award-winning daycare center." In addition to offering generous family leave and child-care assistance, Lotus was also one of the original sponsors of the National Organization for Women's Take Our Daughters to Work Day. Women are well-represented at all levels of management, insiders say, and female employees report being very comfortable in the Lotus workplace.

The company's offices consist of "cubes for the non-managers and offices for the managers. All facilities have a cafeteria/grill." "Lotus has a policy of keeping salaries at the 85th percentile of the industry norm. And in 1997, we got a 4 percent bonus of gross salary," reports one insider. The bonuses are based on company performance. And if a product ships and you have worked on it, you get a bonus based on your contribution. Product bonuses can be quite large." Perks also delight the average Lotus employee. "The vacation is generous. You get two weeks the first year, three weeks the second year," explains one employee. "After five years, you get a month's paid sabbatical. After 10 years, four weeks of vacation. After 15 years, another sabbatical. Also, 12 paid holidays, your birthday (or closest working day), five personal days and five sick days." Both drivers and public transport commuters are rewarded at Lotus. "If you want to commute, Lotus will pay for your pass. Otherwise, there is free parking, which means a lot in Cambridge!" Lotus employees think "Lotus is regaining some of the momentum which was lost to Microsoft. So it's getting more exciting now." Insiders also note that "although Lotus is a wholly-owned subsidiary of IBM, we have our own policies and procedures. Technically, we are IBM employees, but it hardly seems like it."

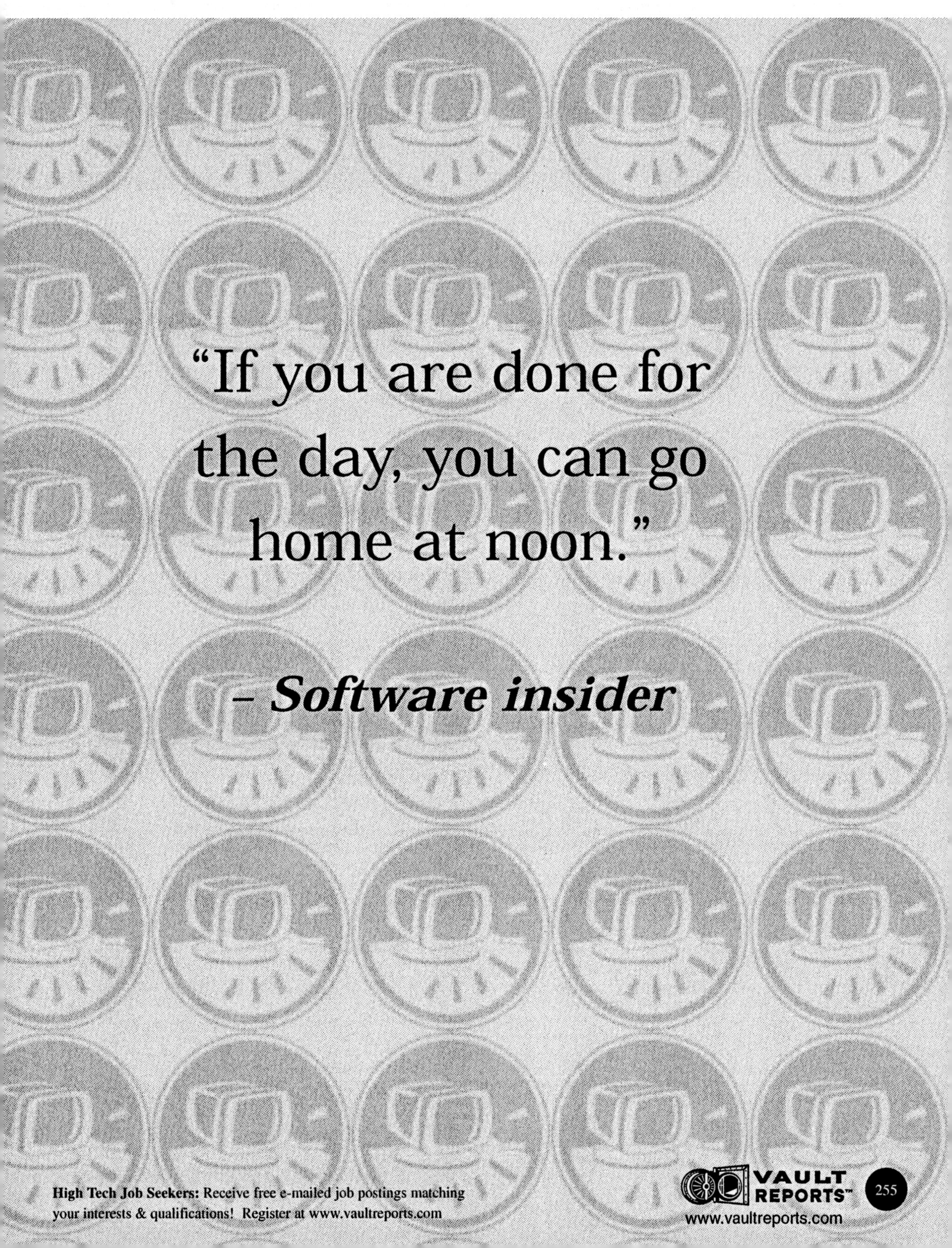
"If you are done for the day, you can go home at noon."
– Software insider

LSI Logic

1551 McCarthy Blvd.
Milpitas, CA 95035
(408) 433-8000
Fax: (408) 954-3345
www.lsilogic.com

LSI LOGIC

LOCATIONS

Milpitas, CA (HQ)
Numerous locations in the U.S. and worldwide

DEPARTMENTS

Design Engineering
Finance & Administration
Operations & Technical Support
Sales & Marketing
Software Engineering

THE STATS

Annual Revenues: $1.3 billion (1997)
No. of Employees: 5,550 (worldwide)
Stock Symbol: LSI (NYSE)
CEO: Wilfred J. Corrigan

UPPERS

- Stock options
- Full tuition reimbursement
- Three weeks vacation
- Good diversity

DOWNERS

- Middling pay
- Intel dominates industry
- Roller coaster industry

KEY COMPETITORS

- Advanced Micro Devices
- Cyrix
- Intel
- NEC
- Rise

EMPLOYMENT CONTACT

College Relations Department
1551 McCarthy Boulvard
MS D-262
Milpitas, CA 95035

Fax: (408) 433-6737
jobs@lsil.com

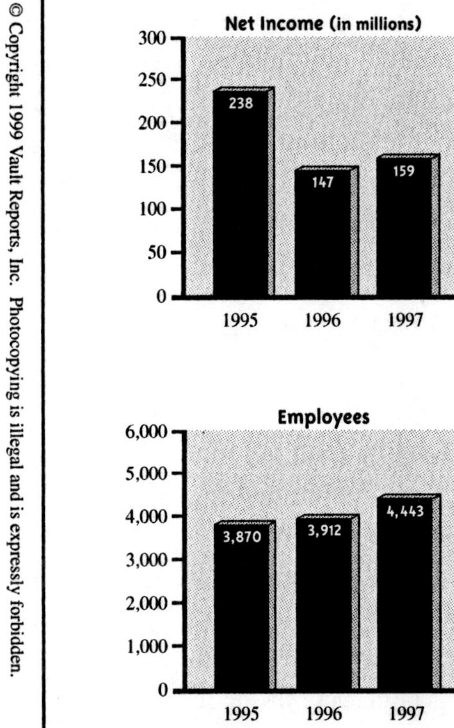

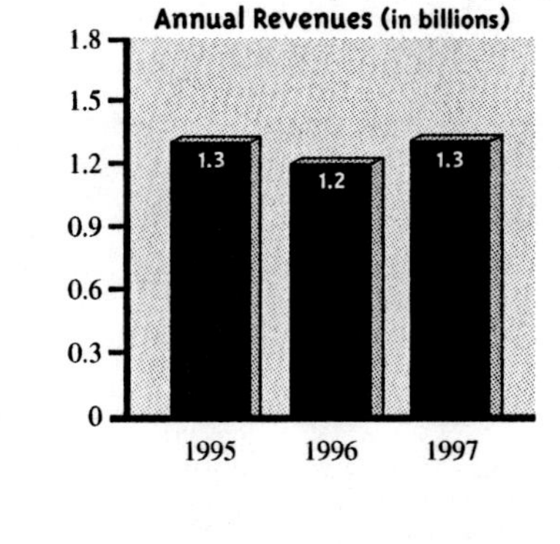

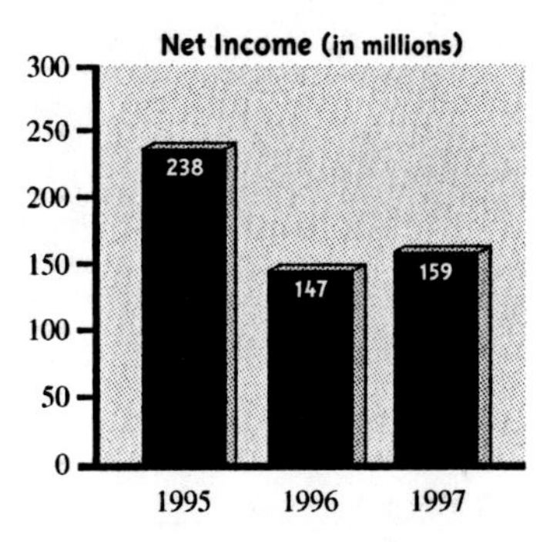

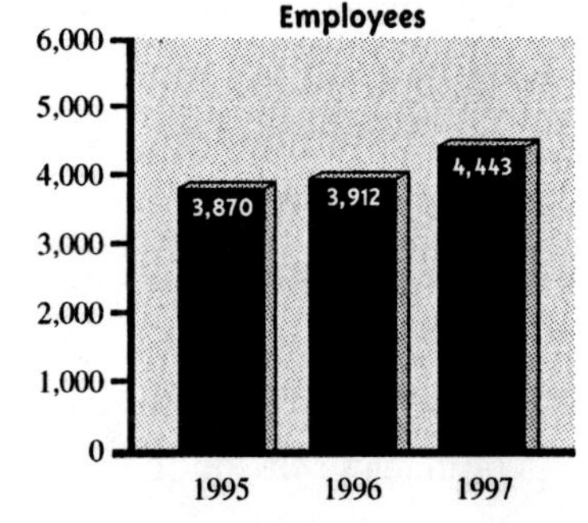

HOURS	PAY	PRESTIGE	DRESS	SATISFACTION
BEST = SHORTEST HOURS	BEST = HIGHEST PAY	BEST = MOST PRESTIGIOUS	BEST = MOST CASUAL	BEST = MOST SATISFIED
5	7	5	7	7

BEST 10 — WORST 1

THE SCOOP

Based in Milpitas, California, LSI Logic Corporation has been climbing its way to the top of the custom microchip pile since 1981. Self-titled as "the system-on-a-chip" company, LSI Logic designs and produces advanced custom semiconductors. One employee terms it a "17-year-old startup." The semiconductors, known as ASICs (application specific integrated circuits), are used in high-demand electronics like video games, cellular phones, and networking equipment. Founder Wilfred Corrigan, an engineer and former CEO of Fairchild Camera & Instrument, named the company LSI for large-scale integration, after a chip with up to 100,000 transistors. Stealthy competition and an overemphasis on commodity chips in the semiconductor industry pushed Corrigan to hone in on the ASICs niche market, a move that proved ingenious.

In 1983, LSI went public with a record-setting $152 million IPO. With skyrocketing sales, the company expanded operations in California and worldwide and debuted regional design centers where customers could design their own chips using company equipment and facilities. The Cold War was kind to LSI Logic: Contracts with big military and aerospace customers pushed company sales to $140 million in 1985, making LSI the US's leading ASIC manufacturer. The company continued to expand in the 1980s, gearing up for a predicted boom in the chip market with hefty investments in manufacturing facilities. When military sales slumped in the late 1980s, however, LSI sales hit a downturn, forcing the company to close factories and slash payrolls. In 1992, LSI suffered a $110 million restructuring charge, its third major loss in four years.

With a narrower focus and slimmer payroll, LSI emerged in 1993 focusing on three key markets: consumer electronics, communications and computers. The company debuted two new high-performance chips, the CMOS ASIC chip and the ATM chip, and a family of Ethernet products dubbed CASCADE. High-profile partnerships helped push LSI to the top again in the 1990s. In 1994, the company kicked off a joint venture with Kawasaki Steel, and Sony debuted its PlayStation video game system, which is based on LSI's CoreWare design. LSI bought out Kawasaki in 1995, and passed the $1 billion revenues mark. The same year, the company announced plans to build a $1 billion chip factory in Oregon. Though an industry slump in 1996 put LSI sales on the decline, the company continued to grow in 1997, forming alliances with fellow chip makers Micron Technology and Sun Microsystems. In October of 1997, LSI Logic announced the development of a new "system on a chip," which reduces the time and manpower required for complex designs by up to 75 percent. LSI's one-chip technology and cheap prices ($35, on average, for a chip) have given it the edge in up-and-coming industries like computer gaming and sub-$1000 PCs that depend on cheap microprocessors. In 1998, LSI announced another technological innovation – smaller, faster chips that can combine the functions of several consumer products, such as cable television, satellite, broadcast and telephone transmissions, in one unit.

The company's operations in Asia remain relatively strong despite the economic turmoil there. In 1998, LSI paid $760 million for Symbios, Inc., the electronics subsidiary of the struggling Korean manufacturer Hyundai. Meanwhile, corporate execs expect revenues from Asia outside Japan to double – had the Asian financial crisis not taken hold, company officials said, sales could have tripled from chips for consumer electronics in the Far East.

GETTING HIRED

Visit the "Employment Opportunities" section of LSI's web site for details on job openings. Positions range from marketing to engineering to finance. Send or fax resumes to human resources. LSI Logic has many opportunities for recent college graduates, mostly for electrical engineers and computer science majors, though business majors can find opportunities as well. LSI Logic recruits at universities with top-notch engineering programs. Resumes may be faxed to (408) 433-6737, or emailed to jobs@lsil.com.

OUR SURVEY SAYS

LSI offers typical high-tech fare, casual dress and hard work, with a twist – a bit more diversity than you might expect from a Silicon Valley company. "This is definitely not a cookie-cutter, uniformed operation in terms of the people," one employee remarks. Another contact says: "The culture is very diverse, in terms of ethnic background, personal backgrounds, and interests." According to yet another: "It's hard to say what a minority is here – we have such a wide range of geographical backgrounds here." Dress tends to vary too – one insider says that the dress code at LSI Logic is "whatever you feel comfortable with. It's jeans to suits around here, and dress-down Fridays are a given." However, one engineer says that "there has never been a tie spotted [at LSI Logic]. Dress ranges from long pants with button-down shirts, to T-shirts." As for treatment of women, one engineer reports: "I have worked in environments where the lab is like a frat house or locker room. It is definitely not like that here." In addition to a diverse work force, the company's family of global subsidiaries means lots of chances to rack up frequent flyer miles. "The best part of working here is keeping my frequent flyer miles," says one insider. "I average over 100K miles per year, and that works out to some serious vacation flying."

An employee at one of LSI's smaller design centers also turns in top ratings for the company. "There's not a lot of hierarchy here, probably more cooperation and less conflict than you find at other companies." Pay is described as "fair, but not the industry's highest." Still, insiders report that LSI tries to stay 5 percent above the industry standard "in all geographies and specialties." Perks include stock options, a 401(k) plan, and a competitive health plan. The 401(k) is especially juicy" The company also offers a tempting stock option purchase plan that "allows employees to buy [stock with] up to 10 percent of their salary at 85 percent of market value."

Hours are fair as well, insiders told us. "Most people work from 9 to 6," says one employee. Though its insiders generally recommended LSI, one contact warns that the semiconductor industry can be unruly: "LSI is a great place to work," the insider says, although he adds: "I would not recommend the semiconductor industry as a great career. It is very high pressure, fast rate of change and quite unstable. But I must admit, it's a helluva ride."

“It’s hard to say what a minority is here – we have such a wide range of geographical backgrounds here.”

– ***Semiconductor insider***

Lucent Technologies

600 Mountain Avenue
Murray Hill, NJ 07974
(908) 582-8500
www.lucent.com

LOCATIONS

Murray Hill, NJ (HQ)
Warren, NJ
As well as 70 offices across New Jersey and others in 90 countries

DEPARTMENTS

Engineering
Finance
Marketing
Sales
Software Development

THE STATS

Annual Revenues: $30.1 billion (1998)
No. of Employees: 136,000 (worldwide)
No. of Offices: 90+ (worldwide)
Stock Symbol: LU (NYSE)
CEO: Richard McGinn

UPPERS

- Flexible scheduling options
- Tuition assistance
- Bonuses
- Profit sharing
- Fitness centers
- Family resource program

DOWNERS

- Slow promotion process
- Unpleasant New Jersey locale

KEY COMPETITORS

- 3Com
- Cabletron
- Cisco Systems
- Nortel
- Siemens

EMPLOYMENT CONTACT

Human Resources
Lucent Technologies
600 Mountain Avenue
Murray Hill, NJ 07974

www.lucent.com/work

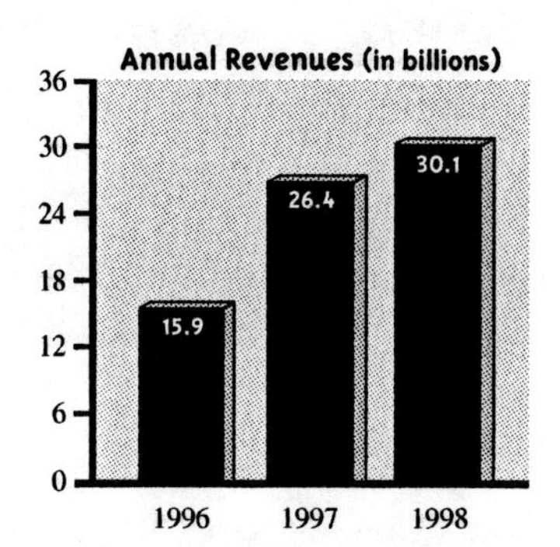

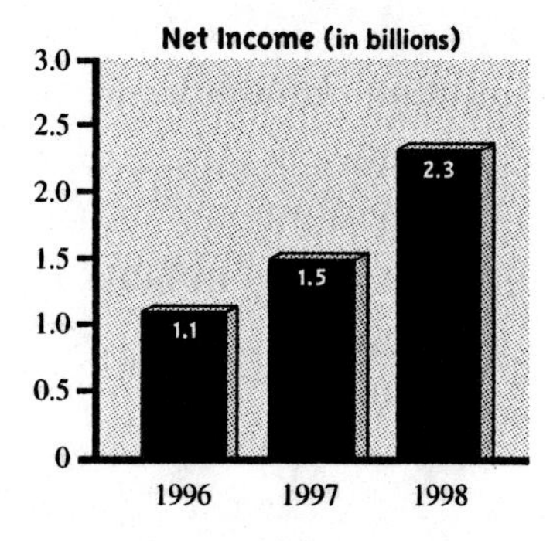

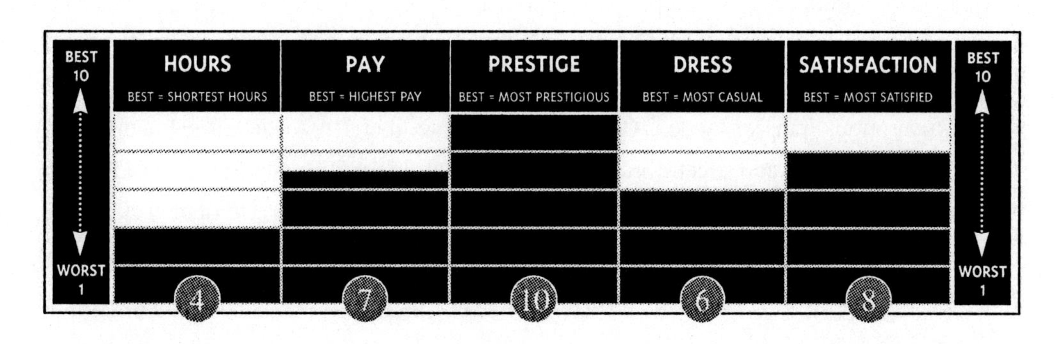

THE SCOOP

Lucent may be little more than a couple of years old, but this is one toddler that already knows how to run. Lucent Technologies is one of the world's leading designers, developers, and manufacturers of telecommunications systems and software. Until 1996, Lucent was part of AT&T; now it is already the nation's leading manufacturer of both the hardware and the software of global communications networks – business communications technology, telephones, wireless networks, and switching equipment. Lucent is now its very own firm, with more than 136,000 employees in 90 countries worldwide. (Twenty-five percent of Lucent employees work outside the United States.) Revenues in 1998 were a very healthy $30.1 billion, and net income reached $2.3 billion. By spring 1998, Lucent's market value actually exceeded AT&T's.

AT&T and Western Electric founded Bell Laboratories in 1925 to conduct research and development. Soon after WWII the company developed the transistor, for which its scientists won a Nobel Prize. The transistor's "switching" ability makes possible the zillions of super-fast calculations needed to run your average computer. That was merely the first in a string of inventions that include the communications satellite, the solar cell, and the UNIX computer operating system. During the 1984 break-up of AT&T, the corporate giant was allowed to keep its research and manufacturing facilities, which came under the name of AT&T Technologies. However, AT&T later decided to spin off Lucent in what was the largest initial public stock offering in U.S. history at the time. The new company consolidated its operations and sold off its non-core business to become even more competitive in this quickly changing industry.

By separating from AT&T in September 1996, Lucent has become free to do business with other telecommunications companies without conflicts of interest. As part of its efforts to boost its core businesses, Lucent acquired the smaller company Agile Networks in October 1996. Agile provides advanced intelligent switching products useful for the Ethernet, as well as asynchronous transfer mode (ATM) tech. In September 1997, Lucent nabbed Octel, which provides voice, fax and electronic messaging technologies that should dovetail nicely with those already offered by Lucent. And the company agreed in October 1997 to acquire Livingston Enterprises, a global provider of Internet equipment. To top off the recent shopping spree, Lucent bought Optimay, a German firm that makes software for cellular phones, for $65 million. The company has also won some high profile contracts, including a three-year $700 million pact with Sprint PCS, a $280 million agreement with AT&T owned Telecorps, and a $100 million deal with Venezuelan carrier Telecel.

To create quicker market response and enhanced customer focus, Lucent has recently reorganized to form smaller customer units. These units, not including Bell Labs, which remains its own discrete unit, are: Microelectronics; Data Networking Systems; Wireless Networks; Business Communications Systems; Optical Networking; Switching and Access Systems; Network Products; Communications Software; New Ventures; and Intellectual Property.

Despite Lucent's success so far, the company is feeling the heat of competition. In June 1998, Lucent sued Cisco, the top networking equipment company that has begun encroaching into Lucent's turf, alleging infringement of eight data-networking technology patents. Earlier that month, the company reached a new contract with 43,000 workers affiliated with the International Brotherhood of Electric Workers and the Communication Workers of America, ending a brief strike that would have crippled Lucent's operations.

GETTING HIRED

Lucent maintains an employment web page named "work@lucent" that can be reached at www.lucent.com/work/work.html. The web page provides access to current job listings, college recruiting information, and the online application program. After applicants complete their resumes online, Lucent keeps them in its "electronic files" and matches them with new job opportunities that arise. The web page also describes a Financial Leadership Development Program, a 24- to 30-month intensive program for those with degrees in accounting, finance, or other business-related fields. Lucent also offers internships and co-ops year round for interested students. Those who are looking for a position in a specific field, take note of an insiders explanation that "by sending a resume to HR, you trigger a process where your resume is scanned and sent through an optical character recognition system. Key words are pulled out, and anyone within the company seeking a matching keyword is sent the resume. For example: ATM, software, object technology, DSP." One insider says: "In the technical areas, college grads come in as STAs and move up to MTS (member of technical staff). Most retain the MTS title throughout much of their career, though about 10 percent become DMTS (distinguished members of technical staff). There is an effort to disguise levels of advancement (hence the common title) to avoid elitism in dealings between colleagues." Though the titles are the same, "salary continues to get both cost-of-living and merit raises. Salary varies by roughly a factor of three between new college grads and veteran DMTS employees, with the biggest raises in the earlier years of employ."

OUR SURVEY SAYS

Because Lucent has only recently become an independent company, it is still making the transition from a "tradition, technocentric view of the world" to a more "customer-focused and entrepreneurial climate." Employees say the company culture is "undergoing rapid change. In Bell Labs, the culture is moving from 'academic' to being high-relevance and having a big impact on Lucent's business." Not only is this shift "creating new opportunities every day," it has also helped make the company more "exciting" and "dynamic." Employees say that morale is currently "higher than ever before" and that they feel "empowered" to tackle the "thorny challenges" that are part of their work, thanks in part to their "universal access" to the "best in current technology." Individual empowerment at Lucent, however, does not mean working alone. Employees praise the "extensive exposure to executives and other upper management officers," the continuous feedback, and the way in which they are treated as "essential team players" from the first day on the job.

One longtime Lucent employee reports that "our staff comes from all over the world. There is a particularly large segment of our staff of Asian background, led by China, India, Bangladesh and Korea. Europeans, Canadians, African Americans, Middle-Easterners and Hispanics are also heavily represented." However, another insider says Lucent is "above average for large technical corporations for minority and female representation at most levels, but there are still few that have advanced to senior level positions. There are only ten Group Presidents, and of those only one is a woman."

The company offers a summer program for women and minority students. "Lucent is a very large company" one employee reports, "so the culture varies quite a bit. Bell Labs is the most relaxed area, very much like a university environment. There are lots of jeans, T-shirts and sandals in the summer, without a necktie or pantsuit in sight. But there are other areas of the company, especially sales, marketing and business areas, with direct contact with customers, where suit and tie for men and dresses for women are expected." The physical ambience "ranges from fair to pretty darn good, depending on whether you are at a 30-year-old building or a new one. Nowhere are we shabby."

Support for employees at Lucent varies somewhat. While department-level managers have secretaries "some male, some female," most employees "type their own reports." However, "everyone has a PC or terminal with shared printers nearby. Office supplies are usually available in a stockroom. We all have access to voice mail on our phone lines. Most have company-owned PCs at home and Lucent will pick up the cost of an extra phone line or long distance charges for network access from home." A Lucent manager tells Vault Reports that "in a recent survey of employees, the question 'Would you recommend Lucent as a good place to work?' got more than a 90 percent 'yes' response. That's very high for any company."

Lycos

400-2 Totten Pond Rd.
Waltham, MA 02154
(781) 370-2700
Fax: (781) 370-3400
www.lycos.com

LOCATIONS

Waltham, MA (HQ)
Chicago, IL
Dallas, TX
Mountain View, CA
Los Angeles, CA
New York, NY
Pittsburgh, PA
San Francisco, CA
Williamstown, MA
International offices in Germany, Italy, France, Japan, the U.K., Spain and The Netherlands

DEPARTMENTS

Accounting
Business Development
Finance
Human Resources
Marketing
Online Media/Editorial
Product Development
Programming
Research
Quality Assurance
Sales

THE STATS

Annual Revenues: $56.1 million (1998)
No. of Employees: 460
No. of Offices: 9
Stock Symbol: LCOS (NASDAQ)
CEO: Robert J. Davis

UPPERS

- Cutting-edge culture
- Good diversity
- Stock options
- Casual work environment

DOWNERS

- Industry uncertainty
- Low pay
- Some tension between technical and business sides

KEY COMPETITORS

- America Online
- Excite
- Infoseek/Walt Disney
- Snap!/NBC
- Yahoo!

EMPLOYMENT CONTACT

Human Resources
Lycos, Inc.
400-2 Totten Pond Rd.
Waltham, MA 02154

Fax: (781) 370-3415
jobs4you@lycos.com

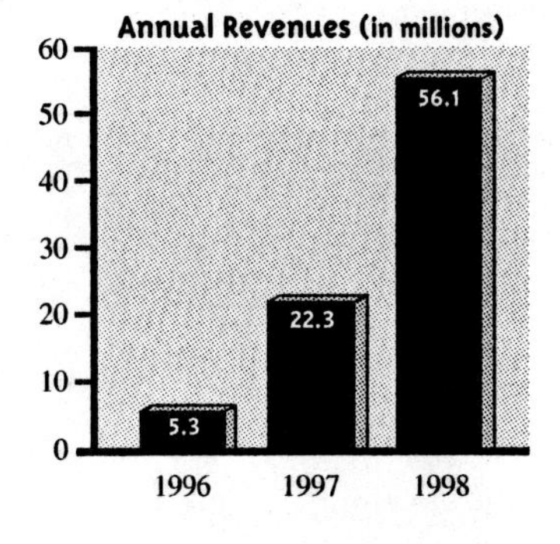

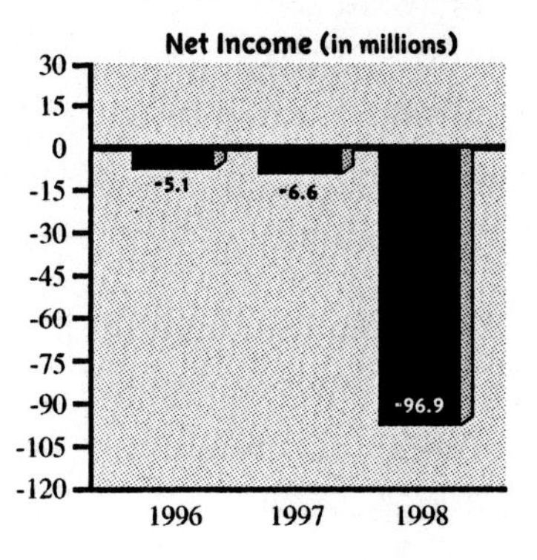

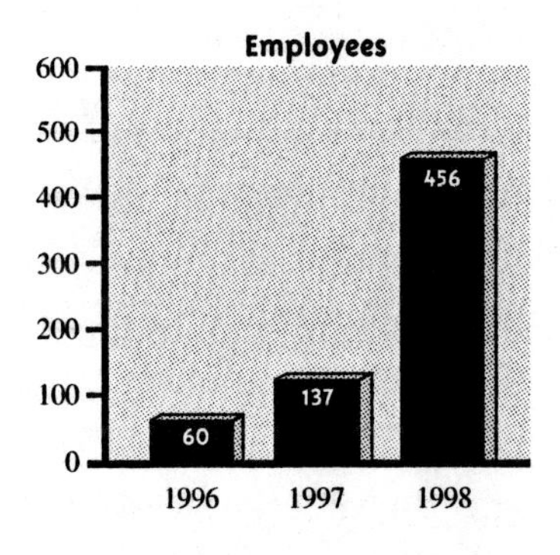

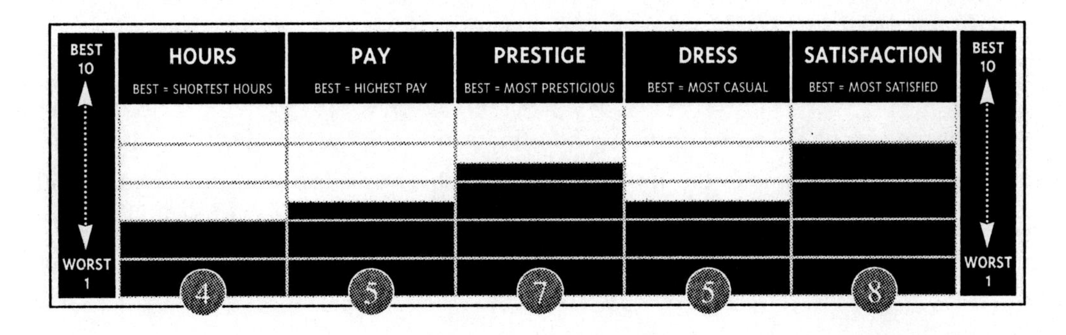

THE SCOOP

You'd need a pretty big black book to catalogue the Web's millions and millions of addresses – good thing there's Lycos. The man responsible is Michael Mauldin, who developed the technology for the Lycos search engine while he was a researcher at Carnegie Mellon's center for Machine Translation. A year later, in 1995, Microsoft and the Library Corp. licensed the technology and in June of that year CMG@Ventures took hold of the exclusive rights to the technology. Lycos, Inc. was formed and Robert Davis was named president and CEO. In April 1996 the company went public with an IPO of $177 million.

Sound like your normal Silicon Valley Internet story? Not quite. Unlike its major search engine-driven web counterparts, which are all headquartered onthe West Coast, Lycos makes its home in chilly Waltham, Massachusetts. And unlike its flashy competitors, Lycos is marked by efficiency, (what some might even call frugality) and an emphasis on the bottom line. This strategy hasn't hurt the firm's traffic, though – Lycos lags behind Microsoft by only 100,000 in monthly page views.

Lycos has used a variety of strategies to keep surfers visiting its pages, including acquisitions. In 1996 the company acquired Point Communications, the publisher of Point Reviews; you might recognize its "Top 5% Web Site" icon, which appears on many web pages. In 1998, the company bought Wired Digital, acquiring Wired's online magazine, *HotWired*, and its search engine, HotBot. Also in 1998, the company acquired Tripod, a site which boasts the largest number of home page builders, with 3.2 million members. The company believes that its focus on communities, which keep surfers at a site rather than sending them elsewhere, gives it a strategic edge over its competitors. And of course, Lycos is not willing to be left out of the advertising game – the company unveiled a $25 million advertising campaign in 1998 featuring a black labrador called Lycos that, guess what, can go get stuff.

As its competitors cut mega-deals – most notably Infoseek's recent partnership with Disney – Lycos remains on the prowl for traffic-boosting partnerships and acquisitions. In 1996, Lycos secured its status among the top four web search companies when it closed a deal with Netscape which would guarantee 700 million referrals a year from the Netscape Site. And Lycos has also cut deals with AT&T and German media giant Bertelsmann. Bertelsmann will create Lycos web sites in Europe; through a deal with AT&T WorldNet, Lycos announced that it would be able to introduce its own Internet access service, Lycos Web.

GETTING HIRED

Click "Lycos Home" on their web site, www.lycos.com, and then go to Employment Opportunites. There, you'll find several job descriptions and the skills you need to qualify. To apply, click "To Apply" at the bottom of the page. One insider at Lycos' Pittsburgh technical center says: "We decide really quick if we want someone or not. We ask difficult technical questions, and don't expect candidates to know all the answers."

OUR SURVEY SAYS

"The pace is fast, the hours are long, the pay is low," says one Lycos insider. Sounds good, eh? There's more to it, though, so don't turn the page. The culture is "cutting edge" and the experience of working in an Internet business is "invaluable." Says a source: "A person's new ideas will be considered and, if good, implemented. This means that your thoughts and ideas can and will help guide the company." The people at Lycos are said to be "very bright and very motivated to get things done." And the work satisfaction is high. "The number of people that use what I develop is incredible, millions of people a day," a Lycos engineer says. "Not many people can say that."

Though the "base pay is low," employees have an "objective bonus system" that makes things "competitive" (e.g., you meet your "objective," you get decent "bonuses"). "Stock options at a very nice price are also a big plus," says one contact. "But sometimes meeting your objective can be time consuming. "The Internet business is very fast-paced; projects and deadlines tend to be very quick," an insider says. "This does have the downside that you do have to put in some longer hours (45 to 55 hours a week) on a regular basis and every once in a while, very long hours to get a project done." One engineer reports working 50 to 60 hours a week. In addition, a contact says, "The suits who run the place are in Boston, but all the real work gets done in Pittsburgh, so right there you've got a major political problem that causes no end of headaches."

The dress code at Lycos is "varied." When you "deal with customers face-to-face, then it's formal business attire." Otherwise it's "casual, though higher-quality button-down shirts, slacks." Lack of diversity isn't an issue, as Lycos employs "many women and others who would claim to be a minority group." Our contacts don't report "any bad treatment" and "there haven't been any complaints about improper behavior." "Opportunities to work across borders" and "travel" are "major perks" of the job. For "lots of cutting-edge, high-tech experience," you can "do no better" than Lycos, says one insider.

"The stock options make me nervous. Either this company will be the biggest thing going in a few years, or it won't even exist."

– Internet insider

Macromedia

600 Townsend Street
San Francisco, CA 94103
(415) 252-2000
Fax: (415) 252-2348
www.macromedia.com

LOCATIONS

San Francisco, CA (HQ)
Berkshire, United Kingdom • Victoria, Australia • Quebec, Canada • Tokyo, Japan • as well other international sites

DEPARTMENTS

Information Technology
Interactive Learning Division
General Administration
Instructional Media
Marketing
Product Development
Product Management
Sales
Technical Support
Web Publishing
Web Services
Web Traffic

THE STATS

Annual Revenues: $113.1 million (1998)
No. of Employees: 550 (worldwide)
No. of Offices: 10 (worldwide)

Stock Symbol: MACR (NASDAQ)
CEO: Robert K. Burgess

UPPERS

- Stock options
- Gym discounts
- Software at cost
- Office pool tables
- Educational reimbursement
- Massive in-office slide

DOWNERS

- Long workdays

KEY COMPETITORS

- Adobe
- Corel
- Micrografx
- Microsoft
- Netscape
- RealNetworks

EMPLOYMENT CONTACT

Lori Duffy
Recruitment Manager
Macromedia
600 Townsend Street
San Francisco, CA 94103

Fax: (415) 626-0554
mmjobs@macromedia.com

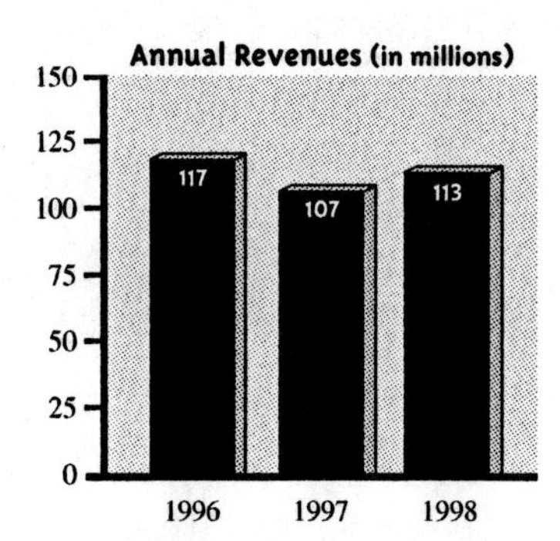

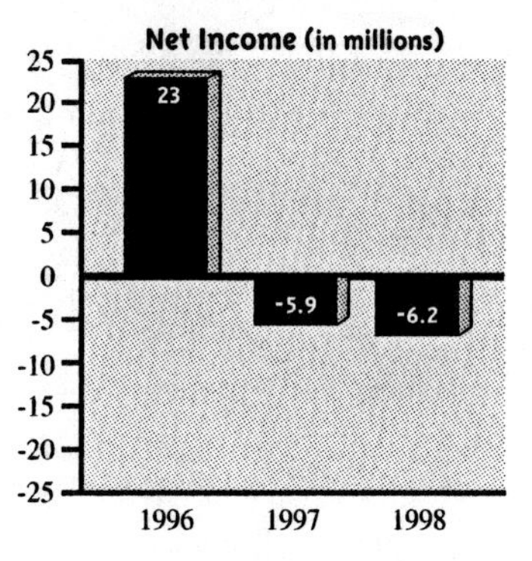

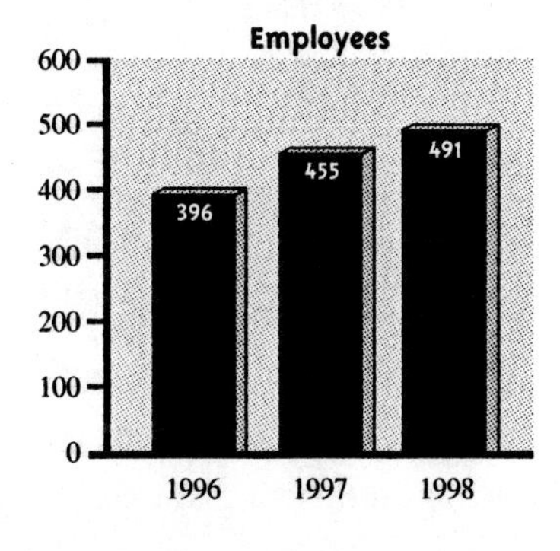

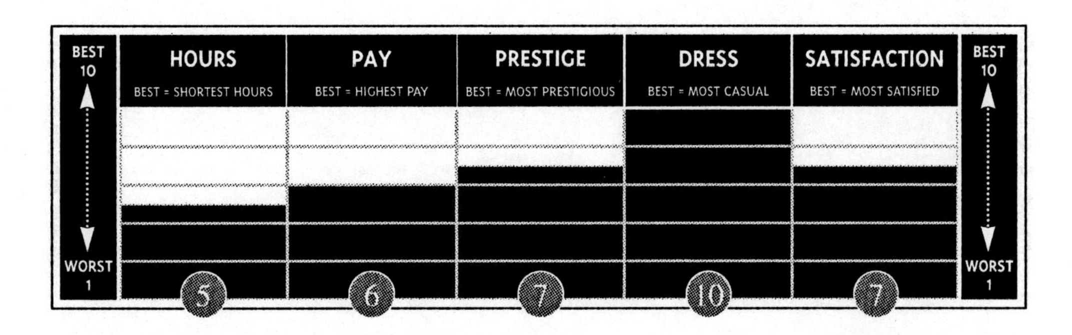

THE SCOOP

An industry leader in the development and marketing of multimedia, graphic, and video software for the World Wide Web, Macromedia is at the forefront of the "interactive" revolution. Formed by the 1992 merger between Authorware and Macromind/Paracomp, Macromedia makes software tools currently employed by more than 500,000 professional developers and more than 2 million business users.

What separates Macromedia apart from many other software players is that its product managers also function like business managers, monitoring the product's movement to the consumers as well as it original production, giving the managers fiscal responsibility for the products they create.

The company's flagship line includes web publishing tools such as Macromedia FreeHand and Macromedia Director, and tools built specifically for the Web, including Flash, Dreamweaver and Fireworks. The company's prominent Shockwave Players allow for stunning web animation and sound to stream smoothly into computer desktops. The software has received increased attention as availability of high-speed modems to support such applications widens. The systems have been used to develop World Wide Web pages for General Motors and Black & Decker. The Shockwave systems are better known for their use in play than in work, though – the systems were used for Tom Cruise's *Mission: Impossible*, and Comedy Central's *South Park* TV show. Teetering on the cutting edge of animation and web design, though, does not ensure a profit – Macromedia lost money in both its 1997 and 1998 fiscal years. Things seem be on an upswing for the multimedia company, however, with the resurgence of working-partner Macintosh, which still has a healthy share of the desktops using Macromedia publishing tools.

GETTING HIRED

"One of the guys I interviewed with was a guy with bright red dreads and skater duds," recalls one insider. For non-sales jobs, insiders advise interviewees "not to wear a suit – everyone wears jeans and T-shirts." This doesn't mean that the interviews are entirely relaxing. Insiders report that some sessions can run "six hours long."

Macromedia accepts resumes by both e-mail and regular mail. The company prefers that applicants sending resumes electronically format them as Microsoft Word files and keep any attached file to under 50KB in size. A leader in web innovation, Macromedia's own employment web site, located at www.macromedia.com, provides regularly updated information on current job opportunities.

OUR SURVEY SAYS

"Macromedia is a company that recently grew from less than a hundred people to around 550 and survived," describes one insider of the company's corporate culture. Pool and ping-pong tables, video games, and monthly free lunches set the tone for the "informal" and "progressive" corporate culture at Macromedia. "The software we create at Macromedia is some of the best around," one contact discloses, "and the company gives us these toys to help keep the creative juices flowing." At times, though, some insiders feel that "the bureaucracy can get in the way." But for most, the environment is stimulating rather than stifling. As one employee says: "There are lots of artists and musicians and the engineering talent is amazing. I've never been around so many brilliant people in my life." There are also "a lot of young people working here." Employees also appreciate the "cultural and intellectual diversity" which includes "everyone from yuppies to punkers to rappers." At Macromedia there are "3 R's" to live by: "Act respectfully, be responsive and get results."

And the dress code? "It sorta depends on which department you work in. Obviously people in marketing have to dress the part" but in development, referred to as "dev," employees are required to wear no clothes "except those required by law." Hours are "like any company doing development." Macromedia employees "usually work normal hours for most of the year." During production time, "they may be as much as 70 hours a week or more." "It can get pretty demanding," one employee admits. As for pay, insiders tell us that stock options "are given to all full-time employees." A recently hired employee comments that she joined the company partly because "Macromedia was willing to pay me exactly what I wanted."

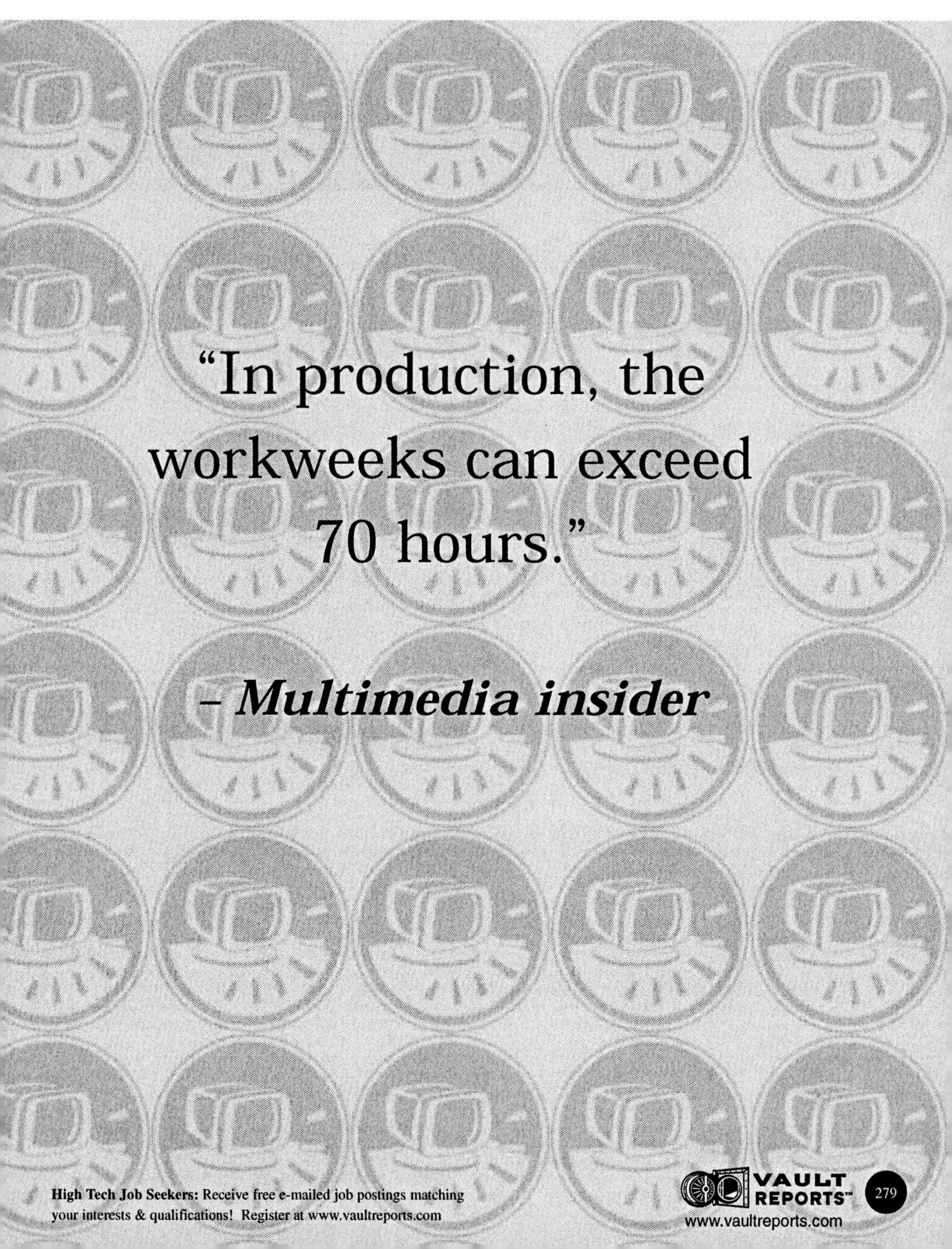
"In production, the workweeks can exceed 70 hours."
– Multimedia insider
High Tech Job Seekers: Receive free e-mailed job postings matching your interests & qualifications! Register at www.vaultreports.com
VAULT REPORTS™
www.vaultreports.com

Micron Technology

8000 South Federal Way
P.O. Box 6
Boise, ID 83707-0006
(800) 9-MICRON
Fax: (208) 368-4435
www.micron.com

MICRON
TECHNOLOGY, INC.

LOCATIONS

Boise, ID (HQ)
Durham, NC
Minneapolis, MN
Nampa, ID
Santa Clara, CA
Penang, Malaysia

DEPARTMENTS

Administration & Clerical
Information Technology
Manufacturing
Purchasing
Retail
Sales & Marketing
Technical

THE STATS

Annual Revenues: $3.0 billion (1998)
No. of Employees: 11,400 (worldwide)
No. of Offices: 5 (worldwide), 4 (U.S)
Stock Symbol: MU (NYSE)
CEO: Steven R. Appleton

UPPERS

- Tuition subsidies
- Profit sharing
- Company gym
- On-site medical care

DOWNERS

- Internal management disputes
- Falling profits
- Below market pay

KEY COMPETITORS

- Acer
- Alliance Semiconductor
- Apple Computer
- Cypress Semiconductor
- Dell Computer
- Gateway
- IBM
- Samsung
- Sun Microsystems

EMPLOYMENT CONTACT

Denise Smith
Micron Technology
8000 South Federal Way
P.O. Box 6
Boise, ID 83707-0006

(208) 893-4425
Fax: (208) 893-7333
denisesmith@micronpc.com

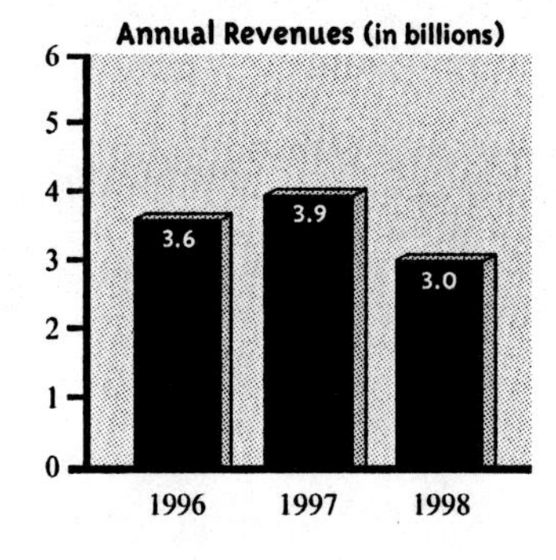

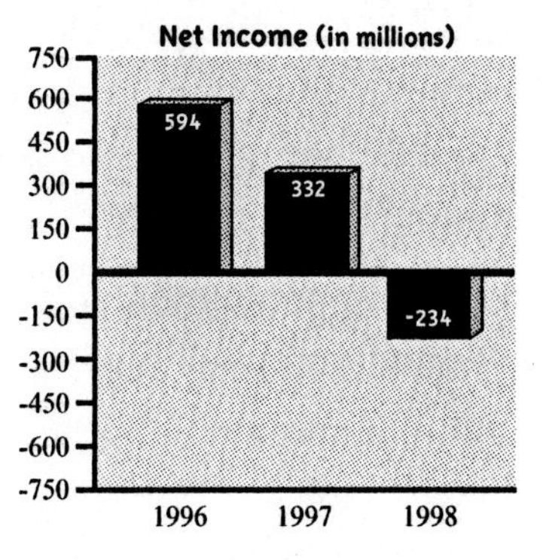

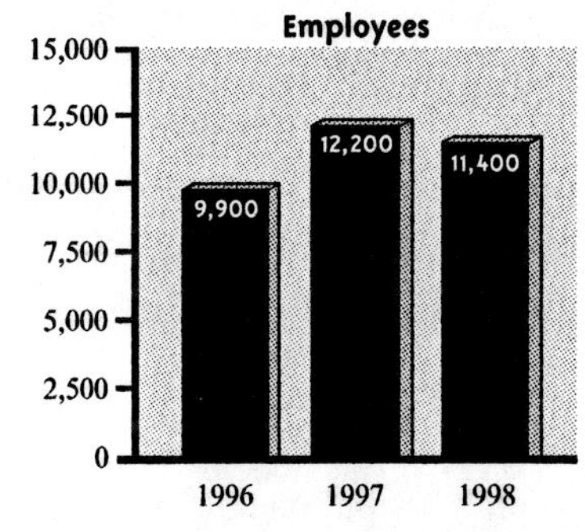

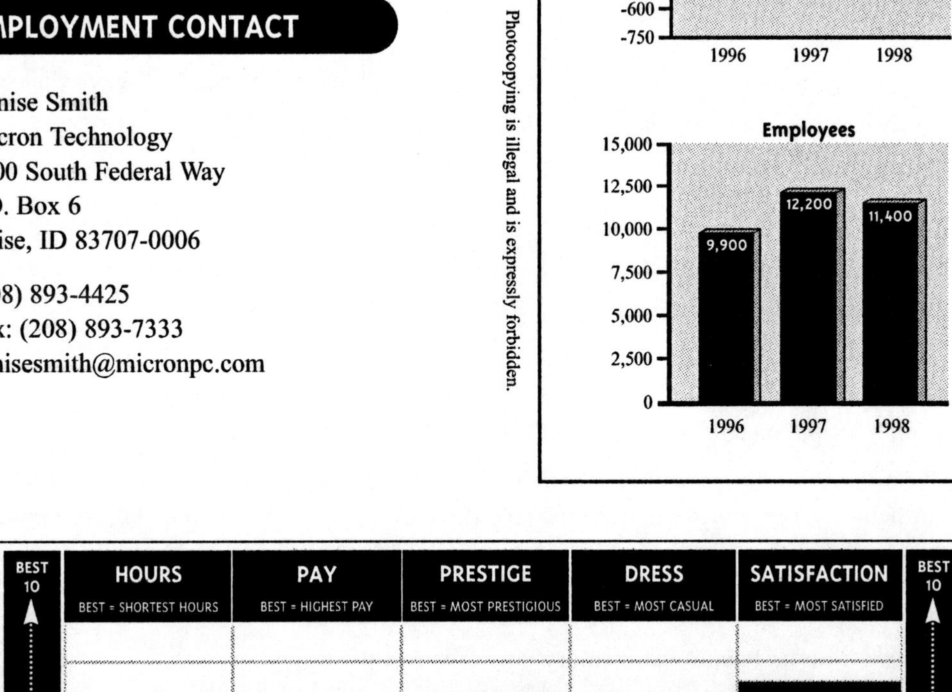

THE SCOOP

Micron Technology is the U.S.'s leading manufacturer of dynamic random-access memory (DRAM), a semiconductor memory chip. Founded as a computer consulting firm in an Idaho dentist's basement, Micron sold its first computer memory chips in 1982, and went public in 1984. In that same year, chip-dumping by Japan (selling chips at below cost in order to force out competitors, a recurring problem that has grown to include Korean firms as well) nearly drove the firm out of business. In the early 1990s, Micron benefited from an increase in chip demand but has diversified and moved into the less volatile PC market.

Micron is affiliated with a myriad of companies, most of which confusingly bear some version of the Micron name. One of the newest and most successful of these is Micron Electronics (MEI), founded in 1995 through a merger of ZEOS International, Micron Computer, and Micron Customer Manufacturing. Micron Electronics, which is majority-owned by Micron Technology, is the No. 3 direct-sales computer maker behind Dell and Gateway. The company, which has grown rapidly, launched a new advertising campaign in August 1998 targeted at mid-sized U.S. businesses.

Recently, Micron has been plagued by internal struggles that resulted in the departure of the company's three top executives. In addition, Micron's core business, has significantly weakened in the face of renewed competition. Demand for DRAM has historically been based on the vagaries of the PC market, though new applications now exist for chips in digital cameras and 3-D games. Awaiting settlement of its internal disputes and a rise in chip demand, Micron has put on hold its new $2.5 billion chip making facility in Lehi, Utah, announcing that "the Company will not outfit or equip the Lehi facility until market conditions warrant." Furthermore, the company, citing the chaotic chip market, refused to make sales predictions. Despite these dire market conditions, employees say the "leadership of Micron is progressive and determined to make the company one of the top five computer marketers in the world."

In 1998, Micron bought the memory-chip business of struggling Texas Instruments for $800 million, well below the market price. The deal, which made Texas Instruments Micron's largest shareholder with a 12 percent stake, pushed Micron past Samsung into the top spot in the memory chip business.

GETTING HIRED

Micron's employment information is listed in what must be one of the most comprehensive career information pages on the Web. Consult www.micron.com for information on Micron's subsidiaries and for links to each subsidiary's job opportunities, benefits, and qualifications. Applicants should note that each subsidiary – as well as the different plants within each subsidiary – has a separate postal and e-mail address to which resumes can be sent. Please note that Micron Electronics and Micron Technology (MTI) have different human resources departments. Employees say that Micron "will accept resumes, but still requires an application for serious consideration."

OUR SURVEY SAYS

Insiders at this "laid-back" and "easy-going" computer company call the benefits "generous" and the pay scale "competitive." "We all get our work done," says a support engineer, "but it is not a Dilbert environment." Micron "works hard" to "guarantee employee satisfaction," and its internal promotion policy provides "ample" advancement opportunities "even from the ground floor." Bonuses and stock options sometimes even "surpass the base salary" and the company "actively encourages" employees to continue their education by paying for it. One employee cites Micron's support of the handicapped as a plus. Others point out that "Micron has a large Hispanic and Asian population." However, Micron pay "is a little under the market, and DRAM dumping has hurt."

Micron's location in bucolic Boise is a plus for some employees. "I like the family environment – low crime, outdoor activities, low population density, friendly people," says one recent hire. Also, "several companies in the area offer Micron employees discounts." Another employee says: "People are honest here – it's an Idaho thing." Better like Idaho a lot if you want to work for Micron, because "anyone who wants to climb the corporate ladder will have to live there." Good perks may alleviate the potato-bound boredom. "There is medical, vision, dental and life insurance. There is a 401(k) plan under which Micron will match contributions up to $1500 a year. There is also profit sharing, which rises and falls with the condition of Micron and the semiconductor market. Two years ago was a tough year, both for Micron and our bonus packet, but 1997 was better." The excellent health facilities include "a new gym, and medical staff. You don't have to pay anything to use the gym, and the medical staff only charges $10 to treat you." Employees in manufacturing may work a compressed workweek, with three 12-hour shifts one week and four such days the next, while management, marketing and engineering employees work traditional 8 to 5 days. "They do try to be fair to the hard workers," says one employee, "and I have even seen a few discretionary bonuses. There is also a good stock purchase plan. Under this plan, each employee gets 100 shares at the lowest point that the stock was in the past year."

The dress code is "pretty basic and reasonable – pants or skirts (no shorts or microminis), no tank tops, and no tattered clothing." Micron will pay for a college or graduate degree "if you want to go for it while you're working, which I'm currently doing," says one studious employee. "Micron keeps paying as long as you keep a C average." While Micron has grown dramatically in the past few years, says one employee, "a sense of entrepreneurialism can still be rewarded. The work environment is moderately casual, and we don't place a lot of emphasis on office status. Our CEO sits in a cubicle."

"We don't place a lot of emphasis on office status. Our CEO sits in a cubicle."

– Hardware insider

Microsoft

One Microsoft Way, STE 303
Redmond, WA 98052-8303
www.microsoft.com

Microsoft®

LOCATIONS

Redmond, WA (HQ)
Houston, TX
Numerous plants and software development centers throughout the world

DEPARTMENTS

Operations
Research & Development
Sales & Support

THE STATS

Annual Revenues: $14.5 billion (1998)
No. of Employees: 22,232 (worldwide)
No. of Offices: 60+ (worldwide)
Stock Symbol: MSFT (NASDAQ)
CEO: William H. Gates

UPPERS

- Amazingly lucrative stock options
- Palatial work campus
- Highest-level work
- Flexible scheduling

DOWNERS

- Fearsome lawsuit
- Little training
- Disliked by other high-tech companies

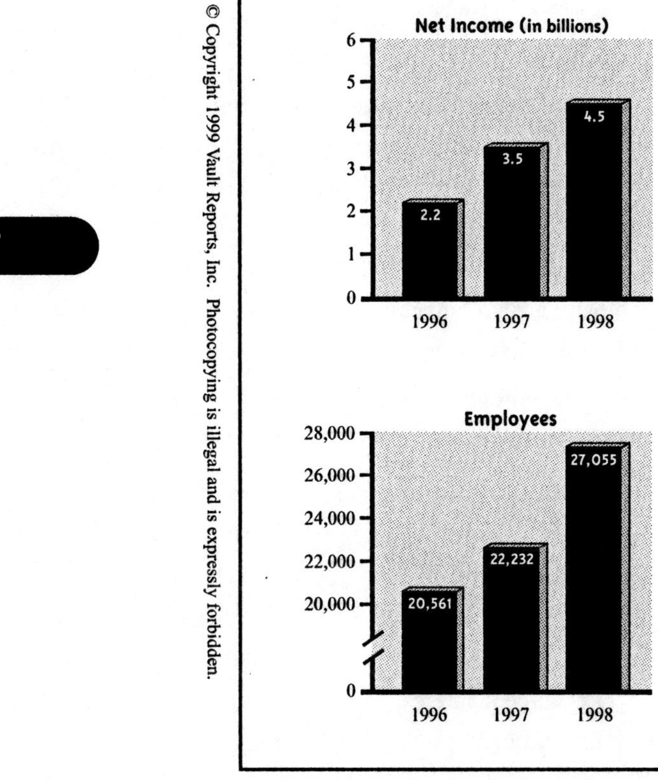

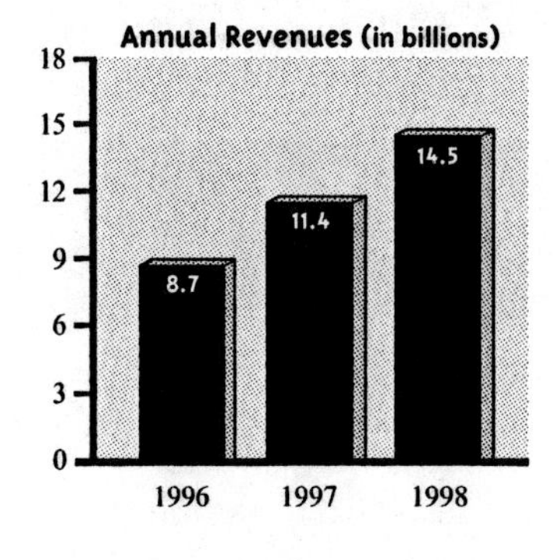

KEY COMPETITORS

- America Online
- Apple Computer
- AT&T
- Hewlett-Packard
- IBM
- Oracle
- Sun Microsystems

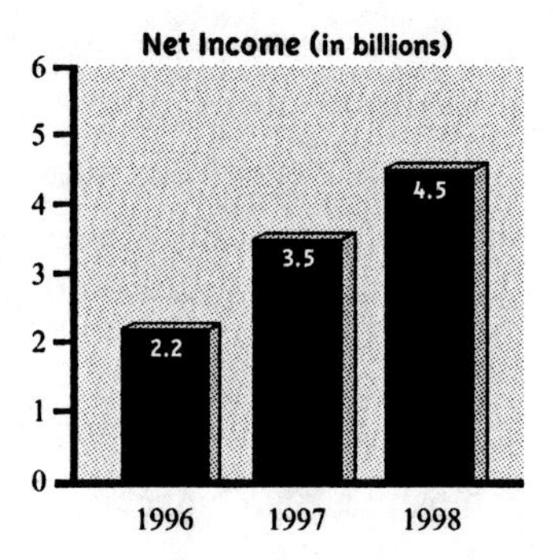

EMPLOYMENT CONTACT

Microsoft Corporation
Attn. Recruiting,
One Microsoft Way, STE 303
Redmond, WA 98052-8303

resume@microsoft.com

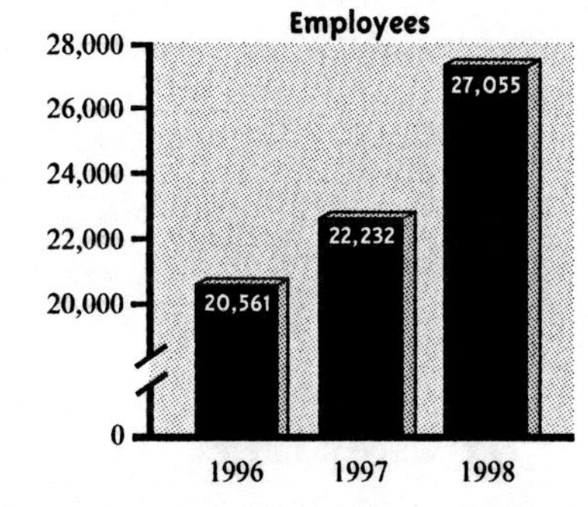

BEST 10 ↑ … ↓ WORST 1	HOURS BEST = SHORTEST HOURS	PAY BEST = HIGHEST PAY	PRESTIGE BEST = MOST PRESTIGIOUS	DRESS BEST = MOST CASUAL	SATISFACTION BEST = MOST SATISFIED	BEST 10 ↑ … ↓ WORST 1
	3	7	10	10	9	

THE SCOOP

Microsoft is undeniably the world's top software company. Microsoft has a wide array of products, from operating systems for personal computers to interactive media programs. Over 90 percent of PCs sold run on Microsoft operating systems. In addition to PC systems Windows 95 and Windows 98, systems like Windows NT run the networks of companies around the globe. From spreadsheets to word processing, Microsoft dominates the applications market as well. Microsoft was the first software company to bundle its applications into a single suite, taking Microsoft Word, Excel, and PowerPoint, and marketing them in the package deal known as Microsoft Office. Not a bad idea – Office now accounts for about 30 percent of the company's revenues. Microsoft plans to release the next version, Office 2000, during the first half of 1999 (though, as is Microsoft's wont, that deadline keeps receding). The company is also branching out into areas in which it traditionally has not had a strong market presence, such as database management. Stepping up its efforts to compete with industry leaders Oracle and Informix, Microsoft has launched the SQL Server 7.0 to boost the already significant BackOffice product line.

One of Microsoft's greatest assets is, well, its assets. With nearly $14 billion in cash and short-term investments at the end of fiscal 1998, and over $22 billion in total assets, Microsoft has a great deal of freedom to pursue new business interests. CEO Bill Gates is now flexing his monetary muscles with an amazing acquisition spree that has spanned the past few years and shows no sign of letting up. Since 1995, Microsoft has dipped into its petty cash drawer to buy an 11.5 percent interest in cable company Comcast, worth $1 billion; a 15 percent stake in UUNet, an Internet service provider; a 5 percent stake in VDONet, an Internet video company; 10 percent of Progressive Networks (now called Real Networks), a web audio company; and WebTV Networks in its entirety, a purchase worth $425 million.

Such impressive figures do not mean, however, that Microsoft has no troubles to overcome. One of the biggest threats to Microsoft's future is the pesky Java Internet software language created by Sun Microsystems. Designed to function with any operating system (or "platform"), Java breaks programs up into small "applets" that allow for swift Internet transmission. This should allow centralized servers to do much of the work currently handled by installed software (a.k.a. Microsoft's bread and butter). Gates may live to rue the day in 1995 when he announced that Microsoft would be Java-compatible, in order to fight the relatively small threat of Netscape, the same way that IBM probably regrets allowing MS-DOS to be licensable to PC-clone producers. Now Sun is suing Microsoft over the company's use of Java in its 4.0 version of Internet Explorer, claiming that Microsoft's program fails its compatibility tests.

To Gates and laissez-faire capitalists everywhere, a far more insidious challenge is coming from the United States government. In May 1998, the Department of Justice and a coalition of 20 state attorneys general charged Microsoft with engaging in monopoly business practices in violation of antitrust laws. By integrating any new technological innovation into Windows, contends the DOJ, Microsoft unfairly limits competition. The suit claims that by incorporating its Office suite into its operating system, and then making it financially impractical for PC manufacturers to include a rival product, Microsoft used its market dominance to crush the competition. And even more important to the future of computing, Microsoft bundled its Internet Explorer with Windows software, thereby damaging rival Netscape Communications Corp. Microsoft argues that the browser is integrated into Windows 98, and that removing it will cause irreperable damage to the operating system. To bolster its case, the DOJ has incorporated evidence intended to establish that Microsoft had engaged in a pattern of anticompetitive practices dating back to well before the Netscape controversy began. Now that the two sides are finally seeing the inside of a courtroom, the eventual outcome (whatever it is) will be sure to send shockwaves throughout the high-tech world, although some suggest that America Online's acquisition of Netscape weakens the government's case.

GETTING HIRED

Even though the company has only 20,000 employees Microsoft receives more than 12,000 resumes a month – this at a company with a high retention rate; many employees stay for over a decade. But Microsoft is also a company that is constantly growing, and that demands regular new talent to keep its products fresh. As top dog, Microsoft hires a great deal of its employees through recruiters (known as Strike Team in Recruiting – they take this seriously), rather than through applications. Gates himself scouts a few graduating seniors every year – the four or five standout computer science undergraduates from around the country. For non-entry level positions, Microsoft human resources can lure the best performers at other software companies with hefty salaries and attractive stock options.

Is there any hope for the unheralded, the resume-bearer with some heavy bond paper and a dream? The answer: Yes, but don't get your hopes up too high. College and MBA students with a year or more of school remaining should really consider applying for an internship. It's the perfect way to get your foot in the door and prove that you can produce for the company. Otherwise, hit the web site or campus recruiters with your resume and hope they call you back. If they call you, simply prove to them that you're an A-1 genius. Microsoft wants to hire brilliant people; part of its recruiting mission statement reads: "The company can teach employees specific skills but it cannot instill intelligence and creativity."

Undergraduates interested in technical positions should be pursuing degrees in computer science or a related discipline, and be highly proficient in C/C++. Internships or work experience in programming are helpful but not necessary. Microsoft's policy of hiring the brightest people it can find means certain sacrifices; Gates' bias is "toward intelligence or smartness over anything else, even, in many cases, experience," insiders say. For MBAs aiming at finance and marketing spots, degrees from highly competitive schools are a must, and real experience in the software industry will get you far. Gates disparages employees who cannot follow the technical side of the business. This does not mean that every marketing whiz is expected to know how to write sharp code, but excellent quantitative and analytical skills are a must.

OUR SURVEY SAYS

"The culture here is focused around doing great work, but it is also a very fun place to be," says one insider. The atmosphere is far from stuffy, leaning, if anything, toward the congenial: "We have a basketball hoop in our hall, which is a great way to blow off steam in the evenings or the middle of the day," says one insider. While average citizens probably would like to write off software programmers as lonely and uncommunicative, Microsoft insiders disagree. One designer says: "The atmosphere is very social – a good-sized chunk of my day is spent in other people's offices or with other people in my office discussing how to accomplish some task or the latest industry news." And while the company has swiftly grown to a $14.5 billion monster, insiders say they "don't feel lost." "We work together in small feature teams, so even though this is a large company, you always feel like an important part of a team," says one Microsoft veteran.

If there's one thing surveyed Microsoft employees universally applaud, it's the "free sodas, all the soda you can drink!" Another popular perk is the "cheap eats" at the subsidized Microsoft cafeteria where, "the food is good and, for the price, great." Other insiders are a bit more serious-minded. One source cites the "package with salary, stock options, 401(k) matching, employee stock purchase plan, and more." Location can be a plus since "living in Seattle is a perk alone. If you're not affected by gray skies, it's a beautiful place." The Microsoft campus itself sports "lots of lawns, and basketball and volleyball courts." If a more structured exercise environment floats your boat, don't worry, Microsoft pays for employee gym memberships. Employees also cherish the "10 to 20 percent discounts on products and services" at businesses around Seattle.

According to Microsoft employees, the dress code is "non-existent." As one insider puts it: "If you're a hot-shot programmer, no one's going to stop and say, 'But he's wearing sandals!' Get real." In fact, the company insists on its web page that it doesn't even matter what you wear to your job interview. Take this, however, with a grain of salt. Dress is still context-specific: "We wear whatever we want to work (shorts, jeans, T-shirts are most common), but this varies according to what job you do. Marketing people and lawyers dress up more than software developers."

Motorola

1303 E. Algonquin Rd.
Schaumburg, IL 60196
(847) 576-5000
www.mot.com

LOCATIONS

Schaumburg, IL (HQ)
Other locations in AZ • FL • GA • NC • NM • NY • TX • and 17 foreign countries

MARKETS

Wireless
Semi-conductors
Advanced Electrical Systems
Components and Servers, including:

- Cellular
- Two-way Radio
- Data Communications
- Personal Communication
- Automotive, Space, Defense Electronics
- Computers

THE STATS

Annual Revenues: $29.8 billion (1997)
No. of Employees: 135,000 (worldwide)
Stock Symbol: MOT (NYSE)
CEO: Christopher Galvin
%Male/%Female: 64/36

UPPERS

- On-site childcare
- Adoption assistance
- Scholarships for employees' children
- Tuition reimbursement
- Flexible scheduling
- Credit union
- Tickets to sporting events

DOWNERS

- Recent poor performance and layoffs

KEY COMPETITORS

- 3Com
- Ericsson
- IBM
- Intel
- Lucent
- Nokia

EMPLOYMENT CONTACT

Human Resources
Motorola
1303 E. Algonquin Rd.
Schaumburg, IL 60196

www.mot.com/employment

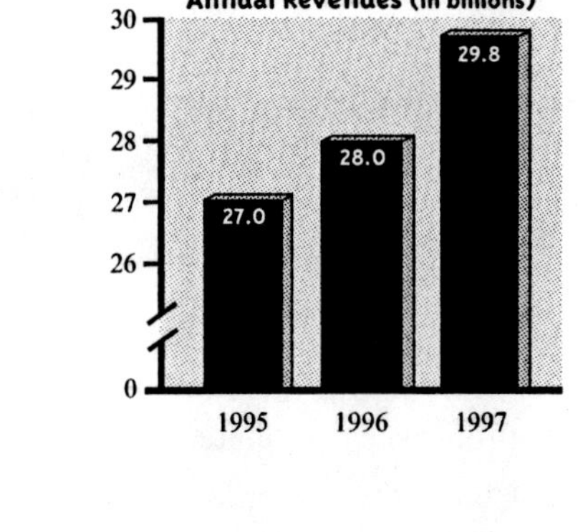

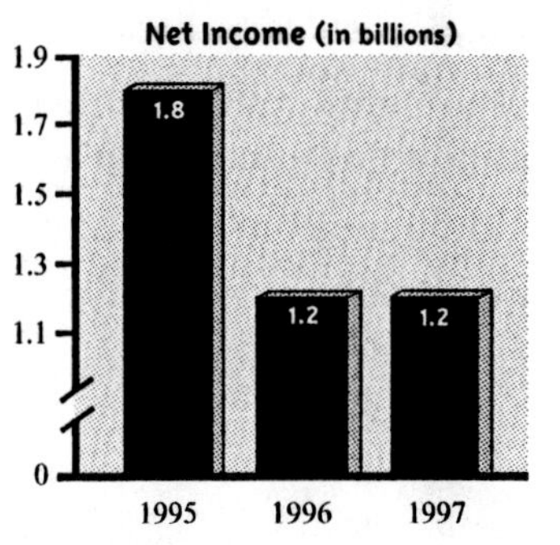

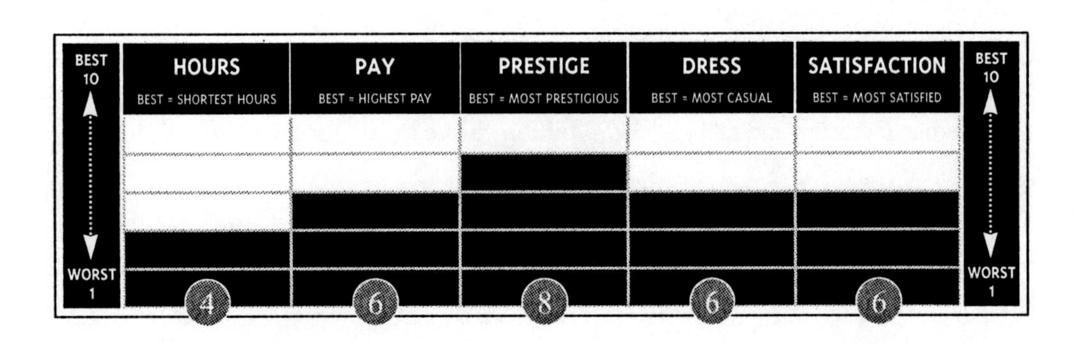

THE SCOOP

Motorola is the world's leading manufacturer of cellular phones, beepers, and two-way radios. Its products are so ubiquitous that its name has become synonymous with telecommunications; in China, for example, cellular phones are simply called "Motorolas." The company that would eventually become Motorola was founded by Paul Galvin in 1928. As the first manufacturer of two-way, hand-held radios, as well as a pioneering maker of semiconductors and car stereo equipment, the Galvin family's business has over the years earned 1016 patents; 403 of these were registered between 1991 and 1995. The company is one of the top three patent holders in the country.

Over the past three years, however, Motorola has not performed well. In February 1998, Prime Communications L.P, a wireless-phone service owned by Bell Atlantic, U.S. West and AirTouch Communications, cancelled a $500 million contract with Motorola after persistent failures with the company's cellular network equipment and software. In one embarrassing episode, PrimeCo's cellular network went down in Chicago as executives from both companies were meeting to discuss the network's problems. PrimeCo's action could have a significant spillover effect: Motorola supplies similar equipment and software to other large companies, including a $1.4 billion order from Sprint PCS and $3.5 billion agreement with two Japanese firms.

The failed PrimeCo contract illustrates a problem that has led to Motorola's dwindling market share and profit margins: the company's technology is simply outdated. Motorola had built its market reputation on its analog cellular technology, only to see the industry and its customers go digital. The company also lacks its own switching and network software like Lucent and Nortel, leaving Motorola unable to compete for telecommunication contracts. As a result, Motorola has seen its worldwide cellular systems market share drop to 15 percent from more than 25 percent. In June 1998, Motorola officials announced a sweeping restructuring and consolidation plan, laying off 15,000 employees (10 percent of its workforce), closing numerous factories, and taking a $1.95 billion charge to pay for the plan.

Despite a weak Asian economy that has contributed to the company's financial woes, Motorola still plans to push full steam ahead in the continent it once dominated, and where it is still a major player. Construction continues on a $750 million plant in China, with several factory upgrades planned for the future. Motorola's aggressive pursuit of Asia, even as the company closes its U.S. plants, reflects CEO Christopher Galvin's gamble that the region will still be the most promising market for cellular technology and other Motorola products despite the global financial crisis.

GETTING HIRED

Motorola's operations are highly decentralized and organized into several different. Motorola's web site, located at www.mot.com, provides information about current opportunities in each of these units. Applicants should be careful to direct their resumes to the appropriate one.

Motorola hires top MBAs primarily into strategy/corporate planning and marketing positions. In marketing, Motorola utilizes the traditional brand management structure, with MBAs starting as assistant marketing managers. Both internships and full-time positions at Motorola can take MBAs around the world. Incoming employees "often have the opportunity to go on rotational programs to get a feel for the different business units before choosing one to work in permanently."

OUR SURVEY SAYS

Motorola is an "engineering and product-oriented" company that encourages "creative solutions to difficult problems." While those in finance and marketing sometimes feel that the technical staff keeps them "at arms length," everyone appreciates the "prestige of working for a company whose name transcends language barriers." Employees say that Motorola's decentralized organization "occasionally causes communication difficulties," but also "allows for a significant degree of departmental autonomy without excessive red tape." The dress code is described as "casual, but more professional than most high-tech firms." There are no dress codes "anywhere in engineering," but "business dress in finance and marketing."

Although one contact describes co-workers as "very cool," and the dress as "very casual," he points out that to advance in such a big company, "one needs to be very mindful of managing his or her career there. The advice to find a mentor is freely and amicably given, but there is no real process to make sure one actually finds a good mentor. Without one, many employees still do fine but find measurable career advancement often unreachable."

MBAs who have spent summer internships at the company report a variety of experiences. "It had its ups and downs," says one, who worked at the company's headquarters. "The work was interesting, but I do not believe I received adequate exposure to upper/middle management and I don't think I was given enough responsibility." Another, who worked overseas reports: "It was very unstructured, so I was able to be creative. There was lots of exposure to the CFO." Another, who worked in Asia, agrees: "It turned out to be a highly unstructured experience – a very steep learning curve and substantial exposure to upper/middle management."

"You should expect that you'll work some overtime, without compensation, and without complaint."

– ***Hardware insider***

Netscape Communications

501 E. Middlefield Road
Mountain View, CA 94043
(650) 254-1900
Fax: (650) 528-4124
home.netscape.com

NETSCAPE

LOCATIONS

Mountain View, CA (HQ)
Offices in every major U.S. city and seven international cities

DEPARTMENTS

Finance & Accounting
Information Systems
Legal
People Department
Marketing
Product Development
Sales & Technical Support

THE STATS

Annual Revenues: $533.9 million (1997)
No. of Employees: 2,385 (U.S.)
No. of Offices: 1 (U.S.)
Stock Symbol: NSCP (NASDAQ)
CEO: James L. Barksdale

UPPERS

- Casual office atmosphere
- Social environment

DOWNERS

- Long workdays
- Uncertainty because of acquisition by America Online

KEY COMPETITORS

- Excite
- Infoseek/Walt Disney
- Lotus
- Lycos
- Microsoft
- Yahoo!

EMPLOYMENT CONTACT

Lamont Monroe
Director of Staffing
Netscape Communications
501 E. Middlefield Road
Mountain View, CA 94043

(650) 937-3473
Fax: (650)528-4135
resumes@netscape.com

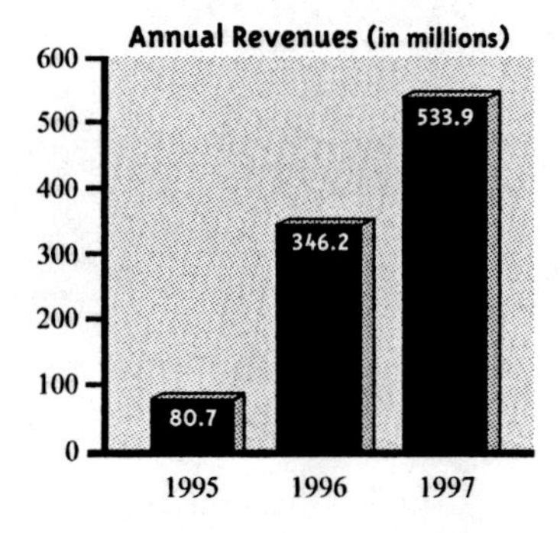

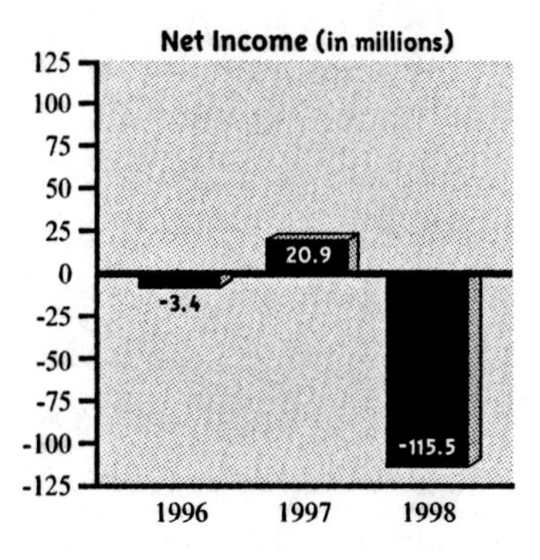

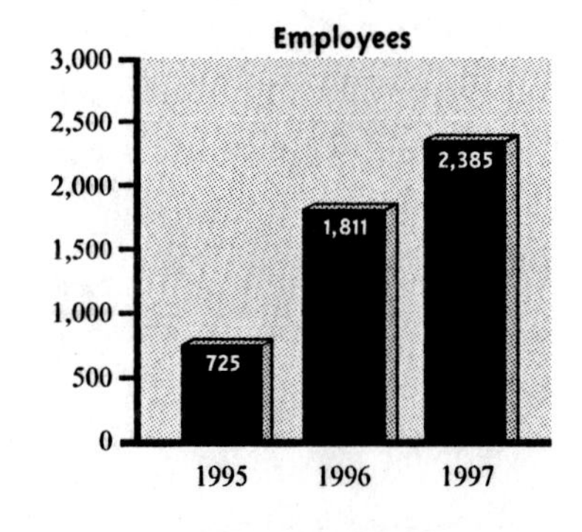

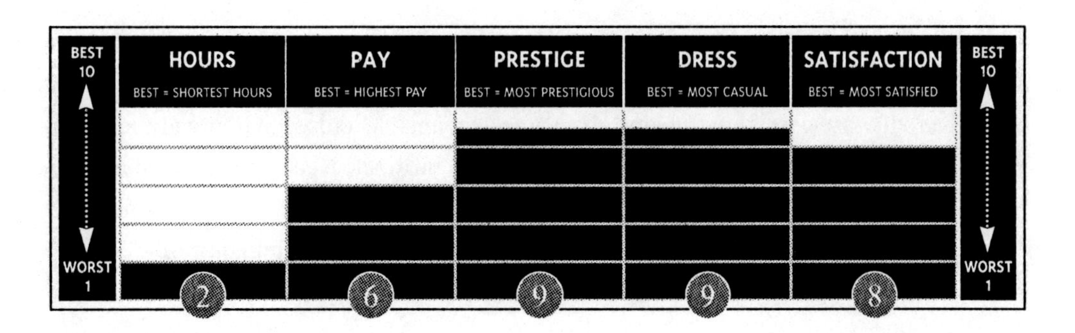

THE SCOOP

By initially distributing its World Wide Web browser, Navigator, free of charge over the Internet, Netscape has become a paramount feature of the Web. In addition to its browser and portal site, the company makes related software products that enable companies to use the Web for commercial purposes, as well as software that integrates web browsing with other applications. Netscape has also been successful with server software for corporate intranets – networks that connect computers within an individual company. Netscape clients have included Chrysler, Lockheed Martin, and Prudential Securities. And in November 1998, the company announced that it would be acquired by America Online for $4.2 billion. That deal, which also involves Sun Microsystems, will not only give the combined company more traffic and "eyeballs," but will take advantage of AOL's media expertise, Netscape's software expertise, and Sun's hardware and support services to move aggressively into what is expected to be the extremely lucrative market of providing e-commerce capabilities and service to businesses.

Netscape's meteoric rise in the technology world began when Marc Andreessen played a key role in developing Mosaic, an Internet software program, at the University of Illinois-Champaign. Andreessen co-founded Mosaic Communications in 1994, but when the University claimed the right to the software's license, Andreessen changed the company's name to Netscape. The company has entered into agreements with other technology companies, such as Sun Microsystems, Oracle, and IBM, to integrate its software into other packages and servers. One of the hottest companies of the 1990s, Netscape is currently developing new software that will transform the Web into an even more exciting – and profitable – medium, and is busy adding features to its own web site on the Internet, transforming it into an Internet "portal" that surfers will keep as their own homepage.

Netscape's major foe? Microsoft – which has bundled a competing web browser, Internet Explorer, with its operating systems software in order to gain control of the Internet market just as it has software operating systems. By some accounts, Microsoft has already seized up to 50 percent of Netscape's share by giving away its web browser; Netscape's income and stock has plummeted. However, Microsoft's strategy has become the subject of a Department of Justice antitrust suit, which charges that Microsoft is unfairly using its dominance of the PC software market to attempt to dominate a new market. Gates and Microsoft contend that Internet

Explorer is an integral part of its Windows software, though Netscape offers uninstalling software on its own web site, saying that getting rid of Explorer has no effect on Windows.

Whether the allegations are true or not, one thing remains clear: Microsoft has taken a huge bite out of Netscape's browser dominance. Netscape, which once controlled 90 percent of the market, has seen its 1997 share drop to 50.5 percent, down from 54.6 percent in 1996. In contrast, Microsoft's Internet Explorer rose from 16.4 to 22.8 percent in the same period. In 1998, Netscape, which after acheiving market dominance, had sold its browser, was forced to admit defeat, and began again to give away its browser for free. The browser had previously represented Netscape's biggest source of revenue.

As a result, Netscape is now shifting gears and focusing on selling software to businesses and beefing up its booming Netcenter web site. In 1998, Netscape announced a partnership with Excite to create a search service on Netcenter, which acts as a "portal," the first site consumers see when they access the Web. The company hopes to leverage its 70 million software users into "the world's biggest media network." Netscape also inked a $15 to 20 million deal with Citibank to build the financial corporation's electronic commerce infrastructure. Such promising developments led Netscape in August 1998 to report a small profit for the second quarter, surprising many analysts who had watched the company post repeated quarterly losses, including a total loss of $44.7 million in 1997.

In November 1998, however, Netscape abruptly folded its cards. America Online, the middlebrow, mass-market Internet service provider, purchased the once-proud company for $4.2 billion of its stock, with a side licensing deal with Sun Microsystems for Netscape's e-commerce software.

GETTING HIRED

Netscape's web site describes current openings, including their responsibilities and requirements. Each position lists a job reference number, which applicants should include with their resumes. Each department has a separate e-mail address for applicants to use when e-mailing their resumes; Netscape accepts faxed resumes as well. The employment web page also describes the company's college recruitment schedule. Netscape recruits computer science students from engineering programs. Netscape also prefers MBAs who have some computer or technical experience. In addition to technical savvy, Netscape looks for lots of personality in its new hires: the official title of the head of recruiting is "Director of Bringing in the Cool People."

The interview process usually begins with a screening conversation by phone. After that, Netscape conducts two or more rounds of face-to-face interviews. Although the interviewers normally dress very informally, the interviewees often (and should) wear suits. Questions are highly dependent on which department one is applying to. On the whole, interview questions are very practical. For instance, marketing applicants are asked simple questions like, "give an example of well-marketed software and why it was successful." However, says one insider, "we don't hire experience as much as we value smart people looking for a challenge."

It's not thought that America Online's acquisition of Netscape will occasion layoffs (except in certain areas, such as the Navigator browser). In fact, AOL CEO Steve Case has offered all 2300 Netscape employees a month of salary bonus if they stay through the merger.

OUR SURVEY SAYS

Netscape employees work "unending days," let their "work take over their lives," but some "love every minute of it." Employees enjoy the "excitement" of a company "on the constant cusp of technological innovation" and like the "underdog cachet" of working for "David in the shadow of Microsoft's Goliath." "Special perks are that you get the weekend off sometimes!" jokes one insider. Still, as one employee says: "This is a fascinating company. Netscape deals with unprecedented challenges in a furiously changing industry."

Recently-hired employees at this "awesome company" comment that they have been encouraged to "dive in immediately" and "quickly acquire significant responsibilities." "The company is driving, demanding, exhausting, but it is also rewarding, nurturing and fulfilling," says one insider. "A company of extremes." Because of Netscape's recent difficulties, however, all is not rosy in Mountain View: Bill Gates as Goliath casts a cold shadow. "The current atmosphere is subdued but optimistic. Otherwise, Netscape has been the best company I have worked at." "Frankly, if there were no Microsoft in the world, this would undoubtedly be the most popular place of employment for anyone beginning in the software field," says another glum Netscaper. "But there is a Microsoft." Many employees characterize Netscape as "busy," "stressful," and a "a very fast-paced company, a very demanding place – no excuses." And, as the recent layoffs indicate, because of competitive pressures, Netscape "is definitely assuming some of the characteristics of larger companies."

A key component of what employees call a "team" atmosphere at Netscape is a "casual, informal" corporate culture that encourages collegiality. "It feels like being a graduate student," one Netscape insider says, pointing to the games of ping-pong, table hockey, and football that he plays near his cubicle. Another has a slightly different take: "Walking around Netscape is like bar-hopping in the city." The company is reportedly very diverse at the lower and middle-management levels, with only "about 20 percent male WASPs." "Netscape is a wonderful collection of eclectic personalities," says another. According to one employee, "women and minorities are treated equally, since, after all, women and minorities are actually the majority at Netscape." However, senior management is "very much white male – not unusual, but not expected," according to one insider. As far as Netscape's dress code is concerned, several employees quote CEO James Barksdale, saying that Netscape requires its hires to "wear something." Summing it up, one product manager says: "It's a very cool place to work. Smart folks, excellent pay, very flexible hours, no dress code pressures, a very diverse mixture of folks."

Norstan

605 N. Hwy. 169, 12th Floor
Plymouth, MN 55441
(612) 513-4500
Fax: (612) 513-4507
www.norstan.com

NORSTAN

LOCATIONS

Plymouth, MN (HQ)
Plus offices in 26 other states and across Canada.

DEPARTMENTS

Applications Development
Business Technology Consulting
Cabling Services
Call Center
Computer Network Systems
Educational Services
Financial Services
Maintenance & Support
Network Services

THE STATS

Annual Revenues: $456.4 million (1998)
No. of Employees: 2,971
No. of Offices: 65 (U.S. and Canada)
Stock Symbol: NRRD (NASDAQ)
CEO: David Richard

UPPERS

- Profit sharing and stock purchase options
- Excellent company training programs
- Community outreach programs

DOWNERS

- Dress code is stricter than at most high-tech companies
- Lower salaries than at other companies
- Few women in management positions

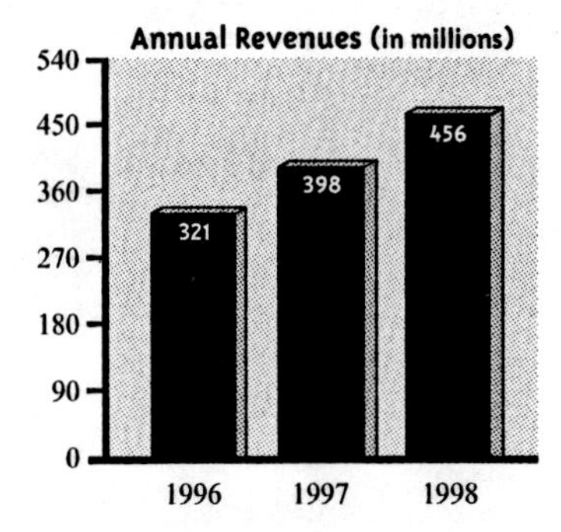

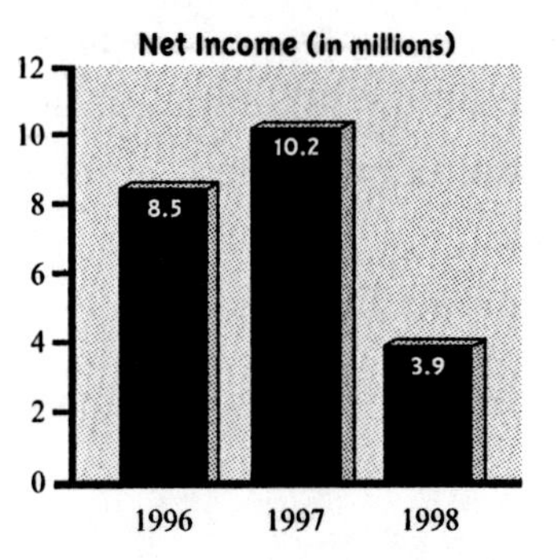

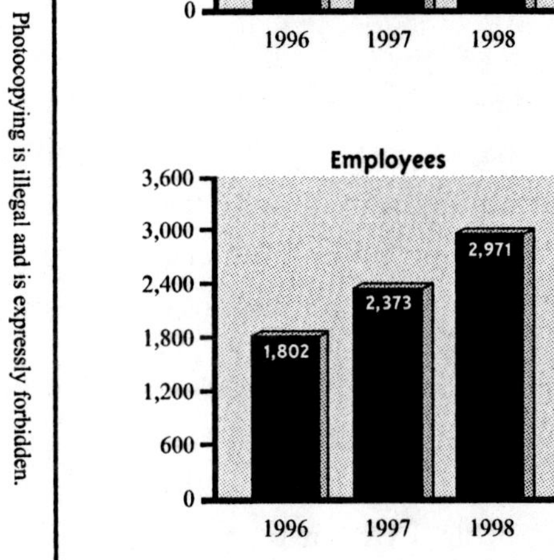

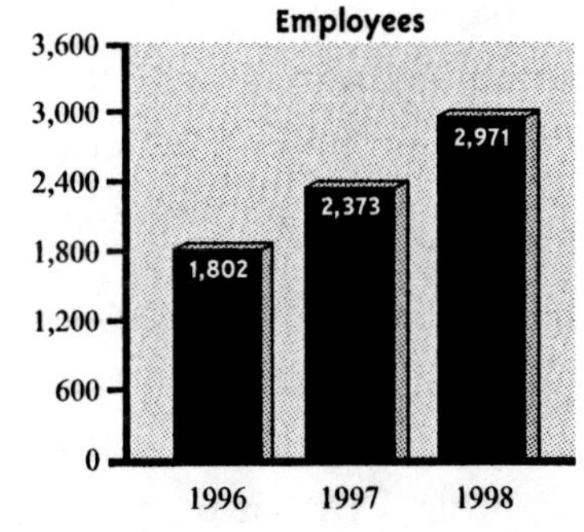

KEY COMPETITORS

- Ameritech
- Andersen Consulting
- Bell Atlantic
- Cap Gemini
- Computer Sciences Corporation
- Deloitte Touche Tohmatsu
- Ericsson
- GTE
- Lucent
- SBC Communications
- Siemens
- Technology Solutions

EMPLOYMENT CONTACT

Corporate Recruiter
Norstan Communications
605 N. Hwy. 169, 12th Floor
Plymouth, MN 55441

Fax: (612) 513-4590
prbannochi@norstan.com

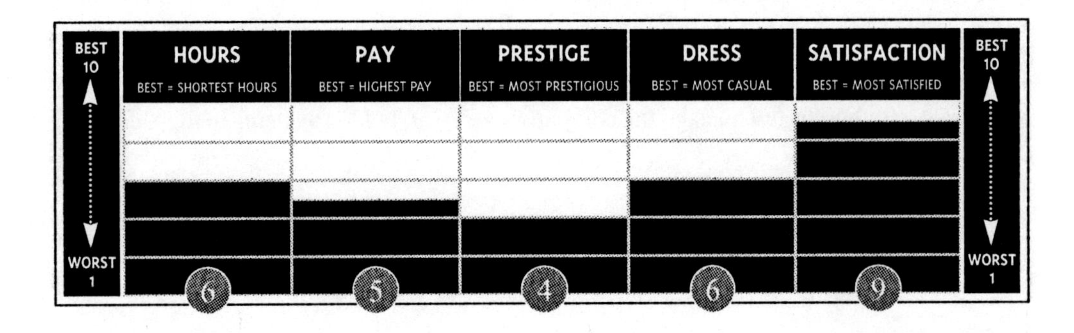

THE SCOOP

As technology options increase, corporations are depending more and more on businesses that can both implement and manage their communications systems – and companies like Norstan are stepping in to capitalize on this growing market. In 1973, Norstan began as a reseller of telephone equipment with a customer base in Canada and the American Midwest. Today, the company offers voice mail, videoconferencing, interactive voice response and high-volume call center systems. As a reseller, the company avoids manufacturing and development costs and has been able to expand and diversify with ease – it sells products made by Siemens, Aspect Telecommunications, Lucent and Sprint.

In 1993, it became clear that there was more money to be made as a service provider than as a mere equipment reseller. So Norstan began the transition from a product-oriented business to a service-oriented organization, specializing in information technology and communications systems integration. Over the next few years, it acquired two technology consulting firms: Prima, based in North Carolina, and Minnesota-based Connect Computer Inc. The two acquisitions were folded into the company's existing professional services arm to form Norstan Consulting. With these purchases, Norstan also added high-quality data communications services to its voice and video business.

In April 1998, Norstan initiated a structural reorganization, which included the integration of its consulting businesses, and the Call Center Solutions Group. The same month, it launched a start-up enterprise, called Connaissance Consulting. In June, Norstan picked up WORLDLINK, a telecom solutions company based in Champaign, Illinois. Norstan will gain 150 new consultants and benefit from WORLDLINK's customer base in the Midwest and on the West Coast. The company plans to further capitalize on its expanded customer base by offering services to their international branch offices. For example, because Norstan has a long-standing relationship with British Petroleum's Cleveland office, it was able to secure a contract to integrate and manage the company's voice and video systems in the U.S. and Europe.

GETTING HIRED

Norstan recruits on college campuses, and does much hiring based on referrals from employees or others in the telecom business. It also advertises openings in newspapers and in the "careers" section of its web page. The interview process starts with a phone screening, followed by a "fairly relaxed," meeting with a supervisor. "Things may get very technical," if you're looking to do something like engineering or system design, but in general, Norstan interviews "are only stressful if you let them be."

OUR SURVEY SAYS

"The company believes in its employees, customers, shareholders and the local community." "The people are excellent," say employees, who feel very strongly about "our corporate culture and values." They say Norstan has "a value system that reflects ethical, responsive and profitable business practices." In addition to its commitment to the customer and the shareholder, the company has a genuine commitment to its employees and its community. Norstan encourages staff to participate in local projects like Habitat For Humanity on company time." And one worker revealed that "the president sends my kid a birthday card with $1 in it every year." "It's small stuff," he says, "but it makes a difference to me."

Though "other companies pay more," Norstan insiders are willing to sacrifice a little cash for "the chance to jump in and basically run your own projects." Our contacts say they are given "as much responsibility as they can handle." In addition, "Norstan is pretty good about recognizing extra efforts," one source says, "and people get promoted as soon as they deserve it." As one of our sources explains, the profit sharing and stock purchase programs are an added incentive: "I will work towards achieving the company goal because I am working towards my own future." The company also encourages personal development through its Education Services arm, which runs training centers around the country. The best part is, "the company will help with most or sometimes all of your tuition." "It's a good place to get that jump on technology if you're right out of college," reports one insider.

Sources say work hours "are what you make them." People are "laid-back for the most part," but at the same time, they "move at a very fast pace." Employees work overtime to get projects done, "but we are by no means a sweatshop." On average, people work "just over 40 hours per week." "A self-starter would do very well in this culture," notes one source. "Initiative is highly prized here," said another, "and if you meet our sometimes strict standards, the rewards are tremendous." The management structure is described as "loose," though there's an implicit understanding that no one is going to "babysit employees." "Dress code depends on what you do, who you do it for, and where you do it." Most wear business casual attire, though the higher your position, the better you dress. "The only real rule is no jeans – except on Friday."

There's "a good mix of men and women" on the lower levels, but you'll find "few women in management positions." However, employees say the company is "changing dramatically" for the better. "The company is pretty white and male," said one insider, "but some departments are getting younger and more diverse. "

“[In interviewing] you may get some hypothetical situations involving the technology. What really matters is the ability to step through a problem in a logical manner.”

– ***IT insider***

Oracle

500 Oracle Parkway,
Box 659202
Redwood Shores, CA 94065
www.oracle.com

ORACLE®

LOCATIONS

Redwood Shores, CA (HQ)
Atlanta, GA • Bethesda, MD • Boston, MA • Chicago, IL • Denver, CO • Honolulu, HI • Nashville, TN • New York, NY • Richmond, VA • Seattle, WA • Beijing, China • Berlin, Germany • Buenos Aires, Argentina • Montreal, Canada • As well as locations around the U.S. and worldwide

DEPARTMENTS

Data Storage Technologies
Internet Products Group
Languages and Relational Technology Group
Oracle Server Product Management Group
Open Systems Integration and Technology
Software Engineering
Systems Management Products Group

THE STATS

Annual Revenues: $7.1 billion (1998)
No. of Employees: 36,802 (worldwide)
Stock Symbol: ORCL (NYSE)
CEO: Lawrence J. Ellison

UPPERS

- Stock purchase plan
- Tuition reimbursement
- Beautiful offices
- Excellent company gym

DOWNERS

- Heavy workload
- Uncertain footing in industry

KEY COMPETITORS

- Baan
- IBM
- Informix
- Microsoft
- Novell
- Peoplesoft
- SAP
- Sybase

EMPLOYMENT CONTACT

Tom Callahan
Human Resources
Oracle
500 Oracle Parkway,
Box 659202
Redwood Shores, CA 94065

Fax: (650) 506-7406

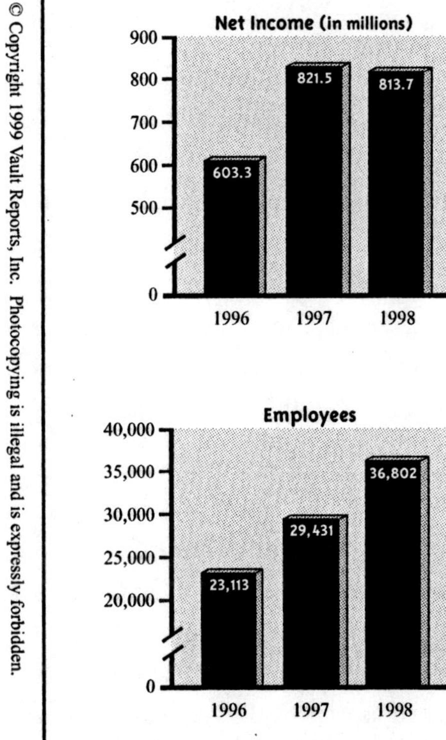

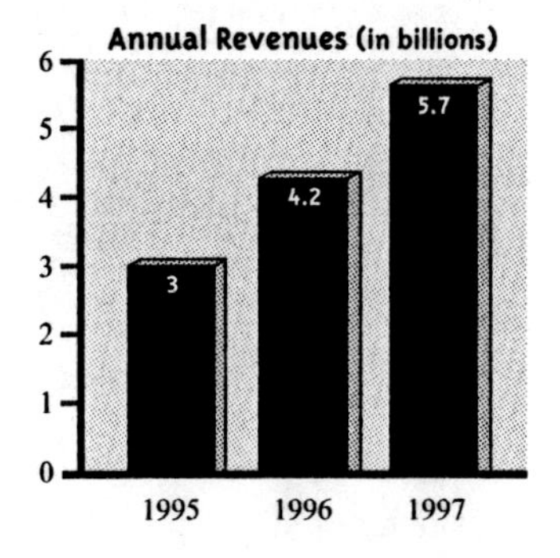

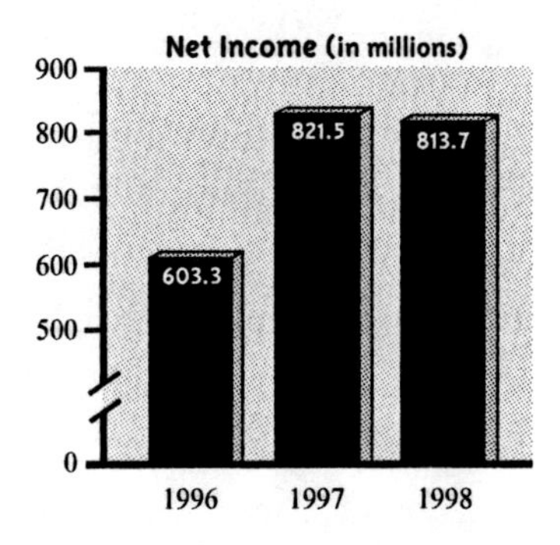

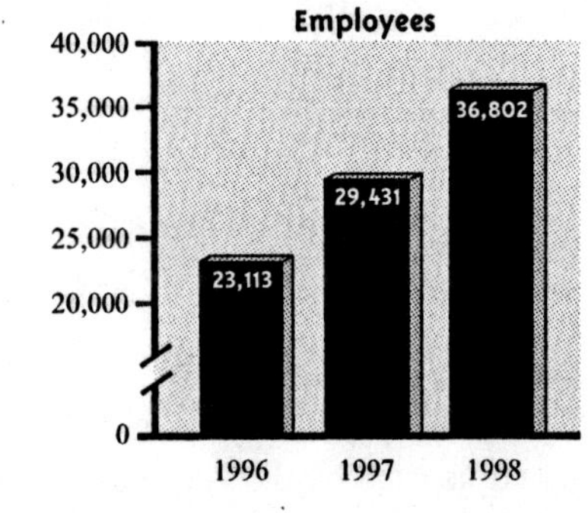

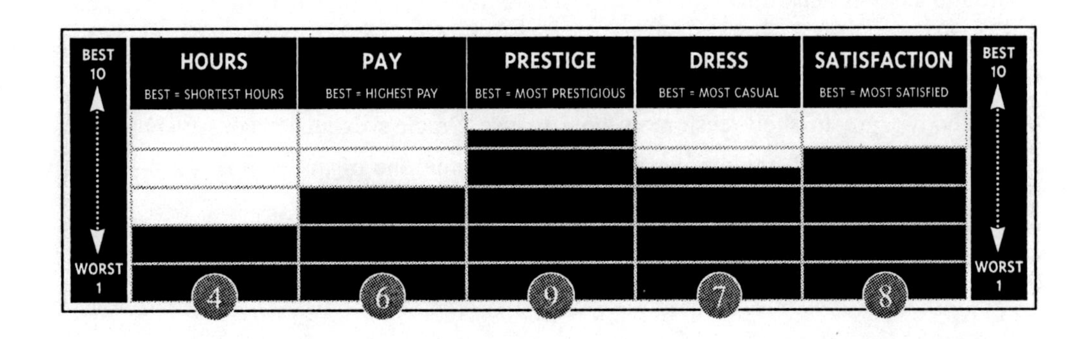

THE SCOOP

Oracle has goals that few Silicon Valley companies would have the vision – or bravado – to consider. First on the agenda is to complete its domination of the database software market. With that coming to fruition, CEO Larry Ellison and his cohorts are setting their sights on wresting control of the world from Bill Gates.

In just over two decades, Ellison has skillfully crafted Oracle into the leading maker of database management software and a fighter capable of taking on the biggest bullies on the block. Founded in 1977 to capitalize on technology first developed – and later fumbled – by IBM, Oracle immediately enjoyed rapid success and growth. However, irregular accounting practices and poor product development during the early 1990s led to concerns that Oracle lacked the ability to keep up with its growth. A shareholder lawsuit forced Ellison to resign as company chairman, but the company managed to rebound. Now, with Ellison recast as a visionary developer of software ideas (his Chief Operating Officer handles the day-to-day business operations), Oracle is ready to take on the 21st century – and the overarching empire of Microsoft.

As one of the largest software companies, Oracle offers several different products, most involving either database management or network connectivity. Oracle software makes it easier for its customers, usually large businesses, to enter, store, or retrieve data like customer product orders, sales records, and personnel information. Oracle's newest program, announced in September 1998, is Oracle 8i, a database capable of managing Internet applications over networks used by thousands of corporations and individuals every day. Oracle has also moved into new product areas to increase sales. To supplement the database software, the company, through its tools department, develops software developing tools that allow users to customize their Oracle software. Oracle applications include over 30 software items for finance and administration, manufacturing, human resources, and other business applications. And somebody's got to show customers how to use Oracle's exquisite constructs – that's why Oracle's service and training department has become one of the company's fastest growing sources of revenue.

GETTING HIRED

Oracle's web site, located at www.oracle.com provides a wealth of information about the company, its products, and its current job openings. The site offers applicants the opportunity to construct and send an on-line resume; the company also accepts resumes via regular mail and e-mail. Oracle employees say contacts within the company can bring a tremendous advantage since a recommendation not only gives the applicant the imprimatur of quality, but offers a chance for Oracle to save on headhunting costs. For snaring technical jobs, however, no recommendation can replace a record of talented software engineering. Resumes are screened by hiring managers, who also conduct the initial interview. In the first interview, Oracle asks applicants about their background and work experience, insiders tell us. Successful interviews are followed by a subsequent round of interviews with personnel from the department in which the position is located. A third round sometimes follows, in which the applicant meets their future sponsor.

OUR SURVEY SAYS

Employees enjoy the "lively and social" atmosphere that prevails at Oracle. They say that the halls brim with "bright and motivated people" who have created a "young and dynamic" corporate culture. Thanks to this enthusiastic work force and to Oracle's recent performance, morale levels are "sky high throughout the company." However, some employees complain that Oracle's "extensive" size and scope makes recognition hard to come by. "This is a huge company, and I think sometimes you have either to sink or swim. You can easily get swept up and passed over in a large corporation like Oracle," says one contact.

"Those technical people – I've heard stories about them," says one financial analyst. "I've heard they have expresso machines in their offices and futons on the floor in case they have to program late into the night. They play Frisbee in the halls and most wear jeans all day long."

Unlike the T-shirt-clad programmers, the managerial staff "always has to dress professionally if we're seeing customers." Yet even this is dressup "only at the khaki level – every day is 'business casual Friday' for us." Those employees not "in the public eye" regularly "wear jeans with a blazer or sweater. It's a much more relaxed corporate environment in Silicon Valley than elsewhere." Other perks include "an impressive, generous stock option plan" to almost all college recruits and a "great cafeteria" that has even "garnered some local press for its tasty fare. Each one has a theme, so if you're in the mood for Chinese, or Indian, or Californian, you always know where to go."

Flexible scheduling allow workers to "avoid rush hour or to get an early work out at the gym." Oracle "has a sexy image in the Valley," employees say, "largely because of the image of the CEO, Larry Ellison," who named the lake on the company campus after himself and who is building a $40 million Japanese mansion to rival Bill Gates' $30 million pleasure palace. Another employee comments, "The reputation of Oracle in the Bay Area is just wonderful. Everybody knows it, and everyone thinks you're lucky to work there." The actual corporate campus is "in a beautiful location" with a "huge lagoon that most of the buildings overlook." "In the local area," says one, "we've had to build new buildings because we've grown so much, and they're all fabulous. They really take care of us." But more satisfying than impressing the neighbors, is Oracle's reputation as an "important, booming company" producing "the industry-leading product."

To a large degree, Oracle employees comment, pay "depends upon the group you join." Employees frequently mention that software development and consulting receive the best packages. Marketing and finance, some say, "get the industry standard," but sales staff receives "slightly lower pay, even though it is bolstered by commissions."

"With our recent financial successes. We've had a lot to party about."

– Software insider

PairGain Technologies

14402 Franklin Avenue
Tustin, CA 92780-7013
(714) 832-9922
Fax: (714) 832-9924
www.pairgain.com

PAIRGAIN

LOCATIONS

Tustin, California (HQ)
Florida • Maryland • North Carolina • Hong Kong • Switzerland

DEPARTMENTS

Corporate Marketing
Engineering
Information Management
Operations
Product Management
Product Marketing
Sales & Customer Service

THE STATS

Annual Revenues: $282.1 million (1997)
No. of Employees: 640
No. of Offices: 3 (U.S.), 2 (worldwide)
Stock symbol: PAIR (NASDAQ)
CEO: Charles Strauch
Year Founded: 1988

UPPERS

- 25-cent soft drinks
- Good reputation for promotions and pay raises
- Tuition reimbursement

DOWNERS

- Dress code is more formal than most high-tech companies

KEY COMPETITORS

- @Home
- ADC Telecommunications
- Adtran
- Alcatel
- Ericsson
- Lucent
- Paradyne
- Raychem
- Siemens
- Tellabs
- Teltrend

EMPLOYMENT CONTACT

Kelly Flake
Senior Human Resources Representative
PairGain Technologies
14402 Franklin Avenue
Tustin, CA 92780-7013

(714) 832-9922, ext. 3211
Fax: (714) 730-3199
hr@pairgain.com

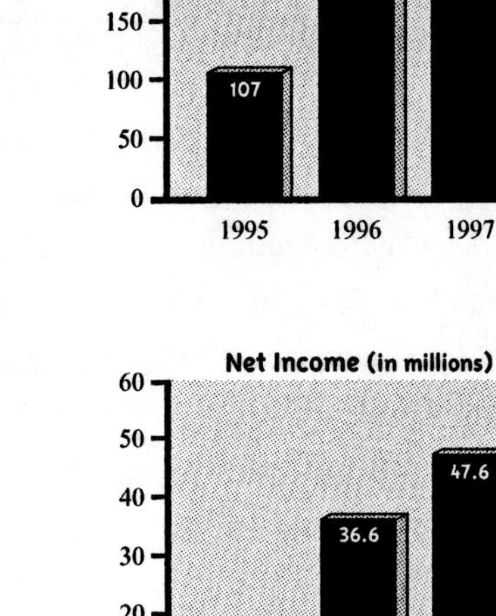

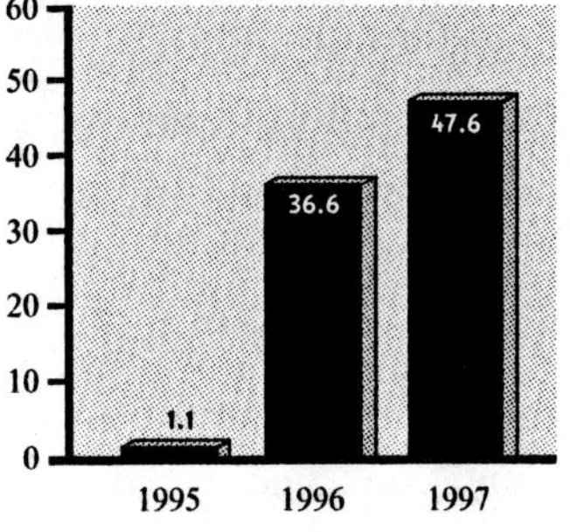

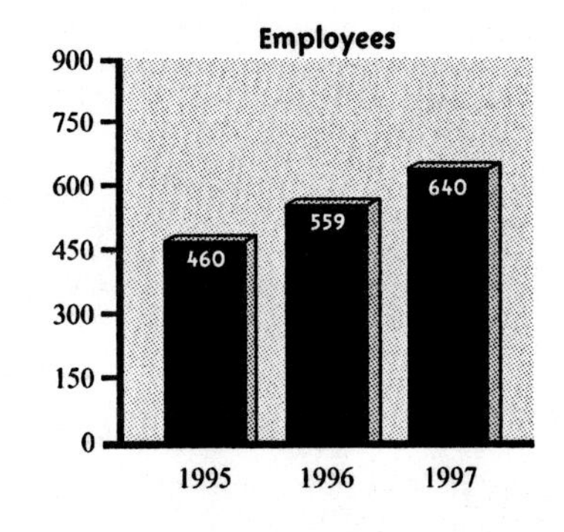

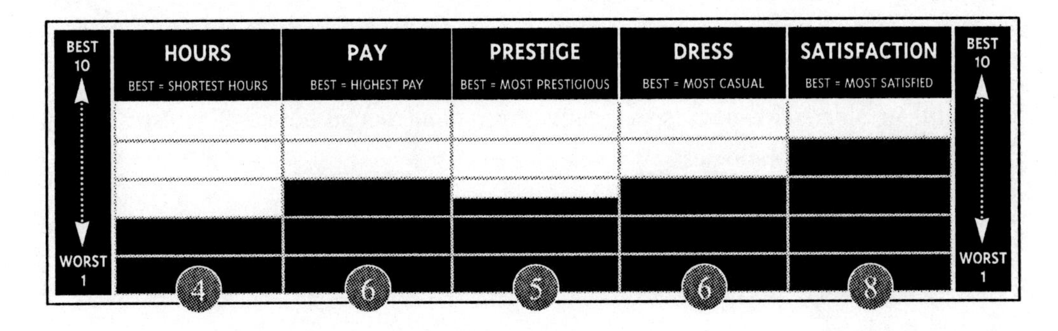

THE SCOOP

In the race to meet the growing demand for multiple phone lines and more bandwidth, PairGain offers a way to maximize capacity for the existing network of copper wires used by telephone companies. PairGain uses a technology called HDSL (High-speed Digital Subscriber Line), which was developed by researchers from the Bell companies (which it now serves). Using digital signal processing chips, HDSL doubles the capacity of the copper wiring and makes data transfer faster and more reliable. With this HDSL, regional phone companies can avoid or postpone the costly conversion to higher-capacity fiber optic technology. They are especially reluctant to make the switch in light of the new challenge posed by cable companies, which now offer high-speed modem access through television coaxial cables; companies like DirectTV and Scientific Atlanta could very likely have a stronghold in this market before most phone companies could have a network of fiber lines installed.

PairGain is one of the fastest growing high-tech companies in the country – sales nearly doubled between 1996 and 1997, and net income grew from $1.1 million to $36.6 million over the same period. The company's success is largely related to the fact that it was the first to offer HDSL technology and because of PairGain's strong relationships with the telephone companies. Since there are a finite number of users, competitors have been hard pressed to break into the market. One of those companies is Adtran, which gained market share by instigating a bitter price war against PairGain in 1998.

Since it was founded in 1988, the company has positioned itself through a number acquisitions and investments in several other companies, including two California-based network solutions makers. In 1997, PairGain acquired AVIDIA Systems, a developer of new networking technology that acts like a high-tech funnel, bundling data into aggregate units that can be transmitted at rapid speeds. More recently, the company agreed on a licensing deal and joint development project with Rockwell Semiconductor. Industry observers say PairGain will soon target the residential market, specifically telecommuters and home-office users who want to utilize the latest technologies.

GETTING HIRED

PairGain periodically hires engineering co-op and interning college students. Those people often get full-time positions when they graduate. Interviews last two or more hours, "depending on how interested [PairGain is]," insiders say. The company posts job openings at its web site, located at www.pairgain.com.

OUR SURVEY SAYS

Employees at PairGain say they work with the "best of the best," and like the fact that everyone "has an open door policy." They say higher-ups are "very approachable, from the CEO to first-level managers. "We don't have fancy offices or a lot of politics," remarked one insider, "teams and teamwork are emphasized." One contact adds: "You'll do well here if you are intelligent, have common sense, are customer service oriented and don't mind working hard."

While salaries at PairGain are "excellent," some employees seem peeved that "we only get two weeks vacation (annually) for the first four years." However, they do get a break for the week between Christmas and New Years and 25-cent sodas year-round (it adds up for frazzled engineers, so don't knock it). In any case, most insiders enjoy all that time spent in the office. They happily report "a great sense of teamwork and camaraderie" in each group, and say their jobs are "challenging and fulfilling." In addition to the basics, workers benefit from "generous" profit sharing and stock purchase plans. To keep employees up to date on the latest developments in the industry, PairGain provides on-site classes and a tuition reimbursement program. It also foots the bills for annual industry conferences. One fun perk: PairGain sends stellar employees on all-expense-paid vacations – most recently to the British Virgin Islands.

The dress code is "anything from business casual to suits," though "our engineers wear jeans a lot." Posted work hours are 8 to 5. While "overtime is not required," employees say that "to do well you need to work at least 50 hours a week." "The main thing is that your project is done on time," explains one source, "and that sometimes does require putting in a lot of extra hours." In addition, the company has a good reputation for promotions and pay raises, as long as employees "put in the hours and work hard to live up to the 'PairGain Quality Commitment.'"

Sources say the company is "very multicultural," and note that the women at PairGain are "definitely making their place," and "many women are in positions of authority." But "the communications industry has been a boys' club for a long time," reports one honest contact, so "it helps if you're not too easily offended."

"I will work towards achieving the company goal because I am working towards my own future."

– ***IT insider***

PeopleSoft

4440 Rosewood Drive
Pleasanton, CA 94588
(925) 225-3000
Fax: (925)-225-3100
www.peoplesoft.com

PEOPLESOFT

LOCATIONS

Pleasanton, CA (HQ)
Argentina • Australia • Brazil • Canada • France • Germany • Japan • Mexico • The Netherlands • New Zealand • Singapore • Spain • United Kingdom

DEPARTMENTS

Account Management • Administration • Business Development • Communication Services • Development • Education Services • Federal/Public Sector • Finance • Global Support Center • Higher Education • Human Resources • Installation Services • Internal Systems • International Positions • ISO 9000 • Legal • Marketing Communications • Middle Market (PeopleSoft Select) • Product Strategy • Professional Services • Red Pepper positions • Release Management • Research & Development • Sales • Technical • Technical Sales Support • Tools Development • Web

THE STATS

Annual Revenues: $815.7 million (1997)
No. of Employees: 5,800
No. of Offices: 23 (worldwide)
Stock Symbol: PSFT (NASDAQ)
CEO: Dave Duffield

UPPERS

- Casual culture
- Shiny happy people
- Booming company

DOWNERS

- Extensive travel
- Fierce competition in field

KEY COMPETITORS

- Baan
- Informix
- J. D. Edwards
- Oracle
- Platinum Software
- SAP

EMPLOYMENT CONTACT

Human Resources
PeopleSoft Corporate Headquarters
4460 Hacienda Drive
Pleasanton, CA 94588

Fax: (925) 694 4444
jobs@peoplesoft.com

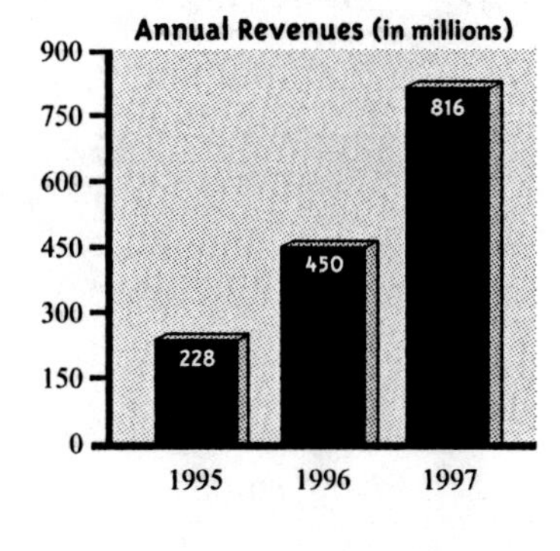

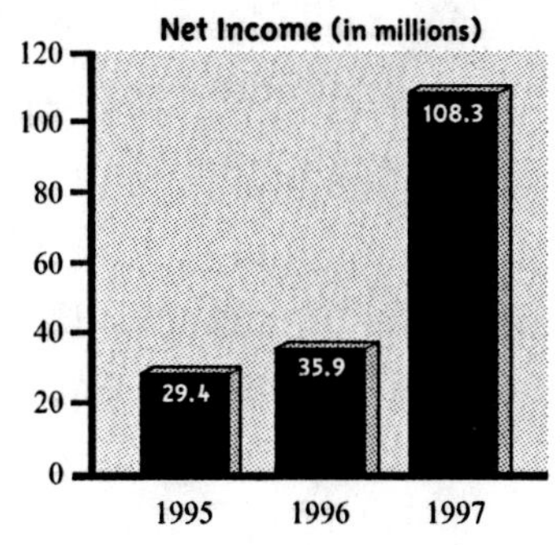

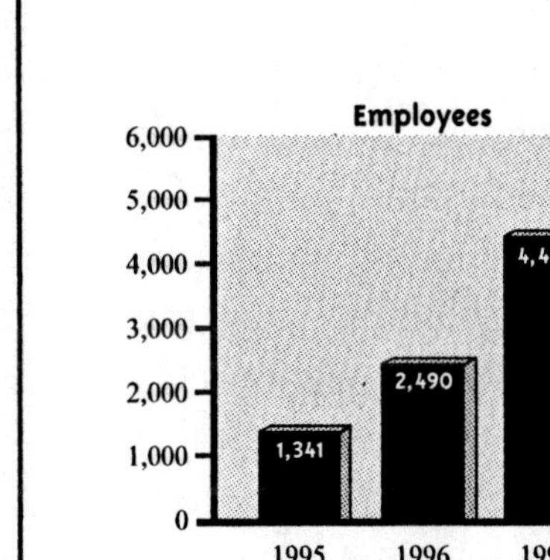

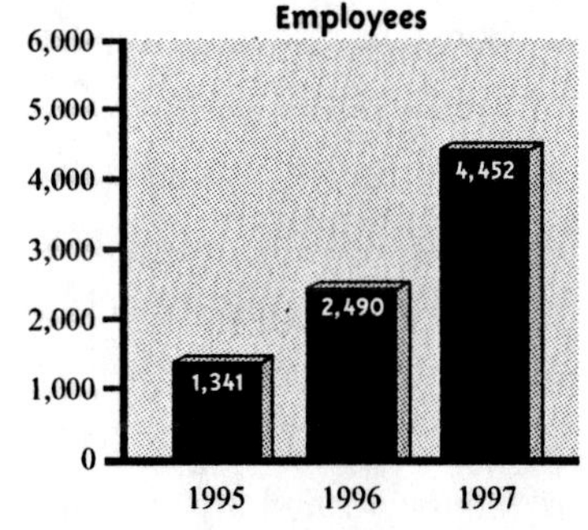

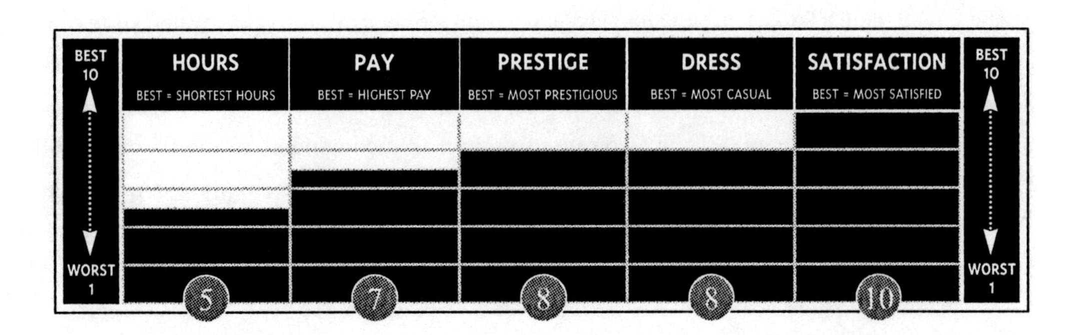

THE SCOOP

"Have fun," "No bullshit," and, oddly, "Keep the bathrooms clean." These are the mantras of Dave Duffield, CEO of PeopleSoft. Duffield's sentiments reflect the quirky philosophy of his brainchild PeopleSoft, the business software firm the former IBM executive founded in 1987. Beginning with human resource application designs, the company has since moved into the dog-eat-dog world of enterprise application software.

What is enterprise application software? Basically, what it does is help corporations run better. The software (otherwise known as Enterprise Resource Planning systems, or ERP) helps companies streamline human resources, organize their books, and smooth out manufacturing processes. The huge, complex (and expensive) systems are normally implemented by armies of systems consultants, driving up costs further; *Fortune* estimates that companies paid $10 billion on ERP systems in 1997.

Such a competitive undertaking seems ill-suited to a feel-good company, but PeopleSoft has prospered nevertheless. Laid-back on the surface, PeopleSoft has moved aggressively to carve out a niche in this profit-rich market. And PeopleSoft has proved to be a real contender in the ERP field, which also hosts such fierce and well-established competitors like SAP and Oracle. PeopleSoft's stock has skyrocketed; it rose 280 percent from April 1996 to April 1998. While PeopleSoft still trails the No. 1 company in the field, SAP of Germany (SAP had 1997 sales of $3.4 billion to PeopleSoft's $812 million), PeopleSoft has shown astonishing growth. PeopleSoft now claims to be No. 2 in the ERP field, passing titans Oracle and Baan, though in an industry ruled by a "bigger is better" mentality, many have accused PeopleSoft of manipulating its financial data to appear larger.

The key to the ERP market, say analysts, is vending a powerful, fully-integrated suite of human resources, financial and manufacturing software. PeopleSoft is the clear leader in the field of human resources software; by some estimates, the firm has 50 percent of that market. Industry observers say that PeopleSoft is also one of the top five contenders in the financial software field. To target the manufacturing field, PeopleSoft acquired the supply chain management company Red Pepper for $228 million of stock in late 1996. PeopleSoft has also been preparing its programs for the global age. Its newest ERP suite, PeopleSoft 7.5, has enhanced manufacturing capacities and support languages, global currencies (including the Euro) and reworked performance measurement skills. PeopleSoft is committed to expanding its client base, and is banking on the idea that others are ready for its advanced organization systems. New customers include the University of Michigan and the Mormons.

By many accounts, PeopleSoft's main selling point may be its "Positively Outrageous Customer Service." CEO Dave Duffield once said "If there's one thing I could have on my gravestone, it would be 'Dave Duffield – he had happy customers.'" Duffield backs up his words with action: the CEO has been known to take customer service calls in his tiny cubicle and to give many clients his personal e-mail address and direct-dial phone number. The cuddly, friendly persona of PeopleSoft has reportedly won over such huge and not particularly snuggly clients as Ford Motor Company. During one negotiation with Corning, the glassware manufacturing client, the deal was reportedly sealed with Corning and PeopleSoft employees hugging. And in a hot high-tech job market, PeopleSoft's fun summer camp atmosphere has added benefits – the firm has one of the highest employee retention rates in Silicon Valley.

GETTING HIRED

Recruiting is done through a number of channels, including headhunters, college campus recruiting, and web-based and other advertising, though at least one employee cites "referrals" as a major source. PeopleSoft seeks cheery candidates who fit its unique, happy culture. Interviews are "almost always informal" and interviewers are said to be "very pleasant" people who "make occasional jokes." Applicants can expect two to three rounds, with four or five 30-minute interviews in each round. For technical jobs, a number of technical questions will be asked, although people skills are also explored. With PeopleSoft's expansion into new markets and new parts of the world, foreign language skills and availability to relocate may be a plus. One of the greatest growth fields: sales. In 1997, PeopleSoft increased its sales force by 80 percent. But you won't get hired as a secretary – PeopleSoft doesn't have them. Duffield claims that secretaries merely filter information and hinder communication. Recent college grads will have better luck. The firm, according to one contact, "just started hiring new graduates from college, and there are a number of internships available."

OUR SURVEY SAYS

The number one point that the company wants to get across is that "PeopleSoft is people." Just like Soylent Green! Employees could not agree more; one proudly states that "our company culture is one that cannot be surpassed." From the casual atmosphere to the air of mutual respect permeating the office, it is simply "an excellent working environment." Reports one happy staffer: "This is a company that will encourage everyone to be their best and it will provide them an environment to be successful in."

The rewards are great: "The benefits [at PeopleSoft] are terrific. Salary-wise everyone is extremely happy. Dress code is casual. There are a lot of activities where everyone participates." Positive energy flows from the top, and Duffield strives to keep everyone smiling and loose. It is working beyond even his wildest imagination; one contact went as far as to say that "when I was hired by PeopleSoft after working as a temp for a year, I felt like I won the lottery." Aside from the external perks, the company provides the chance to branch out into several different areas and locations, but the common denominator is "having fun with very interesting people who really care about each other and our customers." On diversity issues, insiders say PeopleSoft is tops. "The word discrimination does not exist in the PeopleSoft vocabulary," maintains one contact.

PeopleSoft may adhere to a Lennonesque "All you need is love" approach within the office, but there is more going on than is just a huge love-in. The company has also "been doing very well in the stock market. It is a money maker." This joy-filled atmosphere is best summed up by one dedicated insider. "I am always very happy when I get up in the morning to come to work," the employee burbles. "I come with a big smile, full of energy and enthusiasm. I thank God every morning for my job and as long as I am able to work I will dedicate the rest of my life to PeopleSoft." Quite a commercial indeed.

“When I was hired after working as a temp for a year, I felt like I won the lottery.”

– ***Software insider***

Pixar

1001 West Cutting Blvd.
Point Richmond, CA 94804
(510) 236-4000
Fax: (510) 236-0388
www.pixar.com

PIXAR

LOCATIONS

San Francisco

DEPARTMENTS

Film Division
Sequel Division

THE STATS

Annual Revenues: $34.7 million (1997)
No. of Employees: 400
No. of Offices: 1
Stock Symbol: PIXR (NASDAQ)
CEO: Steven Jobs

UPPERS

- Undisputed innovator in its field
- Ultra creative, cute bugs
- Stock options that millionaires are made of

DOWNERS

- Fickleness of show biz
- One location

KEY COMPETITORS

- DreamWorks SKG
- Lucasfilm
- MGM
- Microsoft
- Silicon Graphics

EMPLOYMENT CONTACT

Pixar Animation Studios
Attn: Recruiting
1001 West Cutting Blvd.
Richmond, CA 94804

Fax: (510) 236-0388
hr@pixar.com

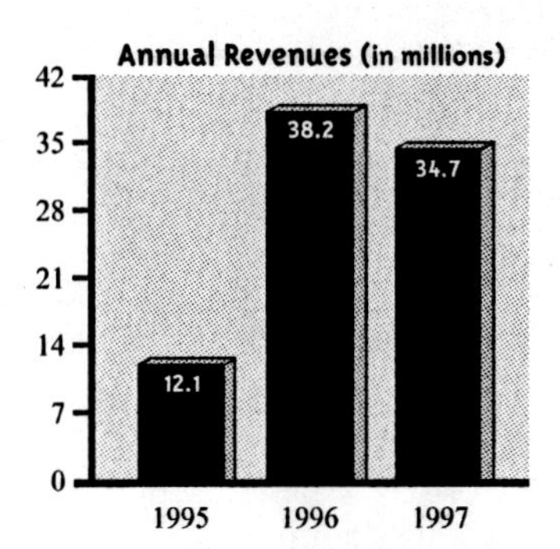

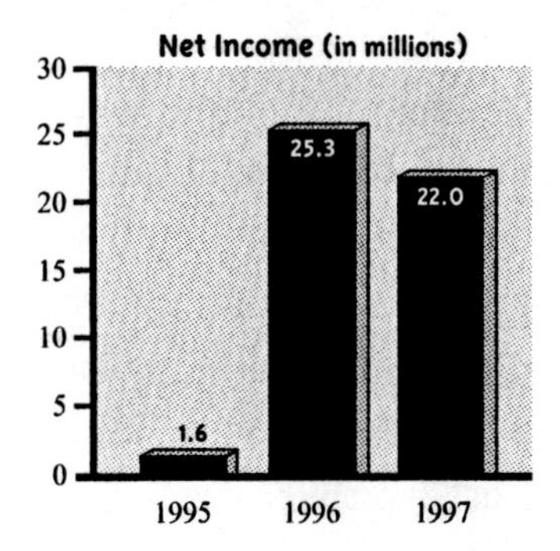

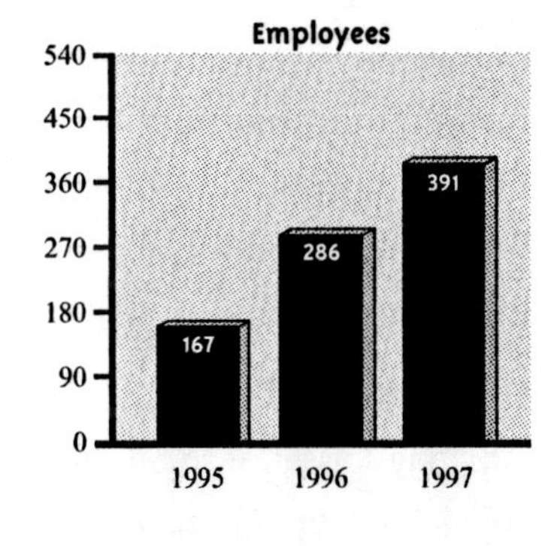

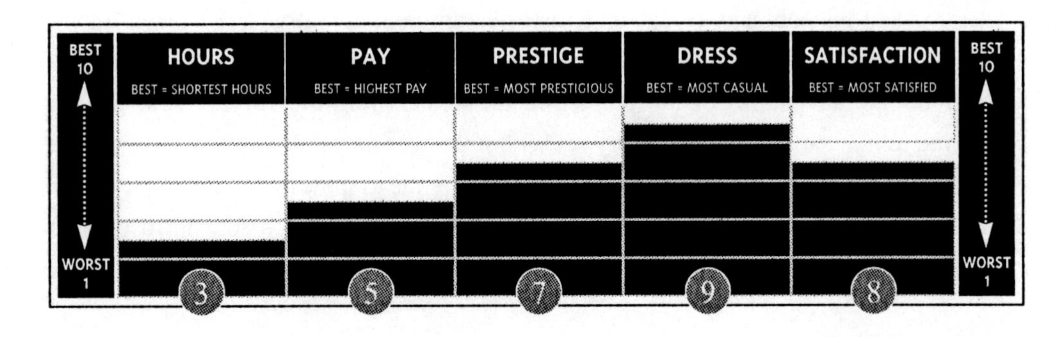

THE SCOOP

Pixar, creators of 1995 mega-hit *Toy Story* and 1998's *A Bug's Life*, is more than just an computer animation studio. The company is poised to become on of the largest entertainment forces of the 21st century. Begun as Industrial Light & Magic ILM, a division of Lucasfilms, the company was spun off in 1986 when Steve Jobs, the founder and on-again, off-again CEO of Apple Computers, acquired the special effects division and renamed it Pixar Animation Studios. Though Jobs served as the well-known figurehead, Pixar was in many ways the vision of one of its senior creators, John Lasseter. Lasseter, who left Disney in the early 80s after becoming disillusioned with the company's lack of innovation, was an early proponent of computer animation, setting the example that Pixar was to follow. The release of his *Luxo Jr.*, a short computer-animated piece about two anthropomorphic lamps, blew away the industry when it premiered at Siggraph, the computer graphics industry's annual convention. Lasseter took computer animation beyond robots and spacemen and redefined the possibilities for the medium, declaring that computer generated images could do for the future of animation what stop-motion photography had for the first 80 years of film. After the premier of*Luxo Jr.*, Pixar focused on honing their burgeoning medium by producing animation for commercials, while Lasseter continued working on short films like *Tin Toy,* which won an Academy Award for Best Animated Short in 1988.

As the potential of computer animation became more understood by the entertainment world, Disney realized its folly in not taking advantage of Lasseter's vision. The Mouse tried tempting Lasseteer to return on three separate occasions, but only succeeded in getting him to collaborate on discrete projects. One such undertaking was the Computer Animation Production System, which replaced hand-coloring. It was 1991 before Disney could convince Lasseter and his company to produce three feature-length computer animated movies. The first of these films, and the first of its kind, was the celebrated *Toy Story*. Originally designed to follow the adventures of the pressed-metal soldier from the *Tin Toy* skit, the *Toy Story* evolved into a completely fresh story set within the world of dolls. The film became the top-grossing movie of 1995. Audiences were wowed by the graphics, but what won the movie the most praise, and several Oscars, was a powerful story line that made full use of computer graphics advantages.

The box-office receipts, and more importantly, the merchandising revenue such as Pixar's Toy Story CD-ROM (the most successful CD-ROM launch ever), gave Pixar a pleasant economic boost – though as Pixar only received 13 percent of *Toy Story* profits. Enter CEO Steve Jobs, who rode to Pixar's financial revenue. While acting as an interim CEO for Apple, Jobs took time to negotiate a better deal with Disney. After a lunch meeting with Disney's Michael Eisner, Jobs walked away with a deal that gave Pixar 50 percent of its box office and merchandising income, as well as a four-movie deal that will provide Pixar with Disney's distribution and market expertise. Steve Jobs' goal for Pixar now is to make it the greatest animation studio of the 21st century, producing a movie a year, starting in 1998 with the hit *A Bug's Life.*

GETTING HIRED

Pixar is a rapidly expanding company that recruits on its web site as well as from several select institutions. "We do check references, they count," says one contact. "The animators have a reel, the artists have a portfolio, the technical production people can point to movies or TV shows/commercials they worked on." Pixar is small enough that "stuff goes through the Human Resource department" and divisions post openings with HR. "[HR does] the initial screening and then forward on viable candidates" to the manager that has the opening.

Because of Pixar's rapid growth and the lack of computer engineers skilled in Pixar's RenderMan imaging software, Pixar actively recruits new employees on their web site, www.pixar.com, where openings are listed regularly. Pixar also recruits eligible candidates from the following institutions: Academy of Art College (San Francisco), Art Center College of Design, California Institute of the Arts (CalArts), Columbus College of Art and, Emily Carr Institute (Vancouver), Fashion Institute of Technology (NY), Kansas City Art Institute, Pratt Institute (NY), Ringling School of Art and Design (FL), Rhode Island School of Design, Rochester Institute of Technology, San Francisco State University (3D Animation and Multimedia, Multimedia/Film, Multimedia Certificate Programs), Savannah College of Art and Design, School of Communication Arts (MN, NC), School of Visual Arts (NY), Sheridan College Schools of Communication Design (Canada), UCLA: Film and Television Dept., University of Southern California: School of CNTV, Vancouver Film School.

OUR SURVEY SAYS

Employees at Pixar find say that though "the atmosphere has shifted a little more towards the corporate side in the past few years" and that the company is beginning to feel much more "like a Hollywood studio," the environment is still "great, really relaxed." "In general, the people are pretty cool" because Pixar is a "flexible employer with excellent employee benefits and a relaxed working atmosphere," although some feel "Pixar has lost some of the feel which made it more attractive in the beginning."

There is "no dress code, or anything silly like that" but "the work hours get long, and Pixar tends to keep the pay scale low," particularly for those in entry-level positions, "just because the name is big." Pixar does offer a social environment in which "there are a number of informal employee-organized physical activities." Pixar employs a full-time Employee Continuing Education Director "who organizes classes on drawing, improvisational comedy, story structure" and various other creative topics.

"In production (actually making a movie), the workweeks can exceed 70 hours" as projects come to completion, but most non-production staff "normally work 40 hours." For those asked to put in extra time there are "perks like free sodas and some snacks, and getting food brought to you." "Part of working at Pixar is understanding production demands, and managing your time accordingly," insiders say. Despite the demanding positions, employees say working at Pixar "is not good, it's great." says one production engineer: "I can't think of a better place to work for."

"Part of working [here] is understanding production demands and managing your time accordingly."

– ***Software insider***

Platinum Software

195 Technology Drive
Irvine, CA 92618
(949) 453-4000
(800) 999-1809
Fax: (949) 453-4091
www.platsoft.com

LOCATIONS

Irvine, CA (HQ)
Foster City, CA • Atlanta, GA • Chicago, IL • Newton, MA • Teaneck, NJ • Dallas, TX • Bellevue, WA • Toronto, Canada • Hong Kong • Singapore • Berkshire, UK • Sydney, Australia

DEPARTMENTS

Channel Marketing
Corporate Communications
Investor Relations
Professional Services
Sales Center
Technical Support

THE STATS

Annual Revenues: $58 million (1997)
No. of Employees: 590 (Worldwide)
No. of Offices: 14 (8 U.S./6 International)
Stock Symbol: PSQL (NASDAQ)
CEO: George Klaus

PLATINUM SOFTWARE

UPPERS

- High morale
- Growing company
- Reasonable hours
- Discounts on soft drinks

DOWNERS

- Prime target for acquisition
- Extensive interview process

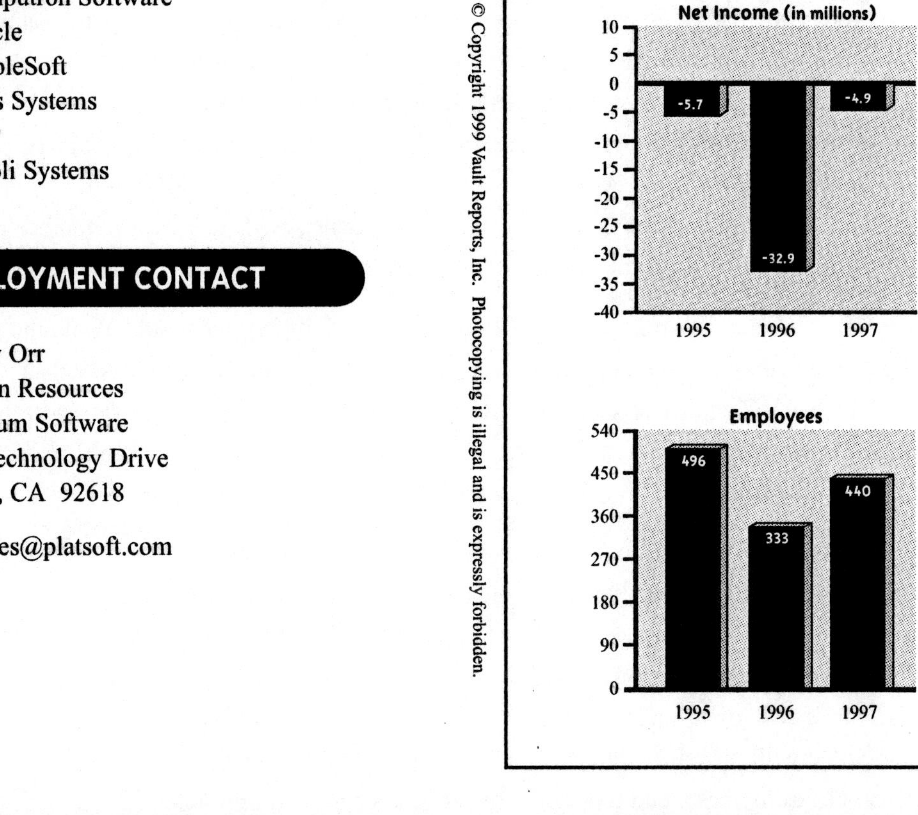

KEY COMPETITORS

- Baan
- Computer Associates
- Computron Software
- Oracle
- PeopleSoft
- Ross Systems
- SAP
- Tivoli Systems

EMPLOYMENT CONTACT

Nancy Orr
Human Resources
Platinum Software
195 Technology Drive
Irvine, CA 92618

resumes@platsoft.com

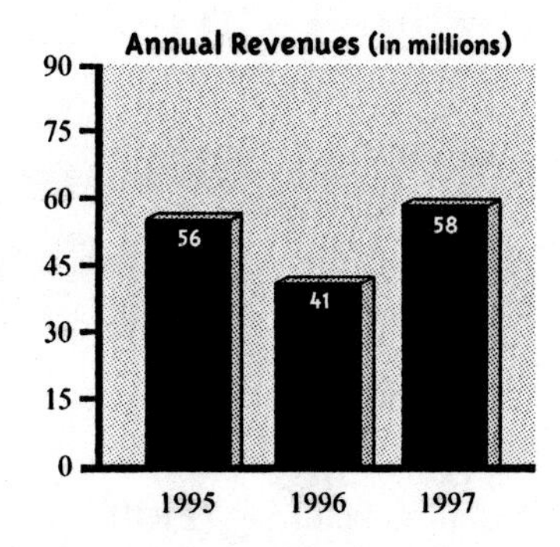

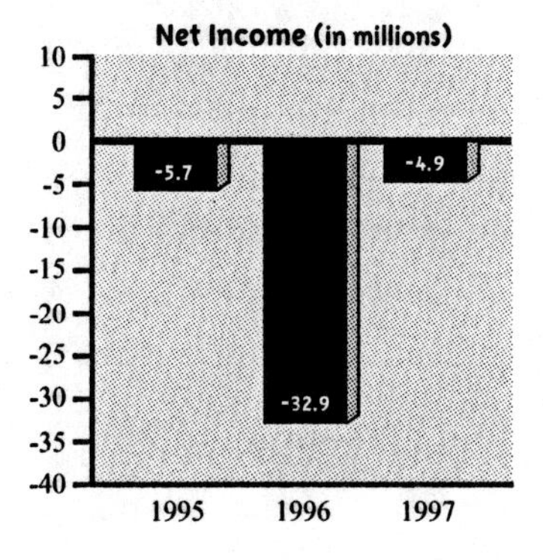

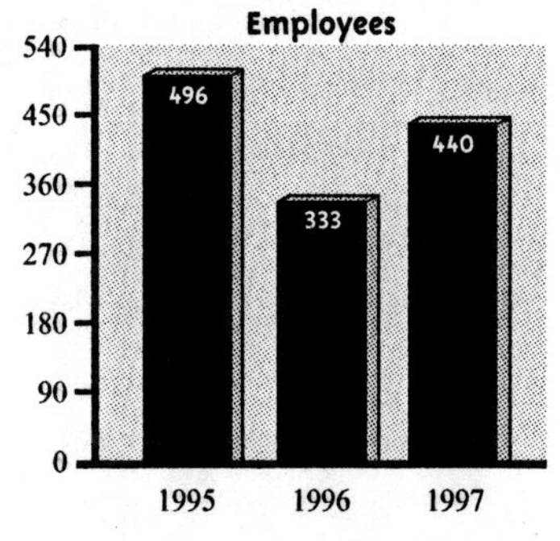

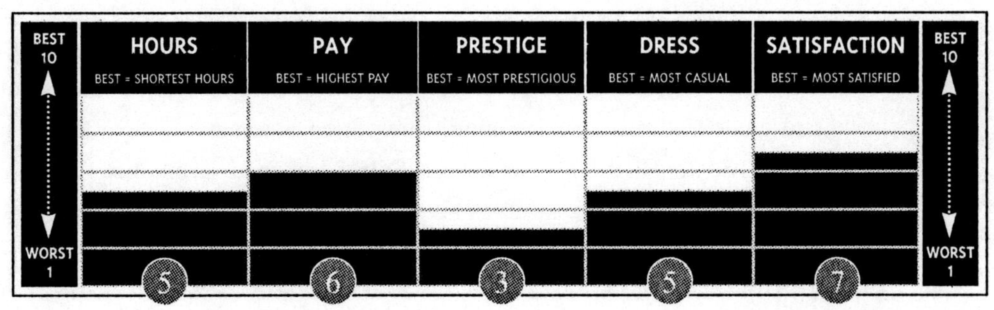

THE SCOOP

Platinum provides companies with a "blueprint" of their businesses with its Enterprise Resource Planning (ERP) software. The program essentially maps out how a company's data and applications fit together on a network. These data items can include purchase orders and inventories to marketing costs, customer service, and foreign currency transactions. The company was founded in 1984 by Gerald Blackie, Timothy McMullen, and Kevin Riegelsberger, who developed the first financial accounting software for multi-user LAN-based environments (The Platinum Series). In 1992, Platinum introduced SQL (often pronounced "sequel"), an accounting application for networks, designed to run with Microsoft products. One of Platinum's most successful products, SQL has won a number of awards, including the 1997 Microsoft Industry Solutions Award for "Best Functionality."

Though consistently praised for the high quality of its software, the Platinum name was slightly tarnished by a bookkeeping scandal in 1994. The SEC discovered that company officials had overstated revenues for two years and charged the corporation with accounting fraud. Without admitting to any transgressions, the company settled the issue and paid shareholders $17 million in cash and stock to end the class-action suit.

In an effort to regroup, the company brought in L. George Klaus as CEO in February 1996. He quickly hired 140 new sales and technical personnel and increased contacts with independent programmers for freelance assignments. Klaus also bolstered the company's sales operations by renewing relationships with a network of independent sales representatives and reinstated Platinum's in-house direct sales force that had been eliminated shortly before his arrival. Klaus is credited with bringing Platinum back to life: by the end of fiscal 1996, Platinum saw a 31 percent sales increase, and reported a much smaller $4.9 million loss.

With a solid foothold in the mid-range market for accounting software, Platinum must now diversify its offerings. In the near future, the company will offer "enterprise-wide solutions" that will include enterprise resource planning, sales force automation, payroll and human resource programs. So far, Platinum has successfully broadened its market focus through partnerships with other companies as well as acquisitions and in-house development. It has partnerships with technology and hardware companies (including Compaq, Hewlett-Packard,

Novell), software companies (including Microsoft, Best Software, and Paradigm Technologies), and related distribution vendors. Through one such alliance with EC Company (a producer of e-commerce software), Platinum is developing an electronic invoice and secure payment-processing system for online shopping.

In June 1997, the company acquired Clientele Software, Inc, which makes customer service applications, such as Help Desk. That year Platinum also purchased FocusSoft, a leading producer of mid-market manufacturing and distribution software. The two companies have become divisions of Platinum Software, and their products will be integrated with its existing marketing, packaging and sales programs. In 1998, Platinum acquired Logic Works, manufacturer of the best-selling data modeling software in the industry. The addition of Logic Works' products makes Platinum a dominant force in the enterprise modeling market. Combined, their applications offer the most comprehensive technology for modeling application components, databases and business processes. This technology is used by application developers, database administrators, and data warehouse managers.

Though it has been doing all the purchasing lately, industry analysts say Platinum is a prime acquisition target itself. This is reinforced by the fact that Klaus has made a name for himself revamping and selling Silicon Valley companies. Prior to joining Platinum, he was called in to fix up San Jose-based Frame Technology Corp., where he led a massive restructuring before the company was snatched up by Adobe Systems in 1995. However, Klaus denies that he has intentions of selling Platinum in the near future.

GETTING HIRED

Platinum posts openings for national and international jobs on its website, www.platsoft.com. Resumes may be faxed, posted, or e-mailed to the Human Resources department at the Irvine, CA headquarters. According to one insider, "they screen you pretty good, though it varies for each department." The insider adds that "referrals definitely help."

OUR SURVEY SAYS

Overall, insiders are thrilled that George Klaus came aboard. Sources say the new CEO "has made a special effort to build a positive and pleasant corporate culture." In addition to improving the company's financial health, Klaus has been instrumental in improving morale, "which helps productivity," noted one employee. Since he arrived, he has instituted an annual company picnic, and throws a party after each quarterly all-company meeting. "With our recent financial successes," adds one insider, "we have had a lot to party about."

Work hours are "40 to 45 hours a week, though people will work overtime and weekends as special needs come up." Employees say "the compensation is good," and are relatively pleased with insurance and other benefits. One source reveals a "small but interesting perk – we can buy soft drinks [in the office] for 25 cents." "It may not seem like much," he continues, "but it's a good indicator of a change in company culture and even financial health."

Dress requirements vary, based on position. Employees with regular client contact (marketing, executives, sales) "tend to dress up – shirts and ties, nice dresses, and business suits for presentations or customer visits." In development and tech support, "people dress down – polo shirts, jeans, and occasionally shorts." Outside of the "regular stress and deadlines," employees say Platinum is "a pleasant place to work," and "the people are great."

"There is an employee free software program that allows all employees to get one free copy of all software made by the company."

– Software insider

Quarterdeck

13160 Mindanao Way
Marina del Rey, CA 90292
(310)309-3700
Fax: (310) 309-4218
www.qdeck.com

LOCATIONS

Marina del Rey, CA (HQ)
Columbia, MO
Dublin, Ireland
Australia
Japan
United Kingdom

DEPARTMENTS

Operational
Sales and Marketing

THE STATS

Annual Revenues: $83.8 million (1997)
No. of Employees: 200 (worldwide)
No. of Offices: 6 (worldwide), 2 (U.S.)
Stock Symbol: QDEK (NASDAQ)
Interim CEO: King R. Lee

UPPERS

- Free software
- Gym discounts

DOWNERS

- Company in chaos
- No job security

KEY COMPETITORS

- Microsoft
- Network Associates
- Oracle
- Symantec

EMPLOYMENT CONTACT

Arla Jordan
Sr. Staffing Consultant
Human Resources
Quarterdeck Corporation
13160 Mindanao Way
Marina del Rey, CA 90292

Fax: (310) 309-3706
ajordan@quarterdeck.com
or HR2empwr@aol.com

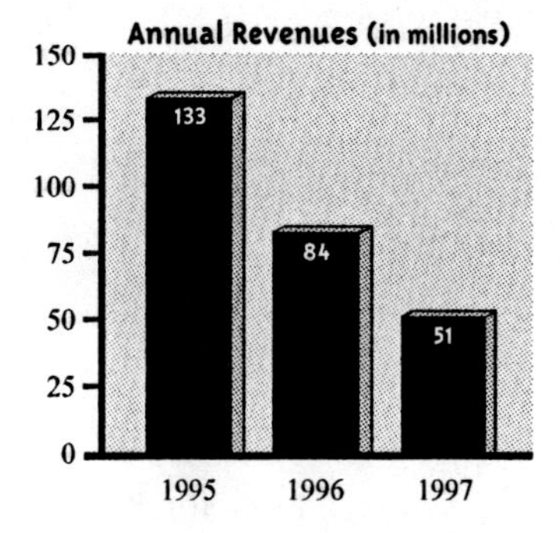

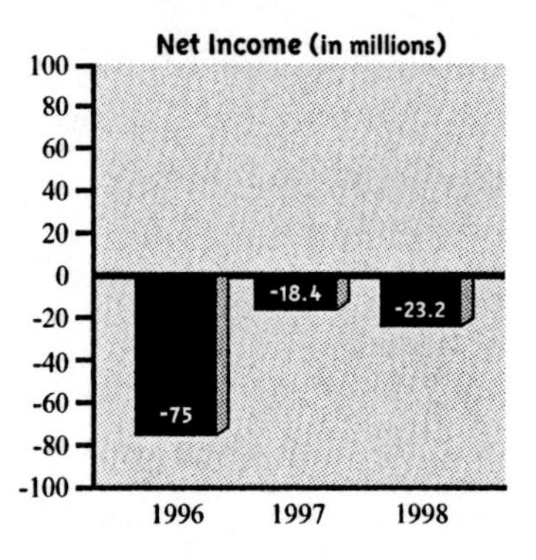

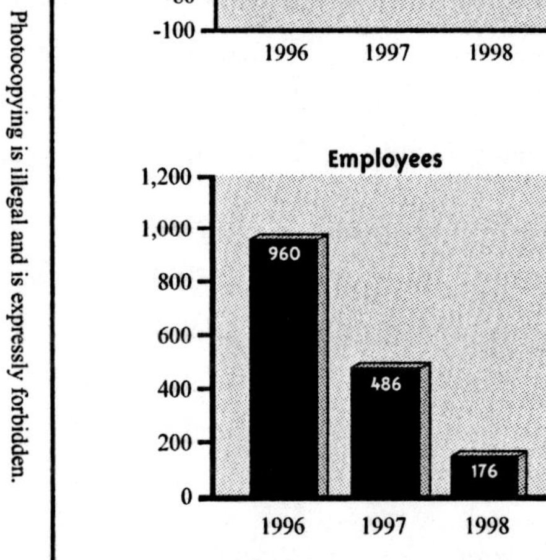

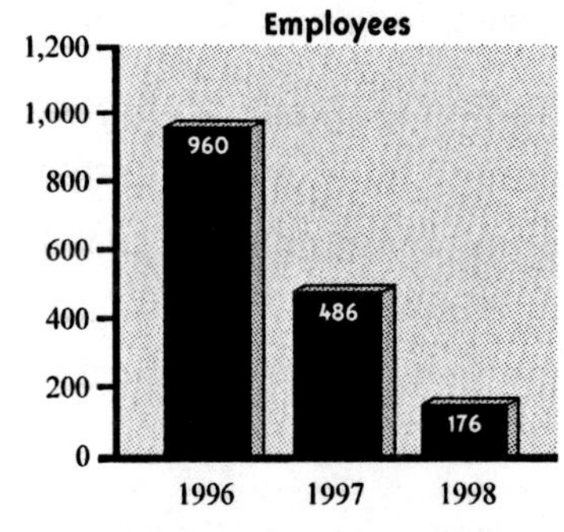

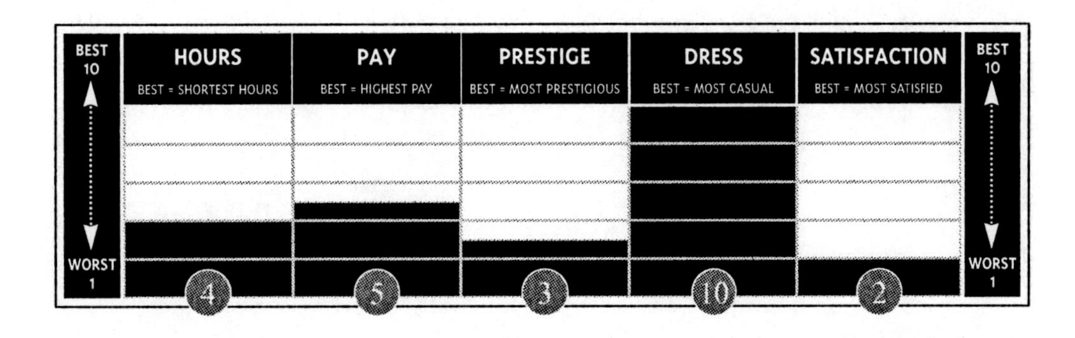

THE SCOOP

Founded in 1983, Quarterdeck Corporation develops, markets, and provides support for software products for desktop PCs, corporate intranets, and the Internet. Its products are mostly disk space and utilities cleanup products, antivirus programs, and information management products such as Procomm Plus 32 and WebCompass. Quarterdeck sells its product lines through thousands of retail outlets in over 35 countries, and also sells directly to consumers, focusing on small businesses and telecommuters. To expand its offerings, Quarterdeck has engaged in a campaign to acquire companies, products, and services. This effort has been accompanied by divestitures of certain underperforming divisions.

The 15-year-old company was founded in a garage by Therese Myers and Gary Pope to develop software to enhance PC performance. Microsoft was both the company's prime competitor as well as Quarterdeck's benefactor. When Windows 3.0 was introduced to the public, Quarterdeck Office Services experienced a boost in sales because its memory management products worked with fewer glitches than Windows.

Sales of Quarterdeck's QEMM software was consistently strong in the early 1990s, and the company went public in 1991. Then its rival, Microsoft, improved Windows's memory management software, and Quarterdeck's sales took a dive. In 1994, King Lee, the head of a consulting firm, took over Myers' position; the next year Gaston Bastiaens, a former manager at Apple computer, replaced Lee as CEO. The company changed its name to Quarterdeck Corp. at that time.

Quarterdeck regrouped by acquiring Internet software developers Internetware, Prospero systems Research (chat software), and StarNine Techonologies (Internet and e-mail). In 1996 Quarterdeck announced plans to buy the communications software firm Datastorm Technologies, the document-sharing-software firm Future Labs, Inc., and the marketing-software firm Vertisoft Systems. In 1997 Quarterdeck entered into an alliance with Marubeni Corp. to boost sales in Japan. The company also entered the antivirus software market that year, purchasing TuneUp.com, an online Windows 95 service center.

Despite these changes, the company continues to struggle, losing $23 million in fiscal year 1998. Quarterdeck has discontinued or "de-emphasized" several categories, including Hijaak, a graphics utility line and products for the Macintosh. "The sale of Hijaak graphics software aggravated some of Quarterdeck's customer base," notes an employee disapprovingly. Curt Hessler, the former I-Net chairman who took over as Quarterdeck's president and CEO in 1997, resigned in 1998 as part of an overall corporate streamlining after a quarter of weak revenue and operating loss (due to low industry-wide sales in PC-utility products). In December 1998, Quarterdeck planned to be acquired by Symnatec, as its stock languished around 50¢ a share.

GETTING HIRED

Quarterdeck posts its job openings, with a full description of each job and instructions on how to apply, on its web site, www.qdeck.com, under Employment Opportunities. The company is not exactly on a hiring binge: The company has laid off 50 percent of its workforce in the past year and a half, says one recent layoff victim. Having a contact in the company is a big plus; Quarterdeck has a "recruitment program" wherein employees receive a bonus for "successfully recruiting new talent," insiders tell us.

OUR SURVEY SAYS

Quarterdeck, its employees say, is on life support. Explains one insider: "The company bought 12 companies in 12 months, and did not do a good job of it, which hurt the firm." "There was no real synergy between management, staff, product lines and corporate identity. The company began to stumble, and the stock went from a high of 39 to around 3, remarks one former employee. Now, the company is trying to "sort itself out" – the staff consists mainly of new employees. Downsizing "over the last year" has resulted in the loss of "people with smarts," though one optimist cites the "chance to rebuild" as a plus.

The dress code is casual "company-wide": "They want you to come in dressed, that's about it." Work hours can be "grueling," due to "too few people chasing after too much work." Because of the chaos, "Quarterdeck offers notoriously poor tech support and customer service and suffers consequently." Lately, however, management has begun to "realize the value of diminishing returns" and the pay is now "acceptable" after having been "lower than what the market required." Perks include "the usual passel" of "medical, dental, and vision" benefits. Starting employees get 10 paid holidays a year plus two weeks vacation, employees tell us. Also, "there is an employee free software program that allows all employees to get one free copy of all software made by Quarterdeck. There's special employee pricing for additional copies." Quarterdeck is located at the end of the 90 freeway ("not the most scenic area in Southern California") across from a gym where the company has arranged for "cheapo memberships." Despite these perks, a contact says, "be aware that Quarterdeck will lay you off in order to salvage their bottom line."

"The mode seems to be we hire people, give them a year and then fire them if they aren't making progress."

– ***Software insider***

RealNetworks

111 Third Ave., Suite 2900
Seattle, WA 98101
(206) 674-2700
www.real.com

REALNETWORKS

LOCATIONS

Seattle, WA (HQ)

DEPARTMENTS

Advertising Sales
Human Resources
Information Systems
Marketing
Product Management
Program Management
Sales
Software Development
Software Testing

THE STATS

Annual Revenues: $32.7 million (1997)
No. of Employees: 326
No. of Offices: 1 (U.S.)
Stock Symbol: RNWK (NASDAQ)
CEO: Rob Glaser

UPPERS

- Flexible work schedule
- Challenging and exciting atmosphere
- Casual dress

DOWNERS

- Potentially long hours
- Company has fast burn rate

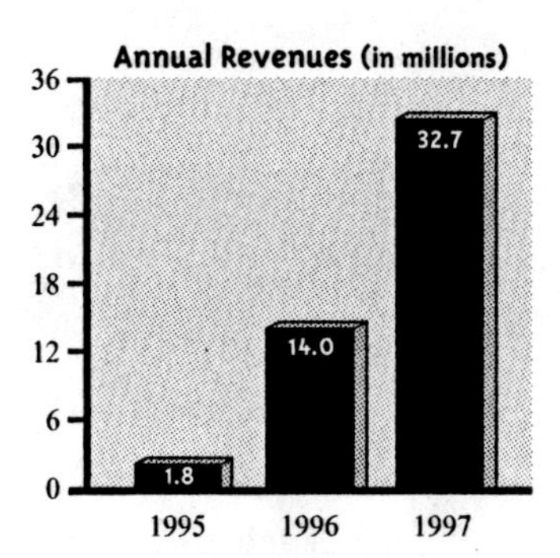

KEY COMPETITORS

- Cisco Systems
- Macromedia
- Microsoft
- Motorola
- Oracle

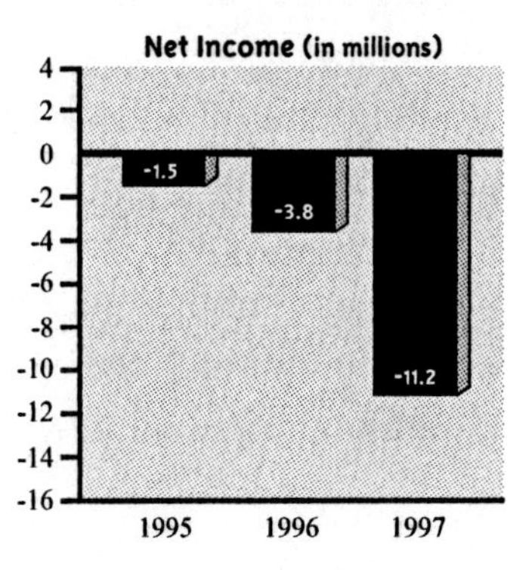

EMPLOYMENT CONTACT

Human Resources
RealNetworks
111 Third Ave., Suite 2900
Seattle, WA 98101

jobs@real.com

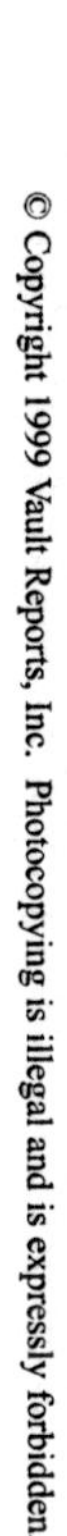

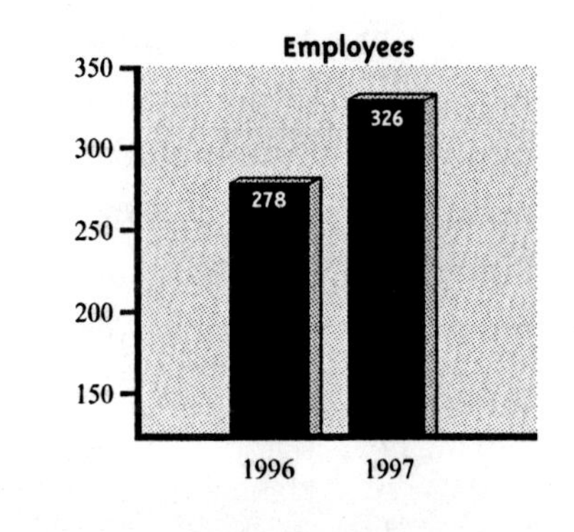

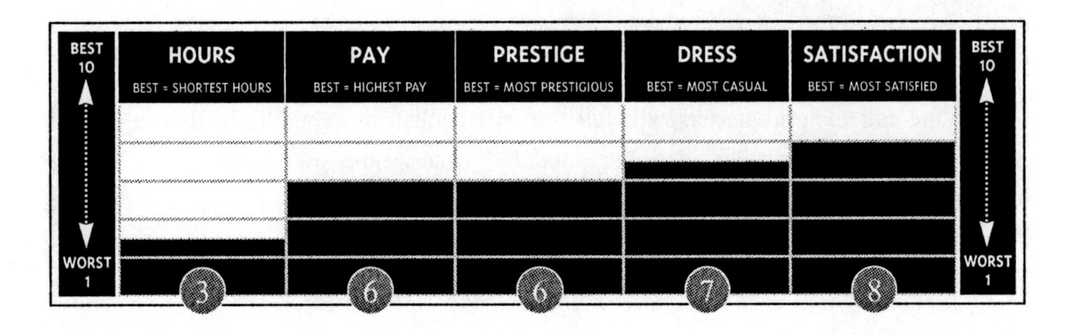

THE SCOOP

The leader and pioneer in the delivery of real-time media over the Internet, RealNetworks lists all the major television networks, the U.S. Senate, Merrill Lynch, and the National Football League among its clients. CEO Rob Glaser left Microsoft, where he had been vice president of multimedia and consumer systems, and founded the company in 1994. The next year, RealNetworks (then called Progressive Networks) introduced the RealAudio system, which allows media providers to create and deliver audio-based streaming multimedia content through the Internet. Downloading the RealPlayer from a company web site lets regular Joes with regular phones and personal computers to play back content on-demand without download delays. More than 32 million players have been distributed; 125,000 are downloaded every day. RN also has created a system to deliver motion video.

The company has teamed up with industry leaders, including Macromedia, Sun Microsystems, Apple, and Netscape to begin building an infrastructure for the fledgling world of real-time media delivery. The company is the creator of the RealMedia Architecture (RMA), which is a framework for the development of streaming multimedia applications. RMA also provides a successful business model, based on Real Network's success, for companies that wish to move into the field. In 1996, the company announced a project called the Real Time Streaming Protocol (RTSP), which would standardize the streaming of multimedia content for clients and servers who use different vendors.

GETTING HIRED

A phone call to headquarters will tell you what jobs are available in this rapidly-growing company. RealNetworks also lists detailed descriptions of openings on its web site (www.real.com), broken down by department. The jobs are surprisingly diverse, ranging from technical writing to managing advertising accounts, and the qualifications are also all over the map: Some positions call for a degree in Computer Science, while for others, an MBA is helpful. Experience in the computer industry is always a plus – many of the key managers at RealNetworks worked for Microsoft. Resumes can be submitted via e-mail.

OUR SURVEY SAYS

Without exception, employees say working for RealNetworks means fairly flexible hours, a casual dress code, and excellent opportunities for women and minorities. One employee, however reports that despite their flexibility, the hours average about 60 a week. Despite formidable job descriptions on the company web site, the average employee age is young, "including many recent grads and interns." Pay is "competitive with other companies in the area." The company culture is described as exciting and energetic, and the employees as especially bright and independent. "We have people here working from all over the world," says one employee. If you manage to land a job with RealNetworks, you'll have a leg up in the computer industry. "The company's name and reputation are so widely known that if you came here, you would be picked up faster than the same person who only worked at Microsoft," one employee says.

SAP

3999 West Chester Pike
Newtown Square, PA 19073
www.sap.com

LOCATIONS

Philadelphia, PA (U.S. HQ)
Waldorf, Germany (World HQ)
Offices in over 50 countries and in 30 U.S.cities.

DEPARTMENTS

Consulting
Customer Support
Development
Sales & Marketing
Technical Support

THE STATS

Annual Revenues: $3.3 billion (1997)
No. of Employees: 16,000(worldwide)
No. of Offices: 80+ (worldwide)
Stock Symbol: SAP (NYSE)
CEO: Hasso Plattner and Henning Kagermann

UPPERS

- Extensive benefits
- Casual dress
- Major perks
- Surfing the wave of Y2K

DOWNERS

- Potentially long hours
- Stiff competition in hiring

KEY COMPETITORS

- Baan
- Computer Associates
- Informix
- J. D. Edwards
- Oracle
- PeopleSoft
- Platinum Software
- System Software Associates
- Tivoli

EMPLOYMENT CONTACT

SAP America, Inc.
Human Resources
701 Lee Road, Suite 200
Wayne, PA 19087

(800) 727-8747
staffing.america@sap-ag.de

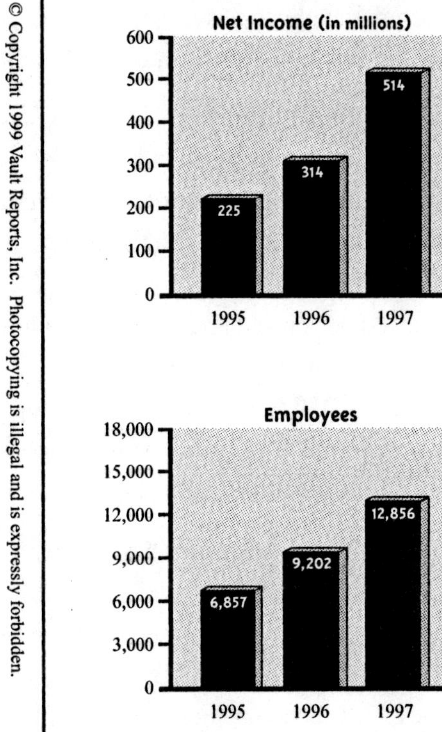

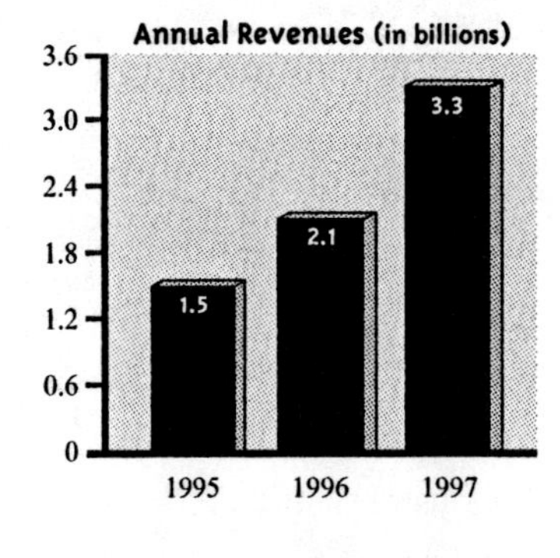

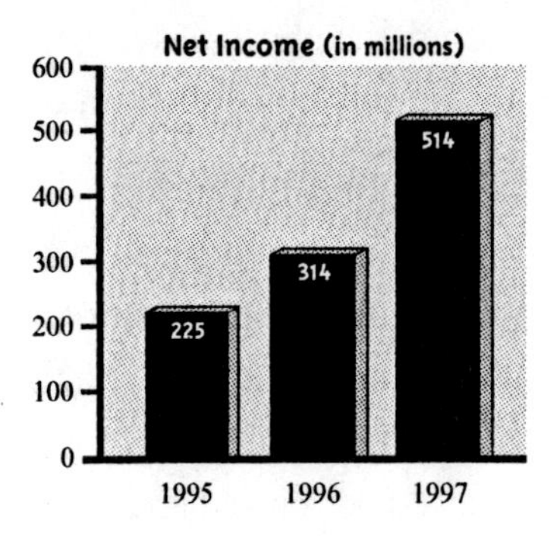

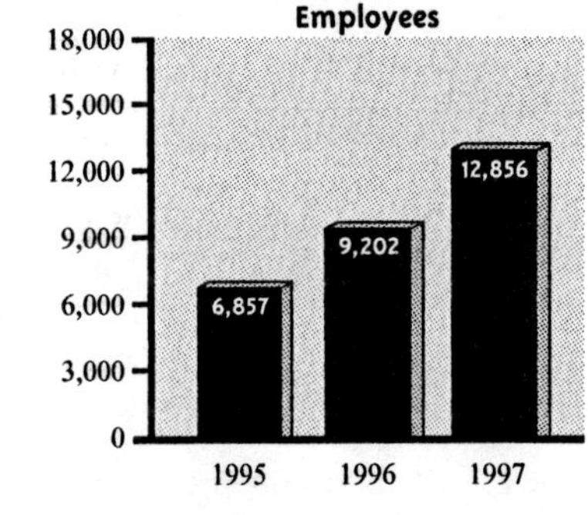

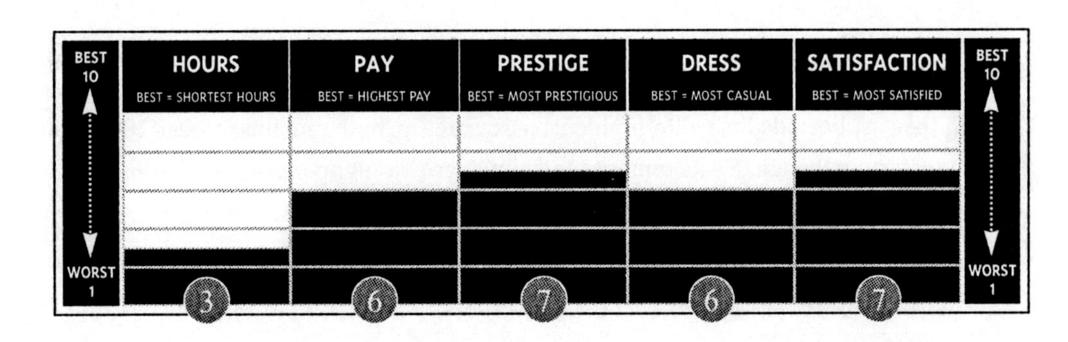

THE SCOOP

After starting as a small business software company in 1972, the German-based SAP (Systems, Applications, and Products in Data Processing) has in the last few years grown by astronomical proportions. SAP is now the fourth-largest software company in the world. Fueled by a unique enterprise application system, R/3, it has become the fourth-largest company in all of Germany and the only non-American concern among the world's five largest software producers.

R/3 may sound like a character from Star Wars, but if anything, it is even more sophisticated and expensive to produce than a droid. The software integrates the computing needs of all the operations of a business, from manufacturing schedules to shipping to accounting, increasing its users' efficiency and reducing costs. For example, Chevron found its system of transactions streamlined significantly by using R/3, thereby cutting costs as much as 25 percent. For its trouble, SAP picked up a cool $160 million. This type of high-fee, high-reward salesmanship has caught on quickly. To date, SAP has gobbled up a staggering one-third of the $10 billion market, including half of the world's top 500 corporations. The company intends to make its system as much an industry standard as Windows is in the realm of personal computing. Listing Microsoft among its clientele, it is clear that such a goal is not terribly farfetched.

Not surprisingly, revenues have grown considerably – the company has posted eye-popping growth figures of 92, 44, 38, and 62 percent in the last four years. This has enabled the company to reinvest in expansion – SAP prides itself on being No. 1 in research and development in the industry. In 1996 alone, SAP apent $2.6 billion on R&D, or about eight times what its competitor Baan spent. This type of investment will eventually allow SAP to tailor future versions of R/3, a software package with a reputation for lack of user-friendliness, to the specific needs of its clients' various fields and to use on the Internet. The public sector appears to be the next major target of the SAP juggernaut, as well as a few select industries that have yet to join the party. Pessimists point to the large proportion of SAP's business that has in the past been dedicated to problems associated with the infamous Year 2000 computer glitch. For now though, SAP continues to enjoy a growing grip on the computing solutions of the global business community.

GETTING HIRED

Recent success has bred an intensely competitive hiring practice. The bandwagon is cruising along, but seating is limited, and getting an interview can be exceedingly difficult. According to one insider at the office, the human resources department receives "about 750 resumes a day, and this year they're only planning on hiring about 200 people." Proficiency with SAP software is a big advantage, even when applying for positions not directly requiring its use. SAP "really needs people to have been in the business world for awhile," so recruiting seems to happen "from just about anywhere except a college campus."

The lucky few are subjected to interviews that are described as extensive and often numerous, but rarely stressful. "The interviewing is very intense," one employee remembers. SAP Canada, for example, "has a four-interview minimum. Further difficulty in landing an offer comes about as a result of the very low turnover rate. Insiders advise, "Don't waste time talking about experience that is not directly related to what you want to do at SAP."

New hires are few and far between relative to the number of applicants, but those who do make the grade are treated to an environment that fosters tremendous loyalty. Most new employees, unless very familiar with SAP's software applications, are put through an eight- to 10-week training process covering most aspects of the particular division, including non-technical skills. For more information on the SAP interviewing process, try the company's web site at www.sap.com.

For application information, SAP can be contacted at staffing.america@sap-ag.de, faxed at 1-800-727-8747 or (1-781-672-6726 from outside the U.S.) or contacted via mail at the addresses listed at the company's web site.

OUR SURVEY SAYS

SAP employees rave about their work environment. The atmosphere is relaxed, from the loose, "business casual dress code" to the "flexible schedule." Despite flexibility, "you can expect a minimum of 50 hours per week and sometimes up to 70." Hours will depend on department, and in most cases, "it all evens out." As far as compensation is concerned, one insider reflects an almost universal sentiment by saying, "the pay is great, the benefits are well above average, and the incentives are terrific!"

Ample vacation time is provided, as well as "the best health insurance I have ever seen and even a pension plan." Additional perks include many free meals for all and a company car for third-year workers from the Germany office. "Without a doubt it is the finest company in terms of work culture," says one contact. But one insider warns, "The mode seems to be we hire people, give them a year and then fire them if they aren't making progress." It seems that SAP is "definitely not shy about letting people go." But while you're there, SAP is very supportive because "the corporate culture is very focused on letting you do your job" and has a very "person-centric" culture. The environment "allows you to express yourself freely, ask for assistance as needed, and keep up-to-date on new product features and company news."

"SAP is a German company, hence there are elements of German influence in many levels of the organization," says on insider. However, xenophobia isn't one of them. Equality of opportunity is not an issue; SAP "has a very diverse culture. There are at least as many women as men in management and many, many minorities are part of the team." "Everyone is treated with equality and with due respect," reports another contact. Seems like the only problem is getting a foot in the door. Once that is accomplished, "the experience is excellent and [SAP] carries a well-recognized brand name." And employees agree that coming to SAP is "a good choice to make" because SAP believes in "performance and spirited work" above all else.

"With some companies, 12- to 14-hour days are expected. Here, you're not expected to work yourself to death."

– ***Software insider***

Scientific-Atlanta

One Technology Pkwy South
Norcross, GA 30092-2967
(770) 903-5000
Fax: (770) 903-4617
www.sciatl.com

SCIENTIFIC-ATLANTA

LOCATIONS

Norcross, GA (HQ)
Tempe, AZ • Menlo Park, CA • San Jose, CA • San Ramon, CA • Santa Fe Springs, CA • Englewood, CO • Kissimmee, FL • Atlanta, GA • Naperville, IL • Braintree, MA • Methuen, MA • Quincy, MA • Laurel, MD • Califon, NJ • Parsippany, NJ • Signal Mountain, TN • Houston TX • Renton, WA

International offices and factories:
Canada • China • Hong Kong • India • Japan • Korea • Taiwan • New Zealand • Singapore • Australia • Spain • Germany • Hungary • Italy • Switzerland • United Kingdom • Argentina • Brazil • Chile • Mexico • Israel • Qatar • South Africa.

DEPARTMENTS

Administration • Business Development • Customer Service • Engineering (Software and Hardware) • Finance/Accounting • Human Resources • Public Relations • Sales & Marketing • Software Development • Technical Services • Training

THE STATS

Annual Revenues: $1.2 billion (1998)
No. of Employees: 5,736
No. of Offices: 6 (U.S.), 4 (worldwide)
Stock Symbol: SFA (NYSE)
CEO: James McDonald
Year Founded: 1951

UPPERS

- Well known in the industry
- Great training programs for engineers
- Culturally diverse workforce
- Free breakfast on Fridays

DOWNERS

- Downsizing
- Excessive politics

KEY COMPETITORS

- ADC Telecommunications
- Andrew Corporation
- ANTEC
- BroadBand
- California Microwave
- General Instrument
- Hughes Electronics
- Lucent Technologies
- NEC
- Tellabs
- Zenith

EMPLOYMENT CONTACT

Brian Koenig
Human Resources
Scientific-Atlanta, Inc.
One Technology Pkwy South
Norcross, GA 30092-2967

Fax: (770) 903-4617
SA.Staffing@SciAtl.COM

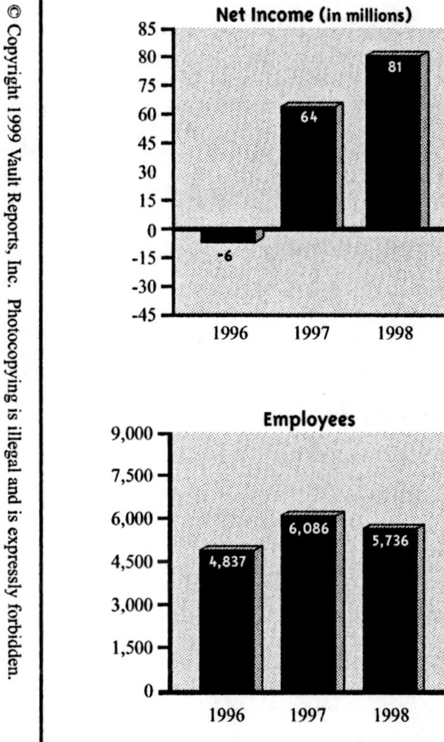
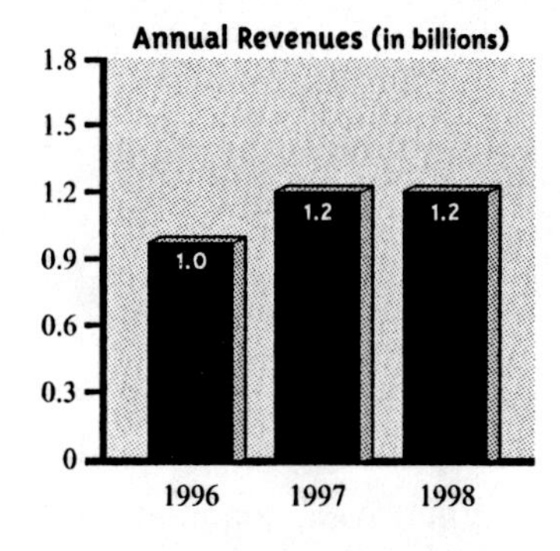

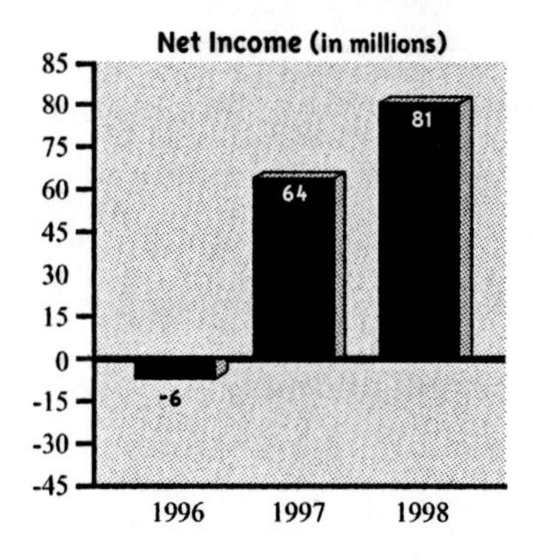

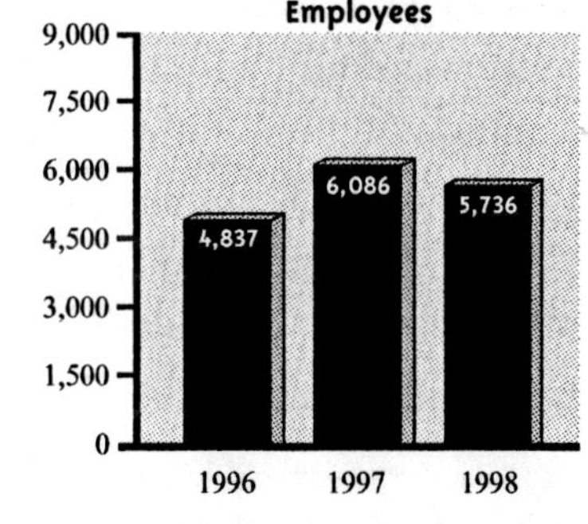

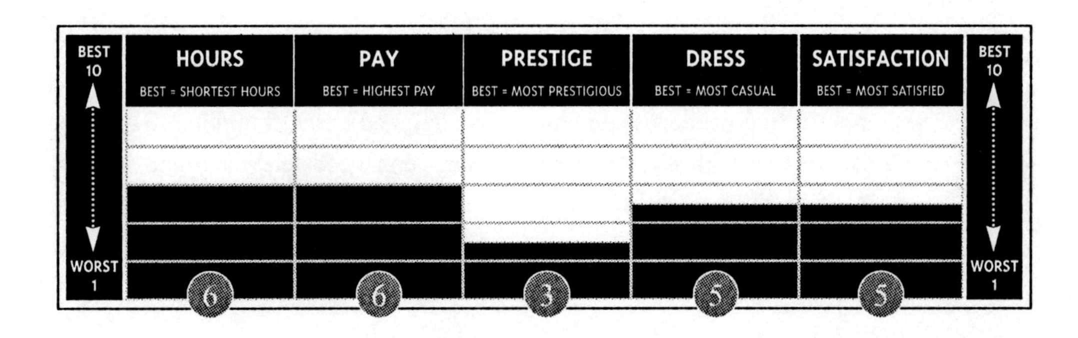

THE SCOOP

Scientific-Atlanta is the industry's leading supplier of broadband communications systems and satellite-based communications networks. Best known for making Home Communications Terminals (those cable boxes on top of your TV set), the company also produces digital video compression products, transmission and distribution equipment, and satellite equipment and systems. SA was formed in 1951 by seven professors from the Georgia Institute of Technology. They started out making electronic test equipment for antennas and performing instrument maintenance for local schools and hospitals. Soon after, they began doing contract development work for private industries and the military.

In 1973, the company launched its first satellite earth station, used to provide cable TV services. It quickly became a leader in the burgeoning industry. In 1978, it acquired Spectral Dynamics, a scientific equipment manufacturer with an established sales network in Europe. Using that network, the company easily transformed itself into an international force. After restructuring and extensive layoffs in the late 1980s, the company set its focus on satellite communications and put a considerable investment in R&D. In 1996 Scientific-Atlanta acquired ATx Telecom Systems, a fiber-optics company. Sales declined after passage of the 1996 Telecommunications Act, so the company strengthened its focus on digital cable and interactive products. It also intensified its international presence with the purchase of Arcodan A/S, a Danish technology firm specializing in optoelectronics and radio distribution equipment. Scientific-Atlanta has also forged strategic alliances with a number of American and international companies, including BT (formerly British Telecom), Siemens Public Communications Network Group, and PowerTV (actually a subsidiary of SA, which develops software for its set-top boxes).

In 1998, the company introduced its Explorer 2000 interactive digital set-top boxes. Equipped with a cable modem, the product allows users to access interactive applications on their TV sets. These include Internet access, video-on-demand, and home shopping. SA also offers a cheaper, analog version with e-mail, web browsing, enhanced television and less extensive video-on-demand capabilities. In April 1998, Scientific-Atlanta announced an agreement with Sun Microsystems to license Sun's PersonalJava application for the Explorer 2000. SA is also in talks with companies like Microsoft, Oracle, and WorldGate about further software applications to run on its new Explorer 2000 box – which is capable of delivering full video on demand. Meanwhile, SA has just introduced a new line of components that increase speed and bandwidth, necessary for high-definition television signals as well as interactive video, data and voice services.

In June 1998, SA announced a restructuring move that will eliminate 275 jobs at its Atlanta HQ, and relocate 150 employees from other company outposts. It will also reduce costs by transferring some production to its Mexican factory. The company is building a $40 million facility that will consolidate its Atlanta operations, which are currently scattered in 18 locations.

GETTING HIRED

The company recruits on college campuses, and posts job listings at www.sciatl.com. Students take note: sources say "the company likes to use co-ops or interns and hires them after graduation." Another important fact: "You have an advantage if you're bilingual." Spanish and Portuguese are especially helpful, as Scientific-Atlanta's "largest production facility is located in Juarez, Mexico and we have an active sales and support staff for Latin America."

Sources say interviews are "pleasant" and "painless," and generally consist of up to four interviews with members of the specific department and an HR representative. You may get some technical questions (depending on the job you're going for), but sources promise "they will not browbeat you." The whole thing will last several hours, and "one of the interviewers will take you out to lunch if it happens to fall in the middle of this process." "Make sure you are familiar with the company's product lines and what we do," advises one insider. At the same time, they don't expect you to be completely familiar with SA's "highly specialized" products – "we often find someone with a favorable technical background and begin a long training period."

OUR SURVEY SAYS

Insiders say "if you can't deal with politics," SA "is not the place to go." "It's a large company," one source says, so it's natural to find "politics between divisions and even smaller groups. Most people just get used to it and accept it." Other contacts add that "workers sometimes get lost in the shuffle," and "highly recommend pursuing a career with a company with a little more stability. At the moment, the company is in the throes of layoffs, and sources say "they are not finished."

Some sources, however, extol "the joy of working for Scientific-Atlanta." They are enthusiastic about their "challenging, very rewarding" jobs, and say "[Scientific-Atlanta is] truly an environment in which you are always learning." One insider says: "It has the benefits of being a large company, but the small company feel of camaraderie and esprit de corps." When traveling, employees enjoy "nice hotels" and "a generous per diem on meals and entertainment." "Scientific-Atlanta has been very good to me and my family," declares one employee, who adds that he has "turned down higher-paying positions because I am truly happy where I am and where I am headed."

Another insider points out that the though the company invests a great deal in training its engineers, it does not do much to promote the development of its other employees – "It seems to be the unwritten policy here." He goes on to add that companies that offer "training, lunchtime seminars and team-building activities have a better workforce. But most insiders at SA find their jobs to be "interesting and challenging." Dress codes vary – formal business dress is required at the corporate headquarters, while employees in other areas wear business casual. Fridays are casual days, and "some people come to work in jeans and tennis shoes." Official work hours are 8 to 5, but "many people come in between 8:30 and 9:00, and leave around 6 p.m." Some stay later if they have a project to finish.

Starting salaries for engineers are "either at or above average, depending on how badly the company wants you." Benefits are "very good," and include health benefits, tuition reimbursement, 401(k) and stock options for some employees. In addition, the company occasionally provides free Odwalla juices, bagels, and pastries for breakfast on Friday mornings. Sources say the company is "very culturally diverse," though the engineering divisions tend to be "male dominated."

Seagate Technology

P.O. Box 66360
Scotts Valley, CA 95067
(831) 438-8111
Fax: (831) 438-3320
www.seagate.com

SEAGATE TECHNOLOGY

LOCATIONS

Scotts Valley, CA (HQ)
Anaheim, CA • Atlanta, GA • Costa Mesa, CA • Dallas, TX • Fremont, CA • Longmont, CO • Minneapolis, MN • Moorpark, CA • Oklahoma City, CA • Pittsburgh, PA • San Jose, CA • Santa Maria, CA • Westford, MA • China • France • Germany • Indonesia • Ireland • Malaysia • Singapore • The Netherlands • United Kingdom

DEPARTMENTS

Administration
Application Engineering
Customer Service
Development Engineering
Equipment Engineering
Finance
Information Systems
Legal
Manufacturing
Materials
Mechanical Design & Drafting
Process
Production
Research & Development
Sales & Marketing

THE STATS

Annual Revenues: $6.8 billion (1998)
No. of Employees: 87,000 (worldwide)
Stock Symbol: SEG (NYSE)
CEO: Stephen Luczo

UPPERS

- Quarterly profit sharing payments
- Bonuses
- Casual dress code

DOWNERS

- Long workdays
- Plenty of competition

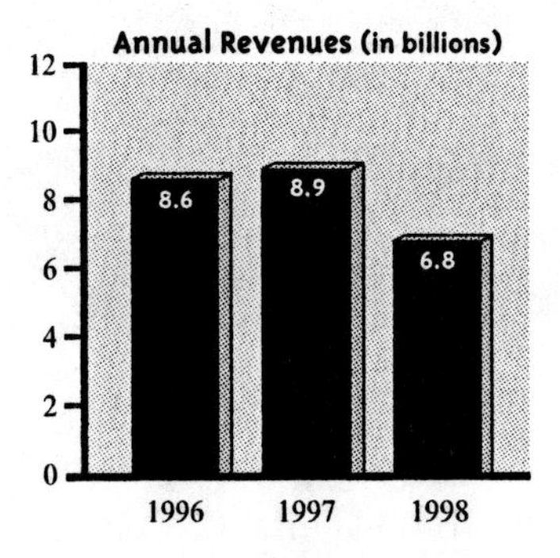

KEY COMPETITORS

- IBM
- Iomega
- NEC
- SyQuest
- Toshiba
- Western Digital

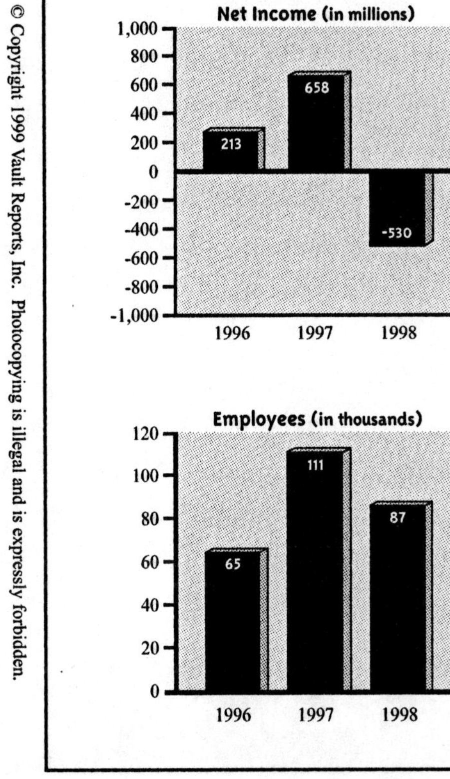

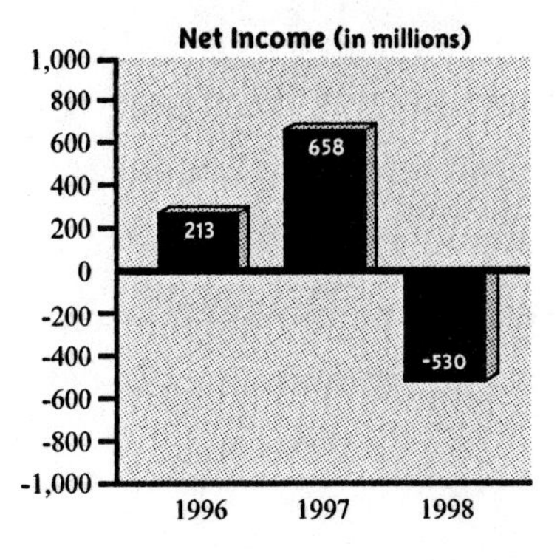

EMPLOYMENT CONTACT

Human Resources
Seagate Technology
P.O. Box 66360
Scotts Valley, CA 95067

(831) 439-2590
Fax: (831) 438-3320
Job hotline: (408) 439-JOBS

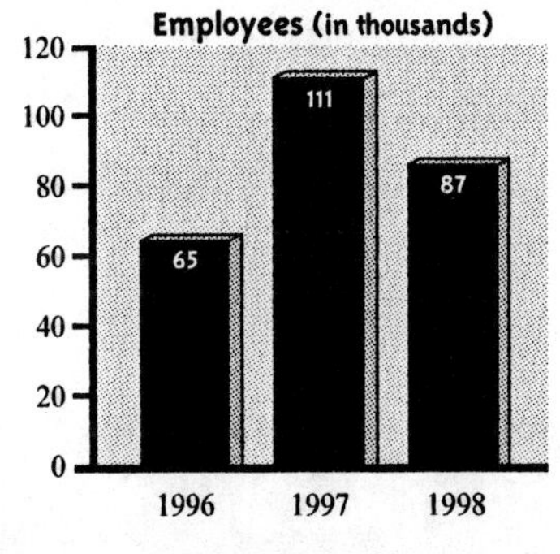

HOURS	PAY	PRESTIGE	DRESS	SATISFACTION
BEST = SHORTEST HOURS	BEST = HIGHEST PAY	BEST = MOST PRESTIGIOUS	BEST = MOST CASUAL	BEST = MOST SATISFIED
4	7	7	6	8

BEST 10 — WORST 1

THE SCOOP

A leading maker of computer disk drives, Seagate's first product, 5.25-inch hard disk drives for IBM mainframe computers, quickly grabbed half of the small disk drive market in the early 1980s. However, Seagate's profits fell along with IBM's in the mid-1980s. To cut costs, Seagate moved its manufacturing facilities to Singapore and began to integrate its production process. By 1986, Seagate had finished its first round of acquisitions and revenues were once again rising. In 1989, Seagate purchased an important rival, Control Data, doubling its size.

In recent years, Seagate's acquisitions have positioned the company as a rising software maker. Its 1996 purchase of Connor Peripherals make Seagate the largest disk drive company in the world, and is only one step towards Seagate's goal to raise $1 billion in software revenue by 1999. However, a British court has recently ruled that Seagate must pay British computer maker Amstrad $153 million for faulty disk drives that led to a recall of over 57,000 Amstrad computers.

An industry slump in 1997, along with growing competition and falling prices in low-end disk drives, led to the ouster of chief executive Alan Shugart, who had founded the company in 1979. Shugart's successor, Stephen Luczo moved in August 1998 to consolidate the company's disk-drive and related operations with a management restructuring.

GETTING HIRED

Seagate's web site, located at www.seagate.com, enables applicants to search through a list of current openings by department, title, or location. Applicants can also post resumes to Seagate through this web page. Resumes must reference specific job openings by using the job code that Seagate provides on the site.

OUR SURVEY SAYS

Seagate's American operations are divided into technical and corporate divisions. Technical employees emphasize Seagate's "top-of-the-line" technological resources and praise the "interactive" environment. Recent hires in technical divisions say that their "senior colleagues are generous with their time" – and that they can be "flexible" in arranging their own. "If I need to take few hours for personal business, I just take it," says one employee. Corporate employees say "the company has an entrepreneurial attitude" that "empowers" them to "take on more responsibility more quickly than we would at other companies." "The corporate culture is uniformly relaxed but the pace is quite fast. In that regard, we're like many other high-tech companies: a relative handful of good people who can wear many hats and juggle many projects." All of Seagate's employees, meanwhile, benefit from the "impressive" profit sharing plan that is a "key component" of the compensation for their "difficult" and "demanding" assignments.

Silicon Graphics, Inc.

2011N. Shoreline Boulevard
Mountain View, CA 94043
(650) 960-1980
www.sgi.com

LOCATIONS

Mountain View, CA (HQ)

Other major U.S. locations in:
Atlanta, GA • Boston, MA • Chicago, IL • Minneapolis, MN • Nashville, TN • New Orleans, LA • New York, NY • Portland, OR • San Diego, CA • Silver Spring, MD

Major international locations in:
Berlin, Germany • Buenos Aires, Argentina • Montreal, Canada • Tokyo, Japan

DEPARTMENTS

Administration • Engineering • Finance • Human Resources • Information Systems • Legal • Manufacturing • Marketing • Sales • Support/Service

THE STATS

Annual Revenues: $3.1 billion (1998)
No. of Employees: 9,900 (worldwide)
No. of Offices: 100 (U.S.), 134 (worldwide)
Stock Symbol: SGI (NYSE)
CEO: Richard Belluzzo

UPPERS

- Cool company
- Historically fun with lots of perks
- High-powered products

DOWNERS

- Recent financial troubles
- Fierce competition

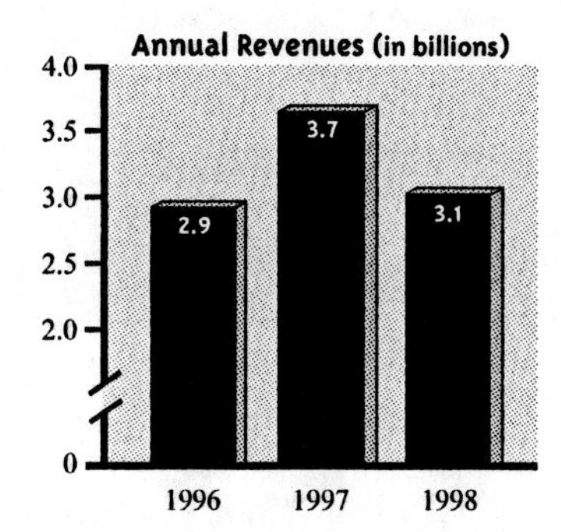

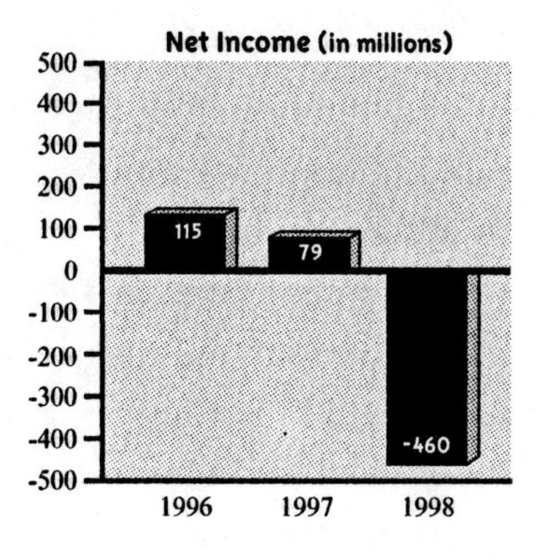

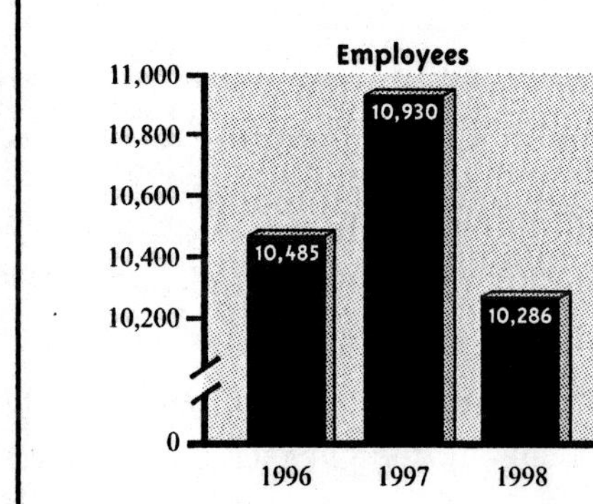

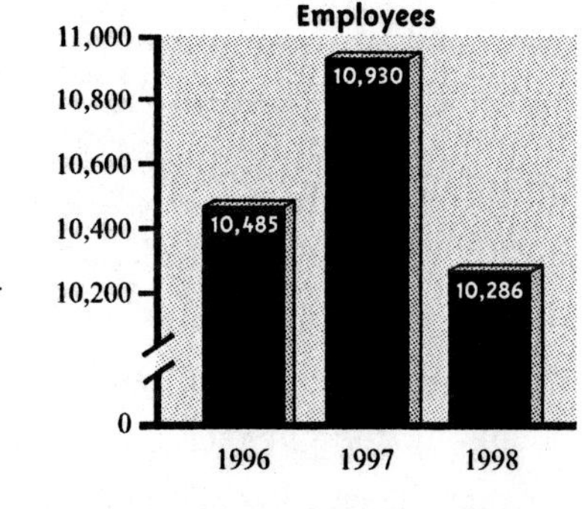

KEY COMPETITORS

- Apple Computer
- Compaq
- Dell Computer
- Hewlett-Packard
- IBM
- Pixar
- Sun Microsystems
- Unisys

EMPLOYMENT CONTACT

Jessie Ikeda
Recruiter for Marketing
Silicon Graphics, Inc.
2011N. Shoreline Boulevard
Mountain View, CA 94043

www.sgi.com/employment

HOURS	PAY	PRESTIGE	DRESS	SATISFACTION
BEST = SHORTEST HOURS	BEST = HIGHEST PAY	BEST = MOST PRESTIGIOUS	BEST = MOST CASUAL	BEST = MOST SATISFIED
4	7	8	8	4

BEST 10 — WORST 1

THE SCOOP

Once the darling of Wall Street and Hollywood, the company whose technology brought you *The Lost World* and *Antz* has fallen on rough times. Riding a microprocessor innovation that allowed the creation of revolutionary 3-D graphics, Silicon Graphics was one of the hottest high-tech companies in the late 1980s and the first half of the 1990s – "the next Apple," as one industry observer all too presciently described the firm. Sales leapt from $86 million in 1987 to nearly $3 billion in 1996. And unlike other high-tech concerns whose revenues were offset by absorbing losses, SGI enjoyed handsome profits nearly every year, peaking with $225 million in net income in 1995. But since 1995, the company has consistently posted poor financial results as competition in its field intensifies. SGI's powerful servers face competition from bigger companies such as IBM and Hewlett-Packard. After its expensive workstations and even its bulwark 3-D graphics unit began losing out to less expensive Windows-based PCs, the company recently began producing Windows NT/Intel workstations. Nevertheless, some analysts believe SGI still has some major assets, namely, its 3-D brand cachet and its stable of skilled employees. Now, say industry observers, SGI must learn to market and sell its wondrous products.

Silicon Graphics, or SGI, was founded in 1982 by former Stanford University computer science professor James Clark (who later left to help found Netscape). Clark and six students came up with new ways to build a 3-D graphics computer, and eventually, a 3-D workstation, which was first marketed in 1984 for $75,000. Their key was to create separate chips and circuits for the graphics that take the burden off a computers' central microprocessor, thus speeding the creation of 3-D models. Previous to the innovation, most workstations could only create 3-D "wireframe" models (outlines that resemble skeletons). Hollywood and advertising agencies soon caught on. Thus was born the SGI era in special effects, one that has made possible the effects in such fine films such as *Mask* and *Twister*, as well as engaging commercials starring morphing faces and dancing crackers. SGI rode its graphics-capable microprocessor to market dominance. Artists, researchers, and engineers were thrilled to bring intensely cool graphics to their desktops. In 1993, when President Bill Clinton visited the company's headquarters to witness a virtual tour of South Central L.A., he declared that the country needed more companies like SGI.

But all high-tech fairy tales, it seems, must come to an end. In the past several years, a combination of strategic mistakes and operations snafus has pushed SGI to the brink of a collapse. The company hit its lowest point (so far) in late 1997, when SGI's stock lost a third of its value in one day, after company officials warned that earnings and sales would fall far short of expectations. In 1997, then-CEO Ed McCracken, who had led the company since 1984, resigned and the company laid off nearly 10 percent of its workforce.

Since January 1998, the new CEO, Richard Belluzzo, has begun to implementing his turn-around plan, focused on a clearer market and situations focus, a redefined product plan, and a new business model. He also announced plans to work with Intel and Microsoft to produce SGI workstations based on Intel's processors and the Microsoft Windows NT operating system and spun off SGI's own chip-making unit, MIPS Technologies. But problems with SGI's Cray division, which makes supercomputers, have slowed progress. By mid-1998, SGI shares had lost two-thirds of their value since 1995.

GETTING HIRED

SGI maintains exhaustive career web pages at www.sgi.com which allow job seekers to search for open positions, submit resumes, and check the company's recruiting schedule at colleges. The site even provides links to a web site that provides figures on the cost of living in different cities, and links to web sites for various diversity organizations of interest to minority candidates

Unfortunately, despite SGI's helpful services, there may be precious few jobs open at the company," insiders say. On the upside, other sources say that SGI takes the long view when considering hiring. "Since things change so quickly, it's often more important to find people who are smart enough to learn new things and get things done in a fast moving environment than it is to get someone who has done the exact job before," says one. "While we always want to fill positions quickly, we'll wait if we have to. Conversely, sometimes we'll get connected with someone and be able to hire them, even though there's not a currently open position, but because they're just the right kind of person, and our upper management will help figure out how to make things happen."

What type of person is the "right kind of person?" Start with SGI's "Purpose and Core Values." SGI's Purpose: "Unleashing the power of human creativity and insight." Its Core Values: "innovation, integrity, passion to excel, fairness and respect, breakthrough results." Schmaltzy, and a bit generic, yes, but a start.

OUR SURVEY SAYS

Like many other high-tech companies, SGI is described by its employees as having a company culture that mixes relaxation and high energy, or, as one insider put it, "laid-back frenzy." "Externally, things look relaxed: many if not most people are dressed casually," reports one employee, "most people don't have to be in the office during specific business hours. Internally, though, it's usually quite hectic. Schedules are as tight as they can possibly be." According to another source: "It's a very casual culture, there is no dress code, and everyone can talk to everyone else. There's an open door policy which people take seriously." At the same time, says an SGI vet, "It can be a pretty intense working environment in spite of all that. Things happen pretty quickly in Silicon Valley, and there's a fast work pace, lots of things going on."

But the laid-back element is equally as strong. The dress code is based on the notion that "if you are comfortable, then you work more efficiently." "Salespeople wear suits, but most everyone else wears casual clothes, all the way down to jeans and T-shirts most days," says an insider at SGI's headquarters. Also, reports one longtime employee: "The work hours are very flexible, and we have many telecommuters." According to a manager at headquarters, "one of my employees comes in at 6:30 a.m., while most of the team comes in between 8 and 9. I arrive between 9 and 9:30 on most days unless I have an early meeting. And in addition to the flexible hours, "every four years you get a paid vacation called a sabbatical when you're supposed to just take time off and 'recharge.'"

Compensation at SGI is described as good but not astounding. “We are very competitive with all the high-tech companies, usually in the top 20 percent of the companies in the Valley,” says one employee. The pay is generally a bit higher than comparable companies in the Valley,” reports another. According to one insider, however, pay is not the main payoff of working with SGI: “Don’t expect to get rich here, but expect to get some great experience that you’ll be able to take anywhere.” Tenure does not correlate with compensation at SGI, according to insiders. “A person is measured against specific performance criteria for placement on a particular pay range,” explains one employee. As far as perks are concerned: “There’s a great cafeteria on (the Mountain View) campus. There are espresso machines in all the buildings.” “Benefits are good,” says one insider, “including on-site gyms, stock options and a stock purchase plan.”

Although employees at SGI are paid relatively well, the mood at Silicon Graphics has been dampened recently by poor fiscal results and subsequent layoffs. “The company has recently gone through growing pains where they had to cut almost 10 percent of its workforce,” says one glum employee. Says one insider: “When the company was flying high, we did lots of fun things – huge parties, free screening for the company of Jurassic Park, Tag Heuer watches for everyone.” Those giddy days, however, have gone the way of dinosaurs: “Now that we are suffering a bit, the company is much more austere.” Says one veteran, “The culture is changing as we emerge from the small-business mentality into a true internationally large company. Some of the fun is gone, but overall it is still a good place to work.” “Think of SGI as a teenager passing through puberty into adulthood,” advises another SGI insider. “That’s where we are. Lots of changes, but a very worthwhile experience overall.”

Sun Microsystems

901 San Antonio Road
Palo Alto, CA 94303
(650) 960-1300
www.sun.com

LOCATIONS

Palo Alto, CA (HQ)
Additional offices in CO • IL • MA • NY • TX • and worldwide

DEPARTMENTS

Customer Service
Engineering
Finance
Hardware Engineering
Information Systems
Network Solutions Consulting
Operations
Software Development
Systems Engineering
Technical Phone Support
Technical Pre-Sales Support

THE STATS

Annual Revenues: $9.8 billion (1998)
No. of Employees: 26,300 (worldwide)
No. of Offices: 160+ (worldwide)
Stock Symbol: SUNW (NASDAQ)
CEO: Scott G. McNealy

UPPERS

- Creative atmosphere
- Excellent job training
- Company café and bistro
- Discounted tickets for sporting and cultural events

DOWNERS

- Long workweek
- Hectic work schedules, especially near project deadlines

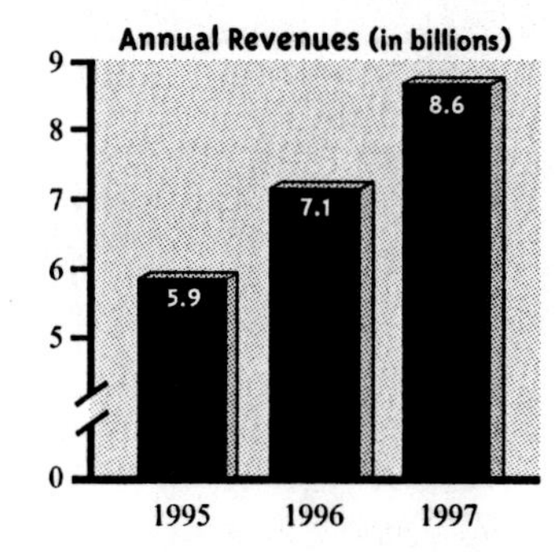

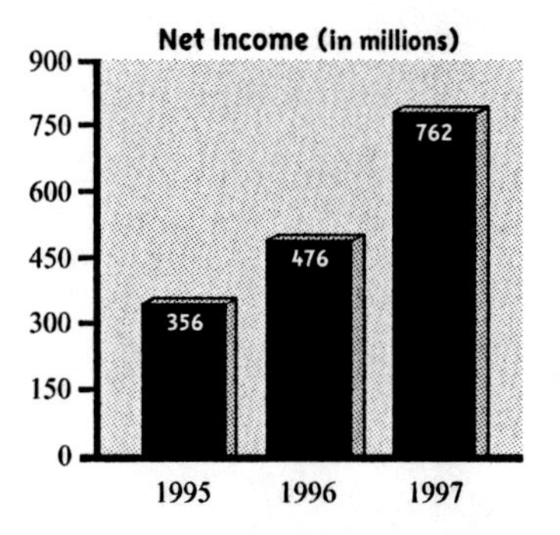

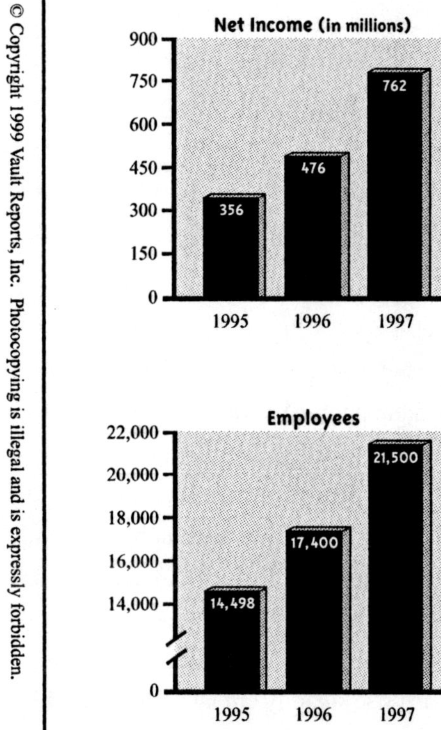

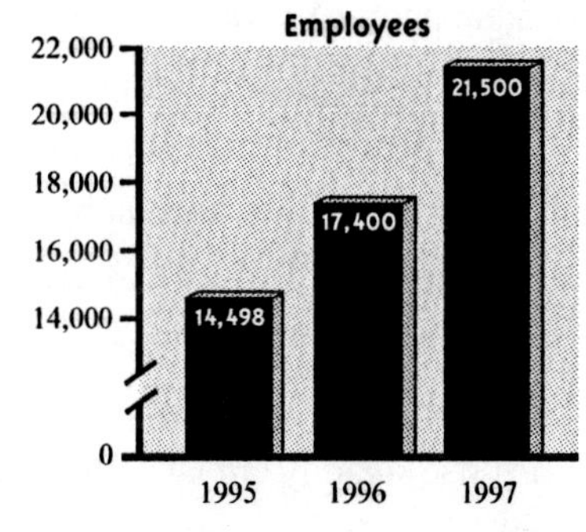

KEY COMPETITORS

- Dell Computer
- Hewlett-Packard
- IBM
- Intel
- Microsoft
- National Semiconductor
- Silicon Graphics
- Tandem Computers
- Unisys

EMPLOYMENT CONTACT

Katherine Hartsell
University Programs Manager
Sun Microsystems
901 San Antonio Road
Palo Alto, CA 94303

(650) 336-0694
katherine.hartsell@sun.com

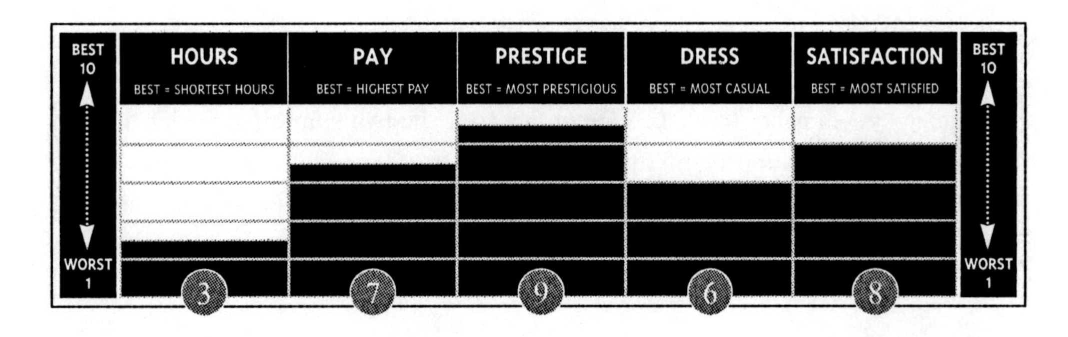

THE SCOOP

Originally a computer workstation builder, Sun was founded in 1982 by four 27-year-olds. Five years later it was a $500 million company, after going public in 1986. Sun succeeded almost immediately by developing operating systems for big-name clients such as AT&T. The company has been a leader in the development of "intranet" networking systems, and Sun's current joint ventures with Oracle, IBM, and AOL have further solidified its position as an industry leader. Just another Silicon Valley success story, you say? Maybe, but while many similar stories are now tales of sputtering organizations caught a step behind emerging technologies, Sun Microsystems has introduced a new technology with the potential to revolutionize computing technology, and has even King Bill of Microsoft worried.

Sun introduced a computer language called Java in May 1995 as a way to animate features on web pages. But in the summer of 1995, industry observers began suggesting that if Java the concept were to prove viable, there would no longer be a need for the operating systems and concomitant software with which Microsoft captivates the computing world. The years following Java's anointing as the Next Big Thing have been heady ones for Sun and its shareholders, with rampant talk of "revolution" and "paradigm shift." Meanwhile, Sun has shunned the Wintel hegemony continuing to produce its own microprocessors (it is the only major computer builder that does not use Intel chips) and developing software for UNIX-based (non Windows NT) systems. Although Windows NT is becoming increasingly popular, Sun is doing well with its sale of Unix servers and workstations, and has seen its earnings and stock price rise steadily. In late 1997, Sun filed suit against Microsoft relating to the software giant's supposed sabotage of Sun's Java language. Sun claimed Java applications ran poorly on Microsoft servers.

It's not as if Sun only has enemies in the high-tech world – opponents of Wintel are cheering on Sun's pugnacious stance. In 1998, Sun and IBM announced a partnership to create an operating system using Java. In June of that year, the two companies initiated a joint effort with the Chinese Ministry of Information Industry to promote Java in the world's most populous country. The Java-based system was released a couple months later. And in November 1998, as part of America Online's blockbuster acquisition of Netscape, Sun announced an alliance with AOL, designed to move both companies into providing e-commerce services to businesses.

GETTING HIRED

Sun recruits at many universities, business schools, and career fairs; a schedule is posted at the company's web site, located at www.sun.com. Also at the web site is a page with which job seekers can search for open positions by function or geographic area. Sun offers internships and a co-op program for both undergraduates and graduate-level students. The summer program offers an annual event with CEO Scott McNealy. For new hires going into sales and systems engineering, there is a one year rotational "Best of the Best" program. The sales positions are for MBAs – about 10 to 12 are hired each year.

Interviews reportedly range from "half-day to full-day schedules" and will include some technical questions, but are less intense and freaky than at other high-tech companies. "My experience is that they are less brainteaser-oriented, and are more to find out what you know about what you say you know," reports one insider.

OUR SURVEY SAYS

Sun Microsystems pushes its employees to keep the company on the cutting edge. In this "aggressive" and "innovative" corporate culture, everyone is "expected to take the initiative, make decisions, and be creative in solving problems," contacts say. There is "no hand-holding" at Sun, and "the flat hierarchy makes it easy to access people throughout the organization." While project deadlines pump up the pressure, Sun employees are also fond of playing practical jokes and of "blowing off steam" at company social events. Advancement opportunities, "generous" pay, and the "top-drawer" job training contribute to an "immensely satisfied" workforce that takes pride in working for a high-tech company with such a "high level of integrity, creativity, and industry prestige."

"Sun is a typical Silicon Valley place – very high pressure," explains one engineer. "Expect a long workweek, though not as crazy as a startup." Another insider says "the work ethic is severe, with many of the movers and shakers in the company working many hours a week." This isn't to say everyone at Sun works like a fiend: "Some people I know work 40, others 70," says one engineer. However, those with their eyes on moving up in the company should note that "those that work 70 are generally recognized and rewarded for it."

Despite the long hours and hectic environment, employees say that "the culture is relaxed, very California." By this they mean low-key when it comes to policies and bureaucracy, dress code, and strict hours. "Dress code?" snorts one employee. "Shorts!" Says one researcher: "Finance and marketing look more conservative." However, "on Friday, you can't tell engineering from finance." Sums up one employee in marketing: "It's business casual for most departments, suits for some positions like sales."

"Work start time is extremely flexible. Most people don't come in until 9:30 to 10 a.m., so that they don't have to sit in traffic. Others I know come in at 7 a.m. to beat the rush and leave at 3:30 p.m.," says one engineer. "It's all up to you, but you do have to try to make sure you don't miss any meetings." "Most of the engineering campus feels a lot like college: You can come and go as you please as long as you get your work done and attend staff meetings," says another. Insiders also report that there is a "strong work-from-home program." Telecommuting is encouraged if it fits your job profile," explains one employee.

“When the company was flying high we did lots of fun things. Now that we are suffering a bit, the company is much more austere.”

– ***Hardware insider***

Sybase

6475 Christie Avenue
Emeryville, CA 94608
(510) 922-3500
Fax: (510) 922-3210
www.sybase.com

LOCATIONS

Emeryville, CA (HQ)
28 offices in the U.S.
54 international offices

DIVISIONS

Channels
Finance
Sales
Customer Service & Support
Marketing
Human Resources
Product Marketing

THE STATS

Annual Revenues: $903.9 million (1997)
No. of Employees: 5,216 (worldwide)
No. of Offices: 82 (worldwide)
Stock Symbol: SYBS (NASDAQ)
CEO: Mitchell Kertzman

UPPERS

- Competitive pay
- Flexible hours
- Stock options

DOWNERS

- Corporate restructuring and downsizing
- Hurt by competition with Oracle

KEY COMPETITORS

- Computer Associates
- IBM
- Informix
- Microsoft
- Oracle

EMPLOYMENT CONTACT

Human Resources
Sybase
6475 Christie Avenue
Emeryville, CA 94608

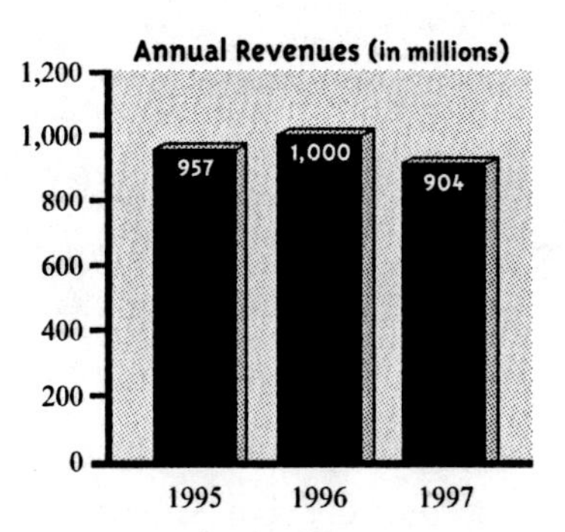

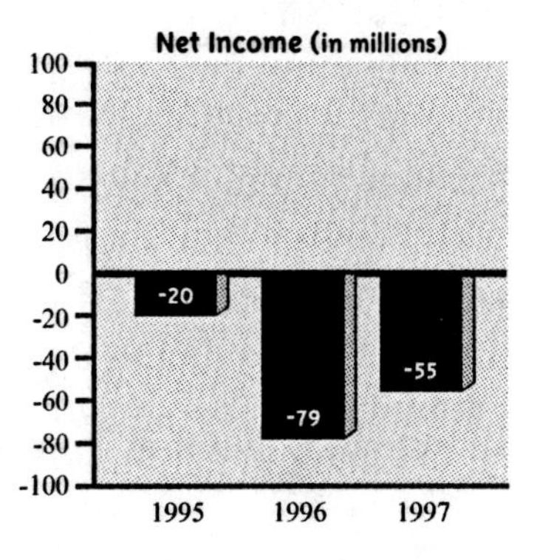

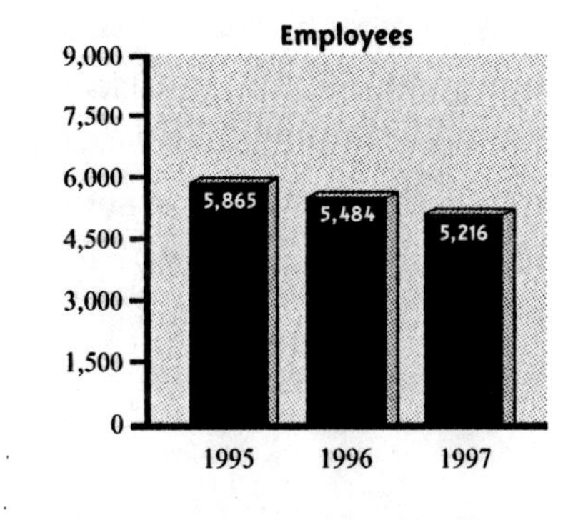

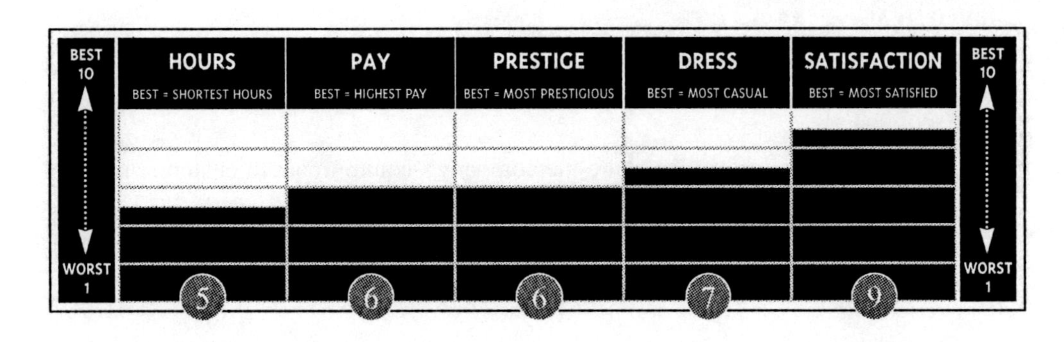

THE SCOOP

Since its 1984 founding by tech-whizzes Mark Hoffman and Robert Epstein, Sybase has grown to become the sixth-largest software company in the world. Based in Emeryville, California, Sybase was the first company to offer high performance relational database management systems, which allow many users to simultaneously access information. Specializing in databases for complex UNIX systems, Sybase raked in big profits when it teamed up with Microsoft in 1994, which co-developed and sold the company's tools. Sybase soon succumbed to cyber-competitors Oracle and Informix, however, and the company was forced to downsize and restructure in 1995. That year, Hoffman was ousted as chairman. The company acquired Powersoft, an application development tool specialist, for $950 million the same year

The company plans to push its Powersoft-branded products, which are focused on the development tools market. Sybase, however, keeps running into trouble. In 1997, the company was forced to lay off 600 workers, and suffered a $60 million loss and a slump in its stock price in January 1998, when it was revealed that Sybase's Japanese subsidiary had misreported earnings considerably. As a silver lining to Sybase's dark troubles, the sudden revelations of President Clinton's alleged intern indiscretions kept Sybase's bad news off the front pages. But don't count out Sybase yet – the firm is still the second-largest database management company in the world after Oracle. Chairman Mitchell Kertzman says Sybase is planning to push strongly into the data warehousing, mobile and embedded computing and e-commerce markets and increase its new-growth sales force.

GETTING HIRED

Sybase looks for applicants who share the company's commitment to customer service. Visit the "Employment Opportunities" section of Sybase's web site for job openings and descriptions. Send or fax resumes to human resources, or e-mail resumes through the web site.

The company offers three- to six-month co-op and internship opportunities for college students. Sybase also posts its college recruitment schedule at the site.

OUR SURVEY SAYS

Sybase's company culture is as laid-back as its California locale, according to insiders. Dress is casual and work hours are flexible. "I've worn a tie twice in six weeks, and both times at trade shows," says one contact. "With some companies, 12- to 14-hour days are expected," one contact notes. "At Sybase, you're not expected to work yourself to death." Nor is "pay stellar," says an employee, who nonetheless "turned down considerably more money and lower-cost-of-living" markets to work at Sybase. Says another insider: "I think the pay is above average, but not the best in the industry." One employee likes the relaxed atmosphere and friendly people so much, "I actually left once for two months, but I came back because I missed the company and culture of Sybase too much!" Insiders agree the company is progressive in its treatment of minorities and women; of the company's eight senior officers, there is one woman. "Women can be found at all management levels of Sybase – no glass ceiling here," says one employee. Another says: "As a man, I am actually a minority within the marketing department." Perks include stock options, stock purchase plans, excellent benefits, and a management level that "genuinely cares about employees."

Symantec

10201 Torre Ave.
Cupertino, CA 95014
(408) 253-9600
Fax: (408) 446-8129
www.symantec.com

SYMANTEC.

LOCATIONS

Numerous offices throughout the U.S • Australia • Brazil • Canada • France • Germany • Holland • Hong Kong • Italy • Japan • Mexico • New Zealand • Russia • Singapore • South Africa • Sweden • Switzerland • Taiwan • The U.K.

DEPARTMENTS

Development
Finance and Accounting
Marketing
Operations
Sales
Services

THE STATS

Annual Revenues: $578.4 million (1998)
No. of Employees: 2,300 (worldwide)
No. of Offices: 22 (U.S)
Stock Symbol: SYMC (NASDAQ)
CEO: Gordon E. Eubanks, Jr.

UPPERS

- Benefits package
- Travel opportunities
- Promotional opportunities

DOWNERS

- Increased competition in industry
- Assimilating mergers troublesome

KEY COMPETITORS

- Microsoft
- Network Associates
- Phoenix Technologies
- Quarterdeck
- Traveling Software

EMPLOYMENT CONTACT

Judy Sugiyama
Manager HR Programs
Symantec Corporation
10201 Torre Ave.
Cupertino, CA 95014

(408) 446-7251
Fax: (408) 366-7439
jsugiyama@symantec.com

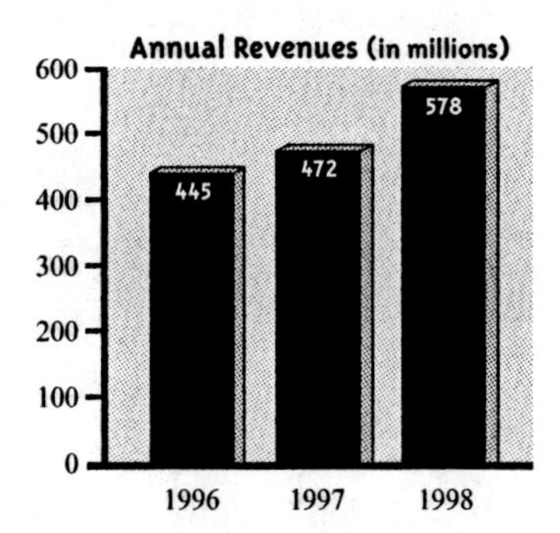

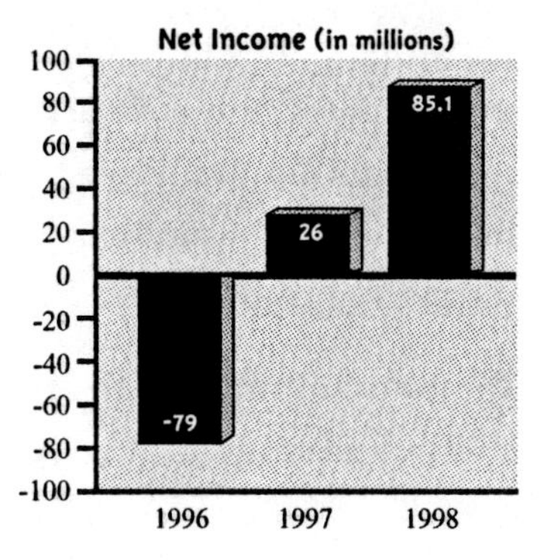

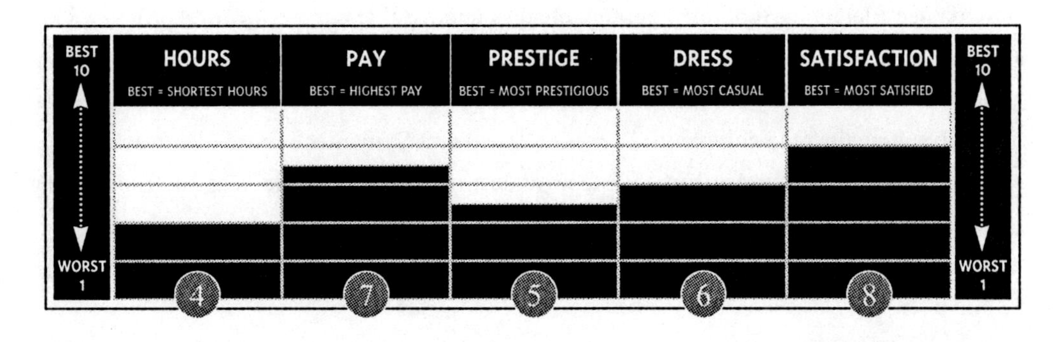

THE SCOOP

When it comes to computers, Symantec does the dirty work. Founded in 1982, the Cupertino, California-based software company has made its name developing utility programs, which perform basic housekeeping tasks like managing and protecting files from computer crashes and viruses. In addition to the Norton product line of utility software, Symantec produces remote connectivity software including the pcANYWHERE, WinFax and ACT! product lines, and Internet development tools with its Café. CEO Gordon Eubanks, who bought the fledgling company from founder Gary Hendrix in 1984, established Symantec's niche markets to escape competition with software giants Microsoft and Lotus.

The company went public in 1989, and merged with Peter Norton Computing, the DOS utilities leader, a year later. Symantec's acquisition spree in the early 1990s has slowed, largely due to the $40 million loss the company suffered as a result of its 1995 acquisition of Delrina, part of the company's failed attempt to cash in on Windows 95 utilities. Recent plans include an increased focus on Internet and Windows NT Systems, and expansion in the Asia/Pacific region. In the past two years, business has rebounded, especially on the strength of the company's Norton utility and anti-virus programs.

Other successful Symnatec products included CrashGuard, a product that prevents computer crashes, introduced in 1997. In June 1998, Symantec announced the $27.5 million purchase of Binary Research Ltd., a New Zealand-based maker of disk-copying software such as the popular program called Ghost which quickly copies PC hard drives. In 1996 Symantec also teamed up with IBM to develop and market new versions of Norton AntiVirus.

Escalating its bitter rivalry with McAfee Associates, also a creator of anti-virus programs. Symantec filed copyright infringement charges against that competitor in April 1996, alleging that its code was copied in several McAfee programs. However, Symantec found itself on both sides of lawsuits in 1998, after users with early versions of Norton AntiVirus filed a class-action suit against the company. The suit alleged that Symantec ignored the warranty given to customers with Norton AntiVirus v.2 to provide an upgrade to fix the Year 2000 software glitch.

GETTING HIRED

Under the "Corporate" portion of Symantec's web site, the "Career Opportunities" section offers an excellent resource for job descriptions and openings – visitors can apply right there. Qualifications and requirements vary by position.

Says one recruiter: "In general, we're looking for highly motivated, high energy, high integrity people who see the value of focusing on customers – and a sense of humor doesn't hurt either." One insider describes the interview process for a technical position in his department as "somewhat stressful. The interviewee will interview with approximately five people. On average, two of the five interviewers will focus on technical questions." When screening applicants, that contact says, "I look for someone who does not respond with 'I don't know,' but instead tries to think of a creative solution – for example, I'm not sure about X, but one possible solution might be…"

OUR SURVEY SAYS

With an excellent benefits package and friendly co-workers, employees describe Symantec as a "wonderful place to work." "It's a pretty casual environment, dress is business casual – it's fun working here," says one insider. "The people are great, very intelligent, high energy." Though Symantec has numerous locations worldwide, employees in finance and marketing work in the Cupertino headquarters. Employees praise the training programs offered to recent college graduates and note that "hiring women and minorities is a priority" "Two of our VPs are women and many of our directors are too."

Our contacts report that the company takes care of its own: "We offer stock options, profit sharing, 401(k) match, great medical coverage with a domestic partner program." Insiders are even more enthused about the little things: "an on-site drycleaning service, on-site car detailing in the summer, an ATM in our lobby, mail services, and parties on our roofdeck."

Trilogy Software

6034 W. Courtyard Drive
Austin, TX 78730
(512) 794-5900
Fax: (512) 794-8900
www.trilogy.com

LOCATIONS

Austin, TX (HQ)

DEPARTMENTS

Artificial Intelligence
Client Services
Consulting
Development
Human Computer Interface
Industry Marketing
Modeling
Pre-Sales
Software Engineering

THE STATS

Annual Revenues: $150 million (1997)
No. of Employees: 400 (est.)
No. of Offices: 1 (U.S.)
A privately-held company
CEO: Joe Liemandt

TRILOGY SOFTWARE

UPPERS

- New cars and exotic vacations for top performers
- Company boats
- Stock options
- Discounted tickets for cultural and sporting events

DOWNERS

- Long workdays

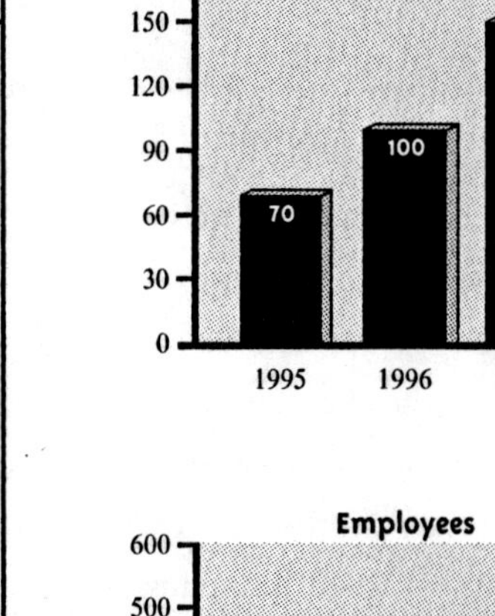

KEY COMPETITORS

- Clarify
- Elcom International
- Firstwave Technologies
- Oracle
- SAP
- Siebel Systems
- Vantive

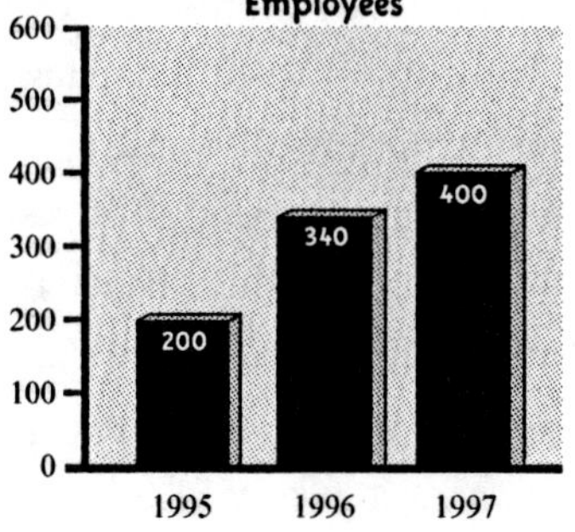

EMPLOYMENT CONTACT

Human Resources
Trilogy Software
6034 W. Courtyard Drive
Austin, TX 78730

recruiting@trilogy.com

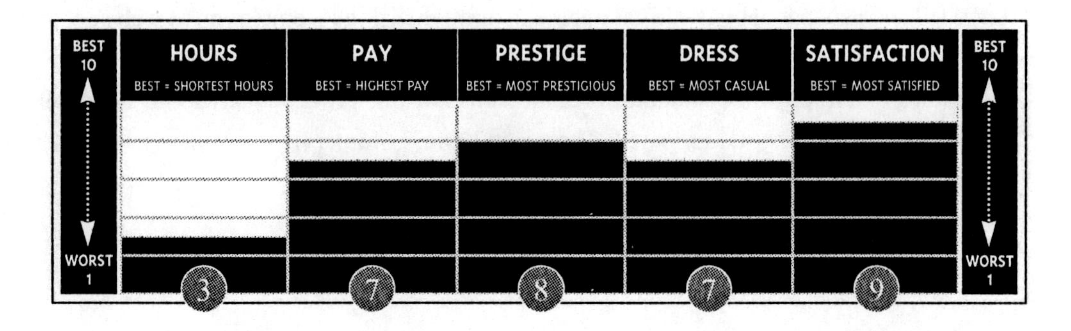

THE SCOOP

Joe Liemandt did what every starting entrepreneur dreams of doing – he caught the big-name competition sleeping. In 1990, Liemandt and his associates dropped out of Stanford University to found what was then called Trilogy Development Group. As a college junior he recognized a market niche that few others were trying to address. Armed with years of research on sales and distribution and the experience of his entrepreneurial father, Liemandt created a first – sales configuration software based on algebraic algorithms. Demand for the product, Sales Chain, skyrocketed immediately, and within two years Trilogy had completed a $3.5 million deal with Hewlett-Packard. More were soon to follow, and Trilogy now provides its software and business support services to major customers like AT&T, IBM, Intel, American Airlines, and Boeing. Trilogy has nearly doubled in size over each year of its history, and, despite the entry of competitors into the lucrative market, Trilogy remains at the top of a heap it created. Not only has Trilogy been called one of the "coolest" companies in America, insiders value the company at nearly $1 billion.

GETTING HIRED

Trilogy Development Group is energetically seeking applicants with computer modeling expertise and computer science PhDs. However, the company's explosive growth has created opportunities in nearly every area. Applicants – who should have extensive computer knowledge – can learn about current openings through Trilogy's employment web site, located at www.trilogy.com. Trilogy accepts resumes submitted via e-mail, regular mail, and fax, and the company will pay for candidates to come to its Austin, TX headquarters for both their first and second interviews. Be prepared: The second round of interviews insiders tell us, is an all-day affair that "may be the same week as the first round."

OUR SURVEY SAYS

Trilogy employees say they share an "unrivaled sprit of camaraderie." Believing that "nothing brings people together better than risk," Trilogy gives new employees "unprecedented amounts of responsibility" and makes them "take chances" in trying to meet the challenges of their work. In fact, a recent *Wall Street Journal* article detailed Trilogy's unconventional training programs, which have included a trip to a casino. While such demands require employees sometimes to "work into the night," they are free to arrange their schedules as they wish. "If you are done for the day," says one employee, "you can go home at noon." Employees praise the "high" salaries and rave about the "unorthodox" perks-such as the company kitchen – and even more unorthodox social events – such as a recent company scavenger hunt.

Unisys

P.O. Box 500
Blue Bell, PA 19424-0001
(215) 986-4011
Fax: (215) 986-6850
www.unisys.com

UNISYS

LOCATIONS

Blue Bell, PA (HQ)
Atlanta, GA • Chicago, IL • Honolulu, HI • Miami, FL • New York, NY • Phoenix, AZ • Sacramento, CA • Burlington, MA • Okemos, MI • As well as numerous other locations nationwide and overseas

DEPARTMENTS

Accounting & Finance
Computer Systems
Customer Services
Information Services
Research & Development
Sales & Marketing

THE STATS

Annual Revenues: $6.6 billion (1997)
No. of Employees: 33,000 (worldwide)
No. of Offices: 80+ (worldwide), 40+ (U.S.)
Stock Symbol: UIS (NYSE)
CEO: Lawrence Weinbach

UPPERS

- Relaxed environment
- Flexible schedule
- Casual dress code
- Abundant overseas opportunities

DOWNERS

- Large bureaucracy
- Scars of past failures
- Expected overtime without compensation
- Sour old-timers

KEY COMPETITORS

- Andersen Consulting
- Compaq
- Computer Sciences
- Decision One
- Electronic Data Systems
- Hewlett-Packard
- IBM
- Inacom
- Sun Microsystems
- Vanstar
- Wang

EMPLOYMENT CONTACT

Human Resources
Unisys
P.O. Box 500
Blue Bell, PA 19424-0001

jobs@unn.unisys.com

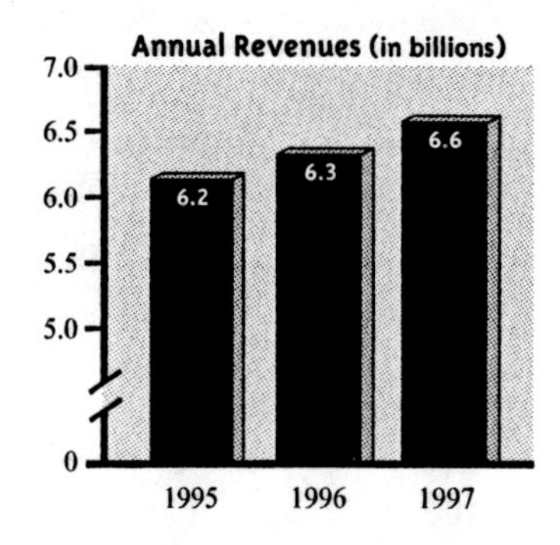

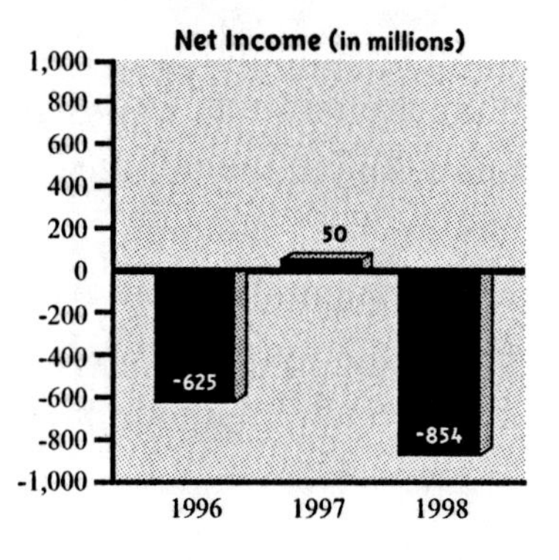

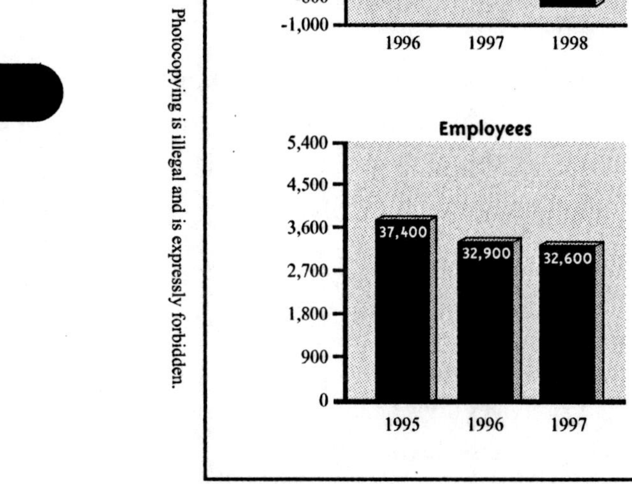

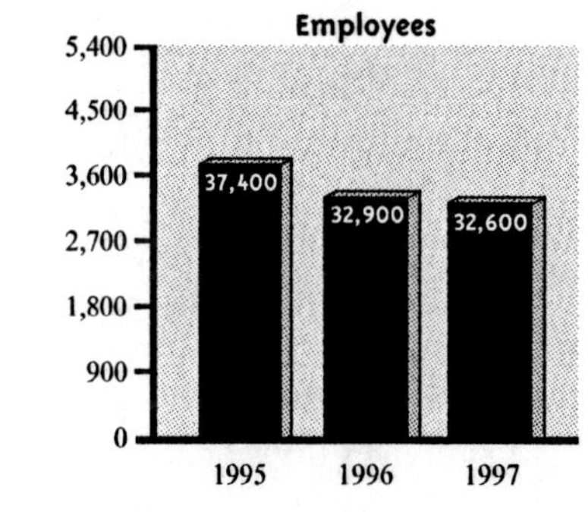

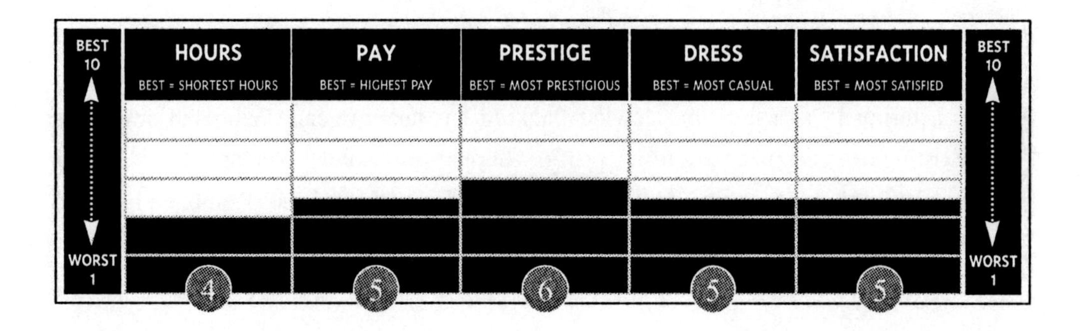

THE SCOOP

In 1986, Burroughs Corporation and Sperry Corporation merged to form what became Unisys. At the time they merged to create Unisys, both were major mainframe computer producers. At first called the Sperry-Burroughs Corporation, the combined firm soon changed its name to Unisys, a synthetic name chosen to reflect the entirely new entity, one of synthetic approach and universal reach. Unisys' then-chairman, W. Michael Blumenthal, claimed that the merger had disproved the "myth that large computer companies could not be successfully merged." But as the 1980s wore on, mainframe computers, Unisys' lifeblood, fell out of favor Mainframes had enjoyed gigantic 50 percent profit margins – and Unisys' cost structure was too fatty for the new PC reality and its slimmer profit magins. And, like other once-mighty computer makers, Unisys missed out on the early part of the boom in personal computers and PC networking.

CEO James Unruh contended with Unisys' myriad problems for seven years, from 1990 to 1997. While the austere Unruh was often assailed for cutting benefits to Unisys retirees and laying off tens of thousands of employees, Unruh managed to wrest the company into some semblance of health, after five separate reorganizations between 1990 and 1997. Under Unruh's tenure, the ranks of employees dwindled to 32,000 from over 75,000. While profits in 1996 were a feeble $49.7 million, that was still an improvement over the $625 million loss of 1995. Unruh's last move was to carve Unisys into three semi-independent business units: the traditional computer hardware operation, a computer maintenance and "help desk" unit, and a services division, designed to sell technology consulting and "vertical market solutions" (customized setups).

The new CEO, Lawrence Weinbach, is bullish on the future of Unisys – and he's embarking on his own transformation of Unisys. Weinbach, previously CEO of professional services giant Andersen Worldwide, has gotten Unisys out of unprofitable businesses, such as manufacturing PCs (a function Hewlett-Packard has taken over). Weinbach now insists that any existing or new businesses turn a profit. He plans to have the firm concentrate on two core businesses: high-end servers and computer consulting and services. Weinbach has also pared down Unisys' debt down from $2.3 billion to $1.2 billion. And he's pursued business partnerships with other high-tech biggies. Recently, Unisys formed a strategic alliance with Microsoft, under which Unisys will pitch Windows NT operating systems to its main customers, governments and financial institutions. Weinbach is also attempting to boost morale by getting rid of penny-pinching decrees, such as the policy that required employees to take the cheapest airfare, even if it included multiple stopovers.

For the future, the company is trying to reduce expenses by $200 million over the next several years by streamlining processes like its purchasing and internal information systems. The firm is also hoping an advertising campaign launched in September 1998, featuring the slogan "We eat, sleep and drink this stuff," will finally help the company shed its image as an old-line mainframe company and develop an identity as a hip firm focused on information services and computer technology.

GETTING HIRED

Applicants should consult Unisys' World Wide Web Visitor's Center, located at www.unisys.com, which provides a lists of current openings sorted by job type and geographical location. Each listing comes with detailed instructions on where and how to submit your cover letter and resume. To be considered for all available positions, applicants should send their materials (in ASCII format) to jobs@unn.unisys.com or mail them to Worldwide Recruiting and Staffing at the corporate headquarters.

OUR SURVEY SAYS

Unisys employees enjoy an "informal," "supportive" corporate culture, a "team-oriented" structure, and the prestige of working for a "dominant" company in the industry. "The people are usually friendly and helpful, not cutthroat," says a longtime employee. "It is a very stable place to work now – there is little fear of layoffs any more," adds an insider. Still, stability does not mean Unisys is stagnant; employees say they enjoy the "constantly changing" nature of their work, reporting working on "new projects" and with "new people" every one to two years. One insider, however, warns that "you should expect that you'll work some overtime, without compensation, and without complaint." Pay is reportedly not stellar at Unisys. "The pay is probably average for the industry," says one insider. Says another: "I won't get rich here, but I'm pretty happy. I've been treated fairly." Employees are happy about other perks, which include "a respectable cafeteria, an active recreation committee, and an excellent library. They are constantly improving the facilities to make it a nicer place to work." However, because of Unisys' financial difficulties, the company's 401(k) program matches only the first 25 cents on each dollar an employee contributes, up to 4 percent of pay. "This is a recent improvement, actually," insiders tell Vault Reports. "The company suspended 401(k) matching some years ago, and only recently reinstated the partial match it now does." While the "challenging assignments" often require "long and intense workweeks," employees feel that the "relaxed" atmosphere – including a "business casual" dress code (unless meeting with a client) – makes it easier to cope with these rigors.

Employees add that Unisys' corporate hierarchy is "minimal," and "everyone is on a first-name basis." The company adds that individuals are encouraged to e-mail the chairman with ideas. "Individuals are empowered to make decisions about general issues rather than wait for the decision to go up and back down the chain of command," says one insider. "In most areas, it is felt that even a bad decision is better than no decision." Another agrees: "I've found that management style is mostly 'hands-off' and usually that means you have to take the initiative. Risk-taking and assertiveness are usually well-rewarded."

Employees praise CEO Lawrence Weinbach, "who has been doing a wonderful job in reducing the debt of Unisys, which gives stockholders – which include many employees – better equity and dividend returns." A contact says of Weinbach, "He is very different from his predecessor. He seems very approachable, and is not bothered by getting his hands dirty." However, one employee sounds a note of caution: "There are a lot of people from the 'old school' who, due to the hard times we've been through, are very unreceptive to change, new ideas, or even a positive attitude."

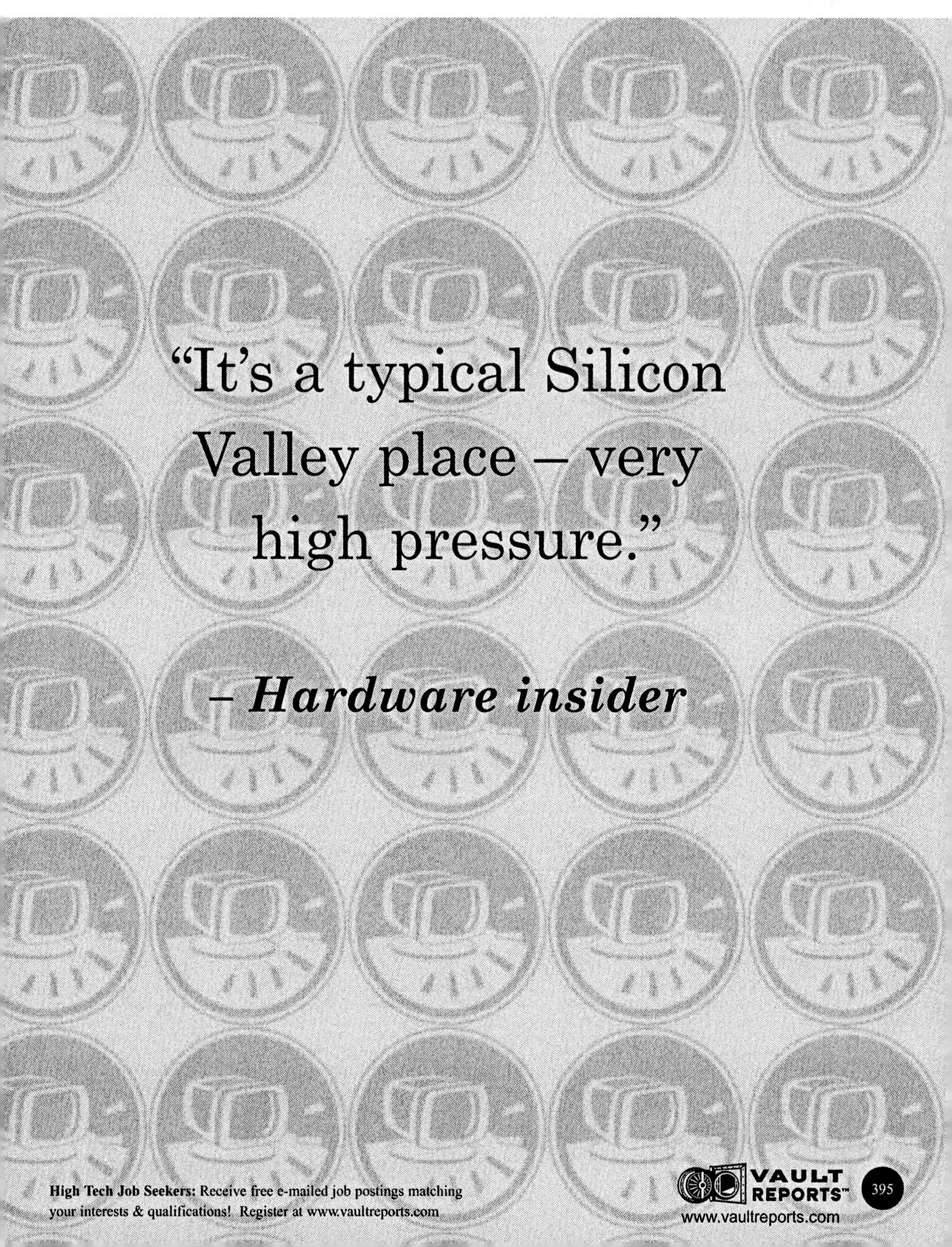
"It's a typical Silicon Valley place – very high pressure."
– Hardware insider

VLSI Technology

1109 McKay Dr.
San Jose, CA 95131
(408) 434-3100
Fax: (408) 263-2511
www.vlsi.com

VLSI TECHNOLOGY

LOCATIONS

San Jose, CA (HQ)
Mesa, AZ • Tempe, AZ • San Diego, CA • Irvine, CA • Duluth, GA • Hinsdale, IL • Hoffman Estates, IL • Burlington, MA • Millersville, MD • Fishkill, NY • Austin, TX • Richardson, TX • San Antonio, TX • Bellevue, WA

DEPARTMENTS

Administrative
Finance
Human Resources
Information Technology
Manufacturing and Quality Engineering
Production and Design Engineering
Sales & Marketing

THE STATS

Annual Revenues: $712.7 million (1997)
No. of Employees: 2,500
Stock Symbol: VLSI (NASDAQ)
CEO: Alfred Stein
Year Founded: 1979

UPPERS

- Competitive salaries
- Attractive incentive programs for product development
- Diverse workforce
- Four-week sabbatical after five years

DOWNERS

- More conservative than other high-tech companies
- "Grueling" interview process
- Recent layoffs/ hiring freeze

KEY COMPETITORS

- Cirrus Logic
- Dallas Semiconductor
- Fujitsu
- IBM
- Intel
- LSI Logic
- Lucent
- Mitsubishi
- Motorola
- NEC
- Oki Electric
- Texas Instruments
- Toshiba

EMPLOYMENT CONTACT

Art Gemmell
Human Resources
VLSI Technology, Inc.
1109 McKay Dr.
San Jose, CA 95131

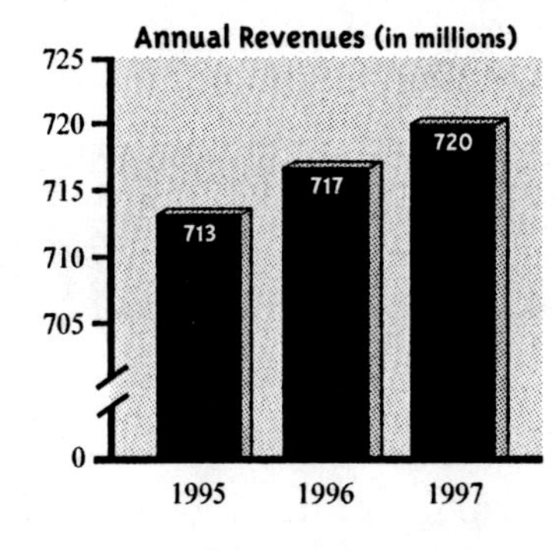

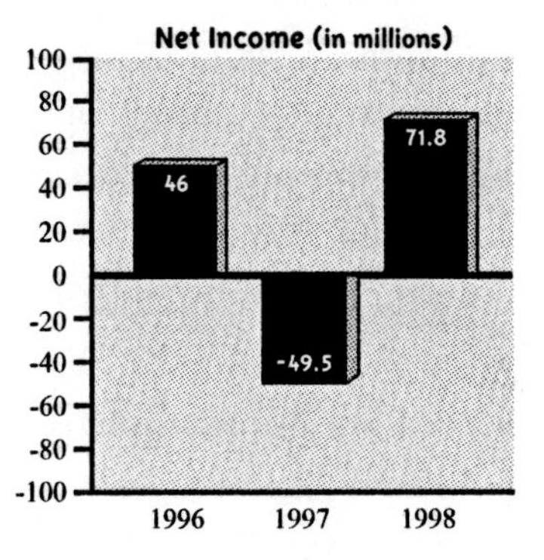

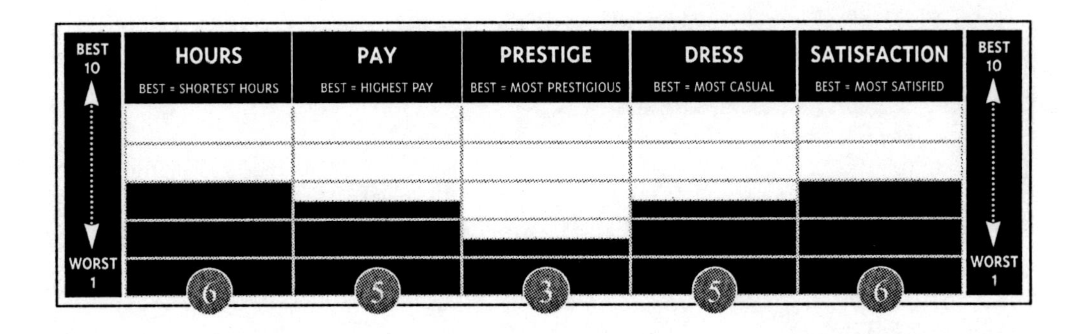

THE SCOOP

VLSI is engineer-speak for "Very Large Scale Integration." It's the process of integrating a multitude of transistors on the tiny silicon memory chips used in computers, mobile phones and video games. Founded in 1979, the company initially supplied the PC market, but over the years, its product line has grown considerably. In 1990 it shifted its focus to the development of custom and semi-custom integrated circuits (ICs). Today, the engineers at VLSI Technology design and manufacture ICs for digital, wireless, networking and advanced computing markets. It provides manufacturers with custom-tailored chips, allowing manufacturers to spend less time developing such technology and thus bringing their products to the market faster. VLSI's customers deal with digital entertainment, wireless technologies, networking, and advanced computing applications. Through strategic alliances, VLSI intends to become "the manufacturing partner of choice" for the leaders in each of these markets. In 1996, engineers from VLSI and Silicon Graphics collaborated to design and produce integrated circuits for a new 3-D visualization supercomputer. In addition, the company has been involved in the development of digital video, cable modems and electronic games.

Losses caused by the faltering semiconductor market in 1996 were compounded by problems at Apple, one of the company's biggest customers. VLSI was forced to shut down its San Jose manufacturing facility and sell off one of its subsidiaries the following year. In 1998, it consolidated its Hong Kong and Korean operations into a single Asian office located in Taipei, Taiwan. After all that cost cutting, the company was able to open a new R&D facility in the "Silicon Valley" of Southern France.

GETTING HIRED

The employment section on VLSI's home page lists available positions and contact information. It also posts a list of its college recruiting schedule in the "University Relations"

section. Resumes may be sent via e-mail or snail mail to the attention of the specific human resources person indicated in each listing. After recent layoffs, the company is in a temporary hiring freeze, but "things are expected to pick up soon," and during normal times, "we are always looking for engineers," reports one source.

Expect "a full day" of hour-long interviews with "a team leader, two or three engineers, a senior-level guy, and someone from HR." The selection process is "very competitive" and "engineering interviews tend to be grueling and technical," but insiders promise that "they don't grill you." "It's not an exam," one interviewee remarks, "so you don't have to know everything." Sources say to "make sure you have a detailed, informational resume that highlights specific projects you have worked on." They stress the importance of "listing exactly what you did, nothing more, nothing less," and specifically warn applicants to be completely honest with their resumes. Interviews will swiftly find out if your "incredible resume" is "mostly bull."

Interviewers "do not follow a standard format, nor do they ask you a fixed set of questions like some companies do." They basically go through your resume and "try to get you talking about your projects." What they are looking for is "insight into how you approach problems and use your analytical skills." In addition, insiders suggest that candidates "show an interest in what the company is doing, and what the work culture is like." "Overall, see that you learn as much about the company as they learn about you."

OUR SURVEY SAYS

It's true that "a lot of people have never heard of VLSI Technology," but insiders say working at VLSI is "a great idea." The corporate culture "is a little bit on the conservative side," but the people are "bright and talented," and "genuinely interested in helping each other" For many, the best part is getting the "insight on what is to come in electronics and its applications." For engineers and the tech community the dress code is business casual, which means "no shorts and sandals" like in other high-tech shops. Sources say the culture is "a little more relaxed" in the company's smaller offices, and more businesslike at the San Jose, San Antonio and Tempe sites.

Pay and benefits are "quite competitive," and "commensurate with your skills." Official hours are "8 to 5 – unless you are working in the assembly area," where they have different 8- or 12-hour shifts. Of course, insiders are "generally expected to work as long as it takes to get the work done," which may mean "30 hours one week, 60 hours the next." In general, however, "very few people work long weeks for extended periods unless they want to." And when they do, "the company notices and rewards you."

VLSI offers the standard benefits, including medical, a 401(k) plan, and stock purchase plans. Two of the company's incentive programs have the engineers raving. For the first, employees receive $5000 for filing a patent and $5000 more if the product is accepted by the U. S. patent office. As part of VLSI's "Impact" program, workers get $200 for relinquishing rights to an idea for development of a new product. Another bonus: "we get a sabbatical of four weeks or longer every five years."

Because VLSI is an electronics firm and it's headquartered in San Jose, "there are probably more minorities here than in your average organization." Women have "fairly decent representation" overall, though "the smaller offices are mostly male, except for sales and administrative staff." Our contacts say they have noticed "an increase in the female workforce over the past few years."

"Once a decision is made, you must back it 100 percent and do what you can to make your project a success."

– ***Semiconductor insider***

Yahoo!

3400 Central Expressway
Suite 201
Santa Clara, CA 95051
(408) 731-3300
Fax: (408) 731-3301
www.yahoo.com

YAHOO!™

LOCATIONS

Santa Clara, CA (HQ)
Boston, MA
Washington, DC

DEPARTMENTS

Client Services
Editorial
Marketing
Product Engineering
Sales
Software Engineering

THE STATS

Annual Revenues: $203.3 million (1998)
No. of Employees: 386
No. of Offices: 3 (U.S.)
Stock Symbol: YHOO (NASDAQ)
CEO: Timothy Koogle

UPPERS

- Stock options
- Rapid growth
- Friendly and casual atmosphere

DOWNERS

- Industry instability

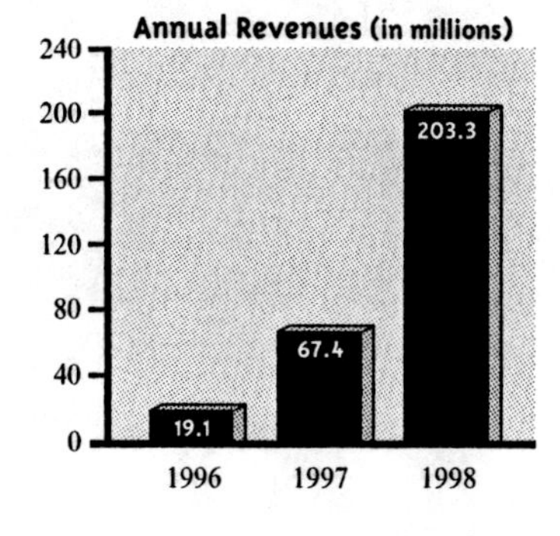

KEY COMPETITORS

- America Online
- Excite
- Infoseek/Walt Disney
- Lycos
- Microsoft
- Snap!/NBC

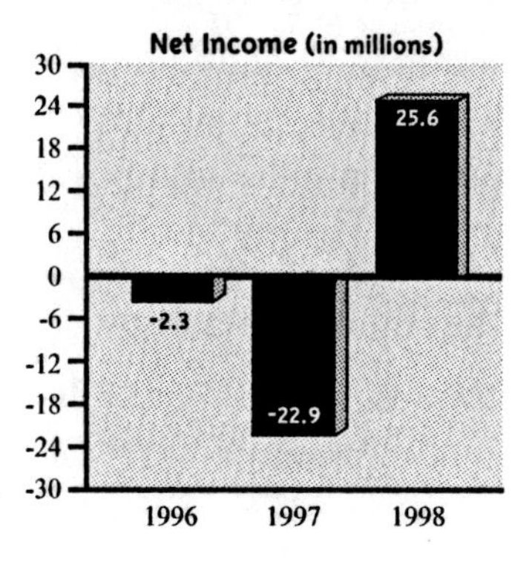

EMPLOYMENT CONTACT

Human Resources
Yahoo!
3400 Central Expressway
Suite 201
Santa Clara, CA 95051

Fax: (408) 731-3301.
hr@yahoo-inc.com

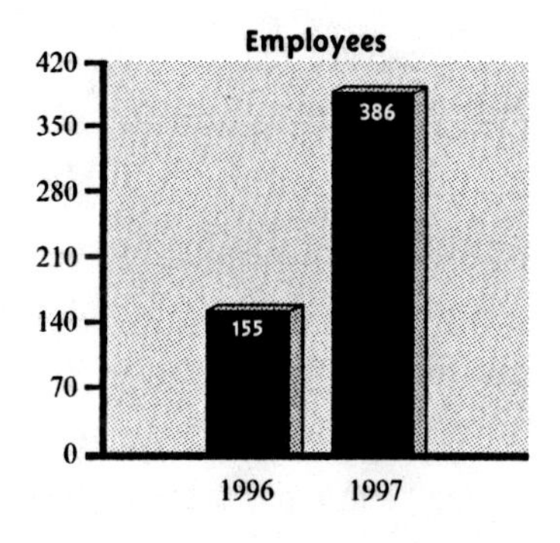

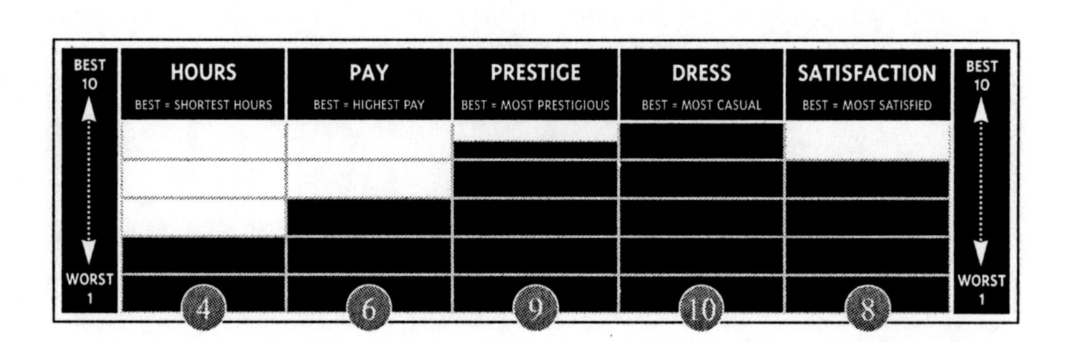

THE SCOOP

Founded in 1994 by David Filo and Jerry Yang – two Stanford graduate students looking for a way to index their favorite World Wide Web sites – Yahoo! met with instant success as an Internet search engine, attracting tens of thousands of web users searching for their favorite pages. After attracting widespread interest and venture capital, Yahoo went public in 1996. Hype about the company pushed its initial public offering to $300 million. The initial $1 million dollar stake put in by venture capital firm Sequoia Capital was worth $560 million at the end of 1997.

The structure of Yahoo's search engine is the product of careful human editorial labor (as opposed to other engines that use algorithms and computers to search the Web), and now the company is capitalizing on this characteristic. Other search engines may be more powerful and may access more sites, but Yahoo's ability to categorize sites has allowed it to experiment with a regional version of Yahoo and customized versions for individuals and children.

The company has seen explosive growth. Its market capitalization hit $9 billion in August 1998 (when the company's stock traded at 23 times its IPO price). The stock went crazy in the last half of 1998; rising 10 percent virtually daily as the company's market cap hit $40 billion. Its stock price is being driven by investors who see the company as the media company of the cyberage. Today, 40 million people visit Yahoo's pages each month. Why do they go? Not just for the search engine. Visitors to Yahoo can get stock quotes, find recipes, chat online about the latest sports news, and order CDs. This one-stop online supermarket strategy has a name: Yahoo and other companies such as Infoseek and Excite are now known as portals (as opposed to search engines). Although Yahoo is the most popular portal, the company doesn't have free sailing into the lucrative e-world of the future – companies like Infoseek are ganging up with traditional media companies like The Walt Disney Company, creating partnerships with deep pockets that they hope will prove to be stiff competition for Yahoo's brand strength.

GETTING HIRED

Like many other cutting-edge multimedia firms, Yahoo's staff is disproportionately small compared to its influence. Still, Yahoo's staff more than tripled from 1995 to 1997, and now applicants can access current job openings through Yahoo's employment web page, located at www.yahoo.com/docs/hr. Applicants can submit a resume either by e-mail or by fax, but Yahoo does not accept mailed resumes or answer phone inquiries.

OUR SURVEY SAYS

Yahoo boasts an "informal atmosphere" and a staff with "substantial industry expertise." A young company, Yahoo offers employees a "dynamic" environment and the promise of "explosive growth," according to insders. A contact praises the "approachable" staff and says that "Filo and Yang are down to earth guys. Everyone likes everyone a lot." Another insider points out that "even the biggies just have plain desks like everyone else." One employee says that she brings her pet parakeet to work "and that nobody cares. Except me and the bird." However, employees point out that Yahoo is "navigating uncharted territory." Employees are confident that Yahoo's "fame among Internet users worldwide" will bring "permanent prestige" to their resumes. Insiders say they love the "kooky job titles and loose atmosphere."

Additional High Tech Companies

Alliance Semiconductor

30399 N. First St., San Jose, CA 95134

(408) 383-4900 • Fax: (408) 383-4999

Makes ultrafast memory chips that retrieve and store data for advanced computer microprocessors. Targets customers in networking, PC, and telecommunications industry.

Analog Devices

One Technology Way, Norwood, MA 02062

(781) 329-4700 • Fax: (781) 326-8703

Integrated circuit (IC) products, which translate pressure and sound into digital signals, account for 60 percent of sales.

Artisoft

5 Cambridge Center, Cambridge, MA 02142

(617) 354-0600 • Fax: (617) 354-7744

Sells computer telephony and PC communications software including LANtastic, a program that allows PC users to share data applications and Internet access. Its products are sold in more than 100 countries.

Asante

821 Fox Ln., San Jose, CA 95131

(408) 435-8388 • Fax: (408) 432-7511

Manufactures hubs, adapters, and switches used in work group computer networks. Has international offices in Canada, Germany, Japan, Taiwan, and the United Kingdom.

Baan

Baron van Nagellstraat 89, 3771 Barneveld, The Netherlands

31-342-428-888 • Fax: 31-342-428-822

A leading provider of enterprise resource planning software. Major markets include auto, technology, aerospace, and process industries.

Banyan Systems

120 Flanders Rd., Westborough, MA 01581

(508) 898-1000 • Fax: (508) 898-1755

Makes networking and Internet software used by Sprint PCS, Siemens, and the U.S. Air Force. Its main product, StreetTalk, allows users to locate resources and other users no matter where they are.

Broadcom

16251 Laguna Canyon Rd., Irvine, CA 92618

(949) 450-8700 • Fax: (949) 450-8710

Specializes in broadband (high speed) data and video transmission technology. Company controls the markets for cable modems and digital set-top boxes.

Cadence Design Systems

2655 Sealy Ave., Bldg. 5, San Jose, CA 95134

(408) 943-1234 • Fax: (408) 943-0513

World's leading provider of electronic design automation (EDA) software used in telephones, fax machines, and computers. Also offers training and consulting services.

Canon

30-2 Shirmomaruko 3-chome, Ohta-ku, Tokyo 146, Japan

81-3-3758-2111 • Fax: 81-3-5482-5135

Products include bubble-jet printers, laser printers, digital cameras, and fax machines. Industry leader in high resolution digital technology.

Check Point Software

3A Jabotinsky St., 24th Fl., Diamond Tower, Ramat Gan 52520, Israel

972-3-753-4555 • Fax: 972-3-575-9256

Develops software that protects corporate networks from unauthorized access. Sun Mircrosystems accounts for 20 percent of sales.

Hitachi

6, Kanda-Surugadai 4-chome, Chiyoda-ku, Tokyo 101-810, Japan

81-3-3258-1111 • Fax: 81-3-3258-5480

Manufactures both electronic components and industrial machinery.
No. 2 maker of workstations after IBM.

Inprise

100 Enterprise Way, Scotts Valley, CA 95066

(831) 431-1000 • Fax: (831) 431-4141

Specializes in programming languages, computer network connectivity tools, and application software. Formerly Borland International. Concentrating mainly on the fast growing Internet and Intranet markets.

Iomega

1821 W. Iomega Way, Roy, UT 84067

(801) 778-1000 • Fax: (801) 778-3190

Sold more than 16 million data storage devices, including zip drives. Plans to release Clik!, a storage disk for handheld computers, cellular phones, and digital cameras.

Mindspring

1430 W. Peachtree, Ste. 400, Atlanta, GA 30309

(404) 815-0770 • Fax: (404) 815-8805

Serves 340,000 subscribers in more than 300 cities nationwide with Internet access.

MRV Communications

8917 Fulbright Ave., Chatsworth, CA 91311

(818) 773-9044 • Fax: (818) 773-0906

Makes hubs, switches, and routers that direct and move data through corporate networks.

National Semiconductor

2900 Semiconductor Dr., Santa Clara, CA 95052

(408) 721-5000 • Fax: (408) 739-9803

Working on a new generation of microchip called system-on-a-chip, which combine a micoprocessor and circuitry into a single unit. About 60 percent of its revenues come from Asia and Europe.

NEC

7-1, Shiba 5-chome, Minato-ku, Tokyo 108-8001 Japan

81-3-3454-1111 • Fax: 81-3-3798-1510

World's second-largest maker of semiconductors behind Intel.
Controls an industry-leading 30 percent PC market share in Japan.

Newbridge Networks

600 March Rd., PO Box 13600, Kanata, Ontario, K2K 2E6, Canada

(613) 591-3600 • Fax: (613) 591-3680

Global leader in ATM switches that route high speed multimedia transmissions over telephone and data networks. AT&T, GTE Wireless, Swiss Com all use Newbridge technology.

ODS Networks

1101 E. Arapaho Rd., Richardson, TX 75081

(972) 234-6400 • Fax: (972) 234-1467

Sells network hubs and switching equipment. Customers include Electronic Data Systems and the U.S. Government.

ONSALE

1350 Willow Rd., Ste. 202, Menlo Park, CA 94025

(650) 470-2400 • Fax: (650) 473-6990

Auctions computer products over the World Wide Web. Has more than 650,000 registered bidders.

Prodigy

44 S. Broadway, White Plains, NY 10601

(914) 448-8000 • Fax: (914) 448-8083

National Internet service provider that offers its 630,000 subscribers Internet access, web hosting, e-mail, and e-commerce. Owned by Mexican telecommunications company Carso Global Telecom.

PSINet

510 Huntmar Park Dr., Herndon, VA 20170

(703) 904-4100 • Fax: (703) 904-4200

Provides corporations with Internet access, web site design and hosting, electronic commerce, and security programs. Has offices in 11 countries and serves more than 25,000 companies.

Quantum

500 McCarthy Blvd., Milpitas, CA 95035

(408) 894-4000 • Fax: (408) 894-4152

Second-largest U.S. maker of diskdrives behind Seagate. Customers include Compaq, Hewlett-Packard, Apple, IBM, and Dell.

Rockwell

600 Anton Blvd., Ste. 700, Costa Mesa, CA 92626l

(714) 424-4200 • Fax: (714) 424-4251

Leading maker of microchips for fax, voice, and data modems. Competitors include 3Com, Honeywell, and Siemens.

Samsung

250, 2-Ga, Taepyung-Ro, Chung-Gu, Seoul, South Korea

82-2-227-7114 • Fax: 82-2-727-7985

World's largest maker of computer-memory chips. Semiconductors account for 46 percent of total sales.

Security Dynamics

20 Crosby Dr., Bedford, MA 01730

(781) 687-7000 • Fax: (781) 687-7010

Leading developer of hardware and software used to protect and manage remote access to computers. Its subsidiary, RSA Data Security, produces data encryption software.

Solectron

777 Gibraltar Dr., Milpitas, CA 95035

(408) 957-8500 • Fax: (408) 956-6075

Builds electronic systems for customers such as Hewlett-Packard, Dell Computer, and Cisco Systems.

Standard Microsystems

80 Arkay Dr., Hauppauge, NY 11788

(516) 435-6000 • Fax: (516) 273-5550

Produces integrated circuits and various non-semiconductor devices.
Over 50 percent of its sales are generated outside of the U.S.

Synopsys

700 E. Middlefield Rd., Mountain View, CA 94043

(650) 962-5000 • Fax: (650) 965-8637

Develops design automation software used in the making of integrated circuits and electronic systems.

Tandy

100 Throckmorton St., Ste. 1800, Fort Worth TX 76102

(817) 415-3700 • Fax: (817) 415-2647

Owns Radio Shack chain, one of the largest electronic retailers in the U.S. Sold Computer City chain to CompUSA.

Tivoli Systems

9442 Capital of Texas, hwy. North, Arboretum, Plaza One, Ste. 500, Austin, TX 78759

(512) 794-9070 • Fax: (512) 418-4992

Makes software for managing different computer networks. Subsidiary of IBM.

Toshiba

1-1, Shibaura 1-chome, Minator-ku, Tokyo, 105-8001, Japan

81-3-3457-2096 • Fax: 81-3-3456-9202

Global leader in laptop computers. Also produces wireless phones and semiconductors.

US West (Media Group Inc.)

188 Inverness Dr. West, Englewood, CO 80112

303-858-3000 • Fax: 303-858-3464

Third-largest U.S. cable operator. Expanding its broadband cable networks to offer integrated high-speed Internet access, cable TV, and telephony services.

VIASOFT

3033 N. 44th St., Phoenix, AZ 85018

(602) 952-0050 • Fax: (602) 840-4068

Sells companies software to correct Y2K bug. Also develops and licences mainframe software for large customers.

ABOUT THE AUTHORS

Marcy Lerner: Marcy is executive editor of Vault Reports. She graduated from the University of Virginia with a BA in history and holds an MA in history from Yale University. Marcy authored articles on career topics for *The Wall Street Journal Interactive Edition* and the *National Business Employment Weekly*.

Nikki Scott: Nikki is news editor of Vault Reports. She graduated from Amherst College in 1997 with a degree in English Literature, with a concentration in Caribbean Literature. After a brief stint in the advertising world, Nikki joined Vault Reports. Nikki has written on career topics for the *Managing Your Career* supplement to the *National Business Employment Weekly.*

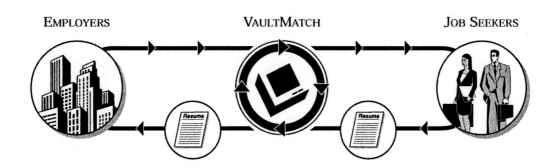

VAULT REPORTS INDUSTRY GUIDES

The first career guides of their kind, Vault Reports' Industry Guides offer detailed evaluations of America's leading employers. Enriched with responses from thousands of insider surveys and interviews, these guides tell it like it is – the good and the bad – about the companies everyone is talking about. Each guide includes a complete industry overview as well as information on the industry's job opportunities, career paths, hiring procedures, culture, pay, and commonly asked interview questions..

Each employer entry includes:

The Scoop: the juicy details on each company's past, present, and future.

Getting Hired: insider advice on what it takes to get that job offer.

Our Survey Says: thousands of employees speak their mind on company culture, satisfaction, pay, prestige, and more.

PRICE: $35 PER GUIDE

GUIDE TO ADVERTISING™

Reviews America's top employers in the advertising industry, including Bozell Worldwide, Grey Advertising, Leo Burnett, Ogilvy & Mather, TBWAChiat/Day, and many more!

100pp.

GUIDE TO FASHION™

Reviews America's top employers in the fashion industry, including Calvin Klein, Donna Karan, Estee Lauder, The Gap, J. Crew, IMG Models, and many more!

100pp.

GUIDE TO HEALTHCARE™

Reviews America's top employers in the healthcare industry, including Amgen, Eli Lilly, Johnson & Johnson, Oxford, Pfizer, Schering-Plough, and many more!

120pp.

GUIDE TO HIGH TECH™

Reviews America's top employers in the high tech industry, including Broderbund, Cisco Systems, Hewlett-Packard, Intel, Microsoft, Sun Microsystems, and many more!

400pp.

GUIDE TO INTERNET AND NEW MEDIA™

Reviews America's top employers in the Internet and new media industry, including Amazon.com, CDNow, DoubleClick, Excite, Netscape, Yahoo!, and many more!

130pp.

GUIDE TO INVESTMENT BANKING™

Reviews America's top employers in the investment banking industry, including Bankers Trust, Goldman Sachs, JP Morgan, Morgan Stanley, and many more!

400pp.

GUIDE TO MANAGEMENT CONSULTING™

Reviews America's top employers in the management consulting industry, including Andersen Consulting, Boston Consulting Group, McKinsey, PricewaterhouseCoopers, and many more!

400pp.

GUIDE TO MARKETING AND BRAND MANAGEMENT™

Reviews America's top employers in the marketing and brand management industry, including General Mills, Procter & Gamble, Nike, Coca-Cola, and many more!

150pp.

GUIDE TO MBA EMPLOYERS™

Reviews America's top employers for MBAs, including Fortune 500 corporations, management consulting firms, investment banks, venture capital and LBO firms, commercial banks, and hedge funds.

500pp.

GUIDE TO MEDIA AND ENTERTAINMENT™

Reviews America's top employers in the media and entertainment industry, including AOL, Blockbuster, CNN, Dreamworks, Gannett, National Public Radio, Time Warner, and many more!

400pp.

9 781581 310436